CONTENTS

PROFESSOR H. B. WHITTINGTON, F.R.S.

drawn by Carolyn Lockett

PROFESSOR H. B. WHITTINGTON, F.R.S.

AFTER seventeen years as Woodwardian Professor of Geology in the University of Cambridge, Harry Blackmore Whittington is retiring in the summer of 1983, and some of his pupils and friends gladly offer this publication as a token of gratitude and respect for his wisdom, published work, inspiration, and friendship.

We are appreciative that his health is good and we recall that during his Cambridge stay he has generally walked to the Sedgwick Museum, and back to his home, some 1½ miles each way; we hope that he may long enjoy physical and mental fitness and that he will be able to continue his fruitful research for many years.

Each Woodwardian Professor, since Adam Sedgwick's 1818 appointment, has worked among Lower Palaeozoic rocks and fossils and Harry Whittington has ably carried on that tradition by biostratigraphical work particularly in North Wales and western Newfoundland, with palaeontological studies over a wider area. His trilobite researches are characterized by precise description and excellent photographic and line illustration and are specially valuable for the clear way in which distinguishing features between taxa are indicated. Thus he has produced a better understanding of many species' exoskeletal morphology, growth-stages, phylogeny, and taxonomy. The stratal and geographical distribution of particular taxa has led him, in co-authorship with C. P. Hughes, to deductions concerning trilobite migrations and to the configuration of palaeozoological provinces (90, 93, 94; numbers refer to publications listed in the appended bibliography); furthermore, he has made correlation between Ordovician strata in Britain and America better founded.

A special feature of Harry's work is his success in achieving orderly publication within a reasonable time of its completion; a notable and easily explained exception being the results of his 1941 fieldwork in Burma, which were delayed because of war.

Though renowned especially for his studies of Ordovician trilobites, he has also described trilobites from other systems (25, 51, 74); additionally, in part as co-author with R. B. Rickards, he has described new graptolites (26, 76, 83, 89). Another diversification of interests arose in 1966 when he started a special study of the well-preserved biota of the Middle Cambrian Burgess Shale of British Columbia.

Harry differs from his Woodwardian predecessors in not having been a member of Cambridge University prior to his appointment as Professor, for he graduated from the University of Birmingham as a student of L. J. Wills. Scholarship awards helped to maintain him when mapping the geology of part of the Berwyn Hills, North Wales. There he showed that a substantial area of ground mapped earlier in the century by H.M. Geological Surveyors as Caradoc is really of Ashgill age; he substantiated some of the biostratigraphical conclusions which had been published privately by B. B. Bancroft in 1933, and he solved some of the nomenclatural problems, posed by Bancroft, concerning certain Caradoc brachiopods (1–3).

These North Welsh researches led to Harry being awarded a Commonwealth Fund Fellowship in 1938, which launched him on what was to become a long series of 'emigrations' to North America. Working under the guidance of Carl Dunbar at the Yale Peabody Museum Harry, with encouragement from P. E. Raymond, described some of the Harvard collection of Bohemian trinucleids (4) and then he turned to North American trinucleids (5) with the objective of comparing and contrasting Ordovician faunas on each side of the Atlantic Ocean. Fortune favoured Whittington further—earlier in the 1930s there had been a resurgence in North America of what E. L. Yochelson, in one of his papers, picturesquely referred to as the opening of 'Pandora's Box of silicified faunas'. Dr. G. A. Cooper of the U.S. National Museum was specially active in dissolving large quantities of limestone and freeing silicified fossils, particularly brachiopods and trilobites, from their rock-matrix; the brachiopods formed his own special study. In Harry's second year at

[Special Papers in Palaeontology No. 30, pp. 1–8]

Yale, Cooper invited him to study at the U.S. National Museum some silicified trilobites which he, in earlier years, had isolated from Ordovician (Trenton) limestones in Virginia. Harry took his first steps in harvesting new information from such exceptionally well-preserved material (6); he creditably revised the descriptions of some time-honoured American species and described a new odontopleurid genus and species. Significantly he also extended his morphogenetical studies, with the advantage that they were of species collected from a single horizon and locality; the photographs, especially of protaspides and meraspides, were indeed praiseworthy.

On the termination of the Yale Fellowship in October 1940, Whittington took a post as lecturer at Judson College, Rangoon, and in the spring of 1941 he worked in the southern Shan States among strata ranging in age from Ordovician to Permian (30). In October of the same year he led a Rangoon student-party to the Lower Salwyn Valley to collect diagnostic fossils from a Permian limestone (31). The advance of the Japanese army into Burma in December 1941 caused the Whittingtons (married in 1940) to take a traumatic journey culminating in their January 1943 arrival in Chengtu (Chengdu), Szechuan Province (Sichuan in new Pinyin spelling); then Harry was appointed lecturer, later professor, in the Ginling Women's College. Whilst there he conducted a geological reconnaisance into the Tibet border mountains of Szechuan (7, 8).

Returning to Birmingham in 1945 as a lecturer, he started geological studies in the Bala area which were completed during later visits from America (58, 73), for that country had called him again in 1946 when G. A. Cooper obtained a grant from the Penrose Fund administered by the Geological Society of America, to enable Harry to study the silicified trilobites which Cooper had isolated from Ordovician limestones (of Llanvirn age) in Virginia in 1935. Thus Harry spent three months at the U.S. National Museum in 1947. This study so fascinated him that he accepted, in 1950, an offer to succeed Preston Cloud as Associate Professor and Curator at Harvard, and in 1957 he was made Professor there.

Before leaving Birmingham, he had inspired his research student to dissolve samples of Welsh Lower Palaeozoic limestones in search of conodonts. That research resulted in a splendid *Philosophical Transactions of the Royal Society of London* paper; its author, F. H. T. Rhodes, is now the President of Cornell University.

During the period from 1950 to 1966 Whittington's research naturally centred for some time around Cooper's Virginian fossils; he satisfied his need to establish more precise knowledge of some of the genera founded many years before and to place their nomenclature on a firmer basis (13, 14). The year 1954 saw his production, with W. R. Evitt, of a classic memoir upon silicified faunas isolated from limestones independently by Cooper and Evitt (23); this gave much new information concerning growth-stages of individual species. That was strengthened by later work (41, 48) and, along with knowledge derived from a restudy of some of the nineteenth-century trilobite morphogeneses (43), formed the framework for his 1957 essay-review (44) which in turn was amplified in the Ontogeny section of the (O) *Arthropoda* (Trilobita) volume of the *Treatise on Invertebrate Paleontology* (50) to which Harry also contributed authoritative descriptions of seventeen families. He is probably the leading world authority on trilobite morphogenesis. He is now compiling with Dr. S. R. A. Kelly a list of the 2,000 or more genera described since the publication of the 1959 trilobite *Treatise* and is organizing a revision of the volume.

Whittington's work with C. H. Kindle in western Newfoundland was undertaken at a time when the usual form of transport to that stormy coast was by boat. The fruits of field seasons in 1955, 1958, and 1961 included the elucidation of the complex stratigraphy of the Cow Head Group (47, 49, 68) and the description of the Table Head carbonates (61). Harry described some of the trilobites in two large and beautifully illustrated bulletins (60, 68); although not silicified, the trilobites include some 'complete' specimens; these works are the standard account of the Whiterockian faunas of the North American Ordovician province.

A story is told that, whilst at Harvard, Harry had humorously but incautiously referred to the large collection there of orthocone cephalopods as 'gaspipes' occupying storage space much needed for his trilobites. This was rewarded by R. H. Flower naming a new orthocone genus and species in the dog-latinized form of 'gaspipe of Whittington' (see appended list of taxa).

During his sixteen years at Harvard, Harry visited Europe at least three times to study in museums and to take part in geological mapping in North Wales (73). One welcome result of these visits was the four-part monograph on the Ordovician trilobites of the Bala area; this was the third of his four post-war publications (14, 34, 58, 72) dealing principally with description of British fossils.

The year 1966 came almost halfway through Harry's professorial career and two events made it specially important for him. The first arose following the Geological Survey of Canada's decision to commission a re-excavation of the Rocky Mountain Burgess Shale exposure discovered in 1909 by C. D. Walcott with its exceptionally well-preserved Middle Cambrian fossils. The G.S.C. invited Harry to be chairman of a group to be chosen by him to restudy the Burgess fauna and flora. In the summer of 1966 he took part in the re-excavation and collection of fossils and started to assemble his team of workers. At first he enlisted the established palaeontologists, D. L. Bruton and C. P. Hughes, and in later years he was to train D. E. G. Briggs and S. Conway Morris and arrange that the workers published in sole authorship. The second event stemmed from the retirement of O. M. B. Bulman from the Woodwardian Professorship in 1966; the Electors for the Chair were glad when Harry accepted their invitation for him to reside again in England. The Cambridge appointment, however, did not curtail his Canadian research. In 1967 he spent five months at Camp Burgess, as part of a larger party than in 1966, collecting a considerable number of fossils; the party improved on collecting methods—adopted earlier in the century—by trying to ensure that part and counterpart were rescued together. In succeeding years, he and his co-workers spent varying periods at the U.S. National Museum studying, photographing, and drawing significant fossils in the older collections made by Walcott, Raymond, and others. The first results of Harry's investigations appeared in 1971 (86). The special lecture, which he was asked to give to the Geologists' Association in 1980, summarizes the interim results of the working party from 1967 to date (105). It shows the great skill exercised in evaluating the method of preservation of this considerable biota and the techniques adopted by all investigators in describing the fossils and illustrating the valuable papers produced under his leadership. This combined research, in part financed by a substantial grant administered by the Natural Environment Research Council, is outstanding among the research contributions which British palaeontologists have made in the last twenty years. Whittington himself has provided substantial studies of the arthropods *Marrella* (86), *Yohoia* and *Plenocaris* (95), *Olenoides* and *Kootenia* (98, 106), *Naraoia* (99), the lobopod *Aysheaia* (101), and the enigmatic animal *Opabinia* (97). With D. L. Bruton he has also studied *Emeraldella* and *Leancholia* (115). The team believes that the Burgess biota lived in an oxygenated environment along the flanks of a submarine escarpment, but that unstable mud slumped downslope in a turbulent suspension carrying the biota down with it, and burying the members in a fine-grained deposit. Whether or not, as I once suggested in a review, thixotropy played a role in this formation is unknown. Among the conclusions of a general interest is Harry's opinion that *Olenoides* had only three pairs of biramous appendages in the cephalon (98) and that *Cruziana* should not indubitably be attibuted to trilobites (106).

No tribute to Harry Whittington would be adequate without reference to Dorothy, his charming and devoted American wife. Dorothy is a botanist and her authoritative knowledge of wild flowers adds zest to any geological excursion which Harry is taking; she is a skilled finder of choice fossils, gladly acknowledged by him in his publications dealing respectively with Newfoundland, and the Welsh Arenig and Bala biostratigraphy; she is a praiseworthy hostess, especially to the many from overseas and Britain who visit the Whittington Cambridge home. This has been a powerful factor in fostering the cause of geology in both her and Harry's homelands.

Harry has a love of teaching. Until 1981 he was responsible for much of the first-year palaeontological instruction at Cambridge; and in nearly every year since 1967, he has been one of the staff on the second-year field trip to south-west Wales, generally accompanied by Dorothy. It happened that Harry usually had his birthday during the trip and many will remember the birthday celebrations organized by Dorothy and others with the active connivance of the hotel managements.

Degrees, awards, fellowships, and trusteeship

1936   B.Sc., 1st class honours Geology, Birmingham.
1938   Ph.D., Birmingham; Commonwealth Fund Fellowship, Yale.
1949   D.Sc., Birmingham.
1950   A.M. (honorary), Harvard.
1956   Guggenheim Fellowship, Europe.
1957   Bigsby Medal, Geological Society of London.
1966   Fellow, Sidney Sussex College, Cambridge.
1971   Fellow, Royal Society, London.
1980   Trustee, British Museum (Natural History).
1983   Leverhulme Emeritus Professorship.
1983   Paleontological Society Medal.

Offices in scientific societies

Paleontological Society (America): Secretary 1956–1962, Vice-President 1963–1964, President 1964–1965.
Palaeontographical Society: Council 1972–1975, Vice-President 1967–1972, 1975–1978.
Palaeontological Association: Council 1967–1969, President 1978–1979.
Cambridge Philosophical Society: President 1979–1982, Vice-President 1983.

Some research students

Birmingham: F. H. T. Rhodes.
Harvard: Z. P. Bowen, L. L. De Mott, A. R. Ormiston, A. S. Hunt, F. C. Shaw.
Cambridge: T. P. Fletcher, R. A. Fortey, D. E. G. Briggs, K. J. McNamara, S. Conway Morris, P. A. Selden, J. Zalasiewicz, D. Campbell, B. D. T. Lynas, J. Almond.

Some taxa related to his name

Trilobita: *Whittingtonia* Prantl and Přibyl, 1949; *W. whittingtoni* Kielan, 1960; *Ceraurus whittingtoni* Evitt, 1953; *Ectenonotus whittingtoni* R. Ross, 1967; *Hibbertia whittingtoni* Tripp, 1965; *Paraharpes whittingtoni* McNamara, 1979.
Brachiopoda: *Eostropheodonta whittingtoni* Bancroft, 1949.
Cephalopoda: *Aethiosolen whittingtoni* Flower, 1966.

There are two further taxa in this volume named in honour of Professor Whittington and one for Mrs Whittington.

BIBLIOGRAPHY

(1)   1938   Fauna of the Lluest Quarry, Llanfyllin (*Wattsella horderleyensis* Superzone), and its correlation. *Proc. Geol. Ass.* **49**, 49–54, pl. 6.
(2)   1938   New Caradocian brachiopods from the Berwyn Hills, North Wales. *Ann. Mag. nat. Hist.* (11) **2**, 241–259, pls. 10, 11.
(3)   1938   The geology of the district around Llansantffraid ym Mechain, Montgomeryshire. *Q. Jl Geol. Soc. Lond.* **94**, 423–457, pls. 38, 39.
(4)   1940   On some Trinucleidae described by Joachim Barrande. *Am. J. Sci.* no. 238 (4), 241–249, 4 pls.
(5)   1941   The Trinucleidae—with special reference to North American genera and species. *J. Paleont.* **15** (1), 21–41, pls. 5, 6.
(6)   1941   Silicified Trenton trilobites. Ibid. (5), 492–522, pls. 72–75.
(7)   1944   Geological reconnaissance between Kuan-hsien and Cho-ke-chi. *Mem. geol. Surv. Szechuan, Chungking, China*, no. 111, 22 pp.
(8)   1946   The physiographical history of Western Szechuan—a review and discussion. *Geol. Mag.* **83**, 141–146.
(9)   1948   A new Lower Ordovician trilobite. *J. Paleont.* **22** (5), 567–572, pls. 82, 83.
(10) 1948   Field meeting at the Forest of Dean and May Hill. *Proc. Geol. Ass.* **59**, 58–61.

(11) 1949 Redescription of the trilobite *Eoharpes* Raymond, 1905. *Q. Jl geol. Soc. Lond.* **104**, 221–228, pls. 11, 12.

(12) 1949 *Dolichoharpes* and the origin of the harpid fringe. *Am. J. Sci.* no. 247 (4), 276–285, 2 pls.

※ (13) 1950 Sixteen Ordovician genotype trilobites. *J. Paleont.* **24** (5), 531–565, pls. 68–75.

(14) 1950 A monograph of the British trilobites of the family Harpidae. *Monogr. palaeontogr. Soc.* London: 1–55, pls. 1–7 (Publ. No. 447, part of Vol. **103** for 1949).

(15) 1950 Swedish Lower Ordovician Harpidae and the genus *Harpides*. *Geol. För. Stockh. Förh.* **72** (3), 301–306, 1 pl.

(16) 1952 A unique remopleuridid trilobite. *Breviora*, **4**, 9 pp., 1 pl.

(17) 1952 The trilobite family Dionididae. *J. Paleont.* **26** (1), 1–11, pls. 1, 2.

(18) 1952 Proposal to suppress the generic name "Polytomurus" Hawle and Corda 1847, and to place the generic name "Dionide" Barrande, 1847 (Class Trilobita) on the "Official List of Generic Names in Zoology". *Bull. zool. Nom.* London, **6**, 157–158.

(19) 1953 [with EVITT, W. R.] The exoskeleton of *Flexicalymene* (Trilobita). *J. Paleont.* **27** (5), 49–55, pls. 9, 10.

(20) 1953 A new Ordovician trilobite from Florida. *Breviora*, **17**, 6 pp., 1 pl.

(21) 1953 A Review. Lower Ordovician trilobites from Western Utah and Eastern Nevada, by Lehi F. Hintze. *Am. J. Sci.* no. 251, 766–767.

(22) 1953 North American Bathyuridae and Leiostegiidae (Trilobita). *J. Paleont.* **27** (5), 647–678, pls. 65–69.

※ (23) 1954 [dated 1953; with EVITT, W. R.] Silicified Middle Ordovician trilobites. *Mem. geol. Soc. Amer.* no. 59, 137 pp., 33 pls.

※ (24) 1954 Correlation of the Ordovician system of Great Britain with that of North America. *Bull. geol. Soc. Amer.* **65**, 258–262.

※ (25) 1954 Two silicified Carboniferous trilobites from West Texas. *Smithson. misc. Collns*, no. 122 (10), 16 pp., 3 pls.

(26) 1954 A new Ordovician graptolite from Oklahoma. *J. Paleont.* **28** (5), 613–621, pl. 63.

※ (27) 1954 Status of Invertebrate Paleontology, 1953. VI Arthropoda: Trilobita. *Bull. Mus. comp. Zool. Harv.* no. 112 (3), 193–200.

(28) 1954 *Onnia* (Trilobita) from Venezuela. *Breviora*, **38**, 5 pp., 1 pl.

(29) 1954 Ordovician trilobites from Silliman's Fossil Mount. Pp. 119–141, pls. 59–63. *In* MILLER, A. K. *et al.* Ordovician cephalopod fauna of Baffin Island. *Mem. geol. Soc. Amer.* no. 62, 234 pp.

(30) 1954 New evidence for the Permian age of the Moulmein System. *In* CLEGG, E. L. G. *Records geol. Surv. India*, 78 (2), 193–194.

(31) 1954 Geological reconnaissance between Loilem and Ke-hsi Mansam, Southern Shan States. Ibid. **78** (2), 203–216, pl. 9 (map).

(32) 1955 Additional new Ordovician graptolites and a chitinozoan from Oklahoma. *J. Paleont.* **29** (5), 837–851, pls. 83, 84.

(33) 1955 Proposed use of the plenary powers to suppress the generic name "Phillipsella" Oehlert, 1886, and proposed addition of the name "Phillipsinella" Novak, 1886 (Class Trilobita) to the "Official List of Generic Names in Zoology". *Bull. zool. Nom.* **11** (9), 283–284.

(34) 1955 [with WILLIAMS, A.] The fauna of the Derfel Limestone of the Arenig district, North Wales. *Phil. Trans. R. Soc.* B **238** (658), 397–430, pls. 38–40.

(35) 1956 Beecher's supposed odontopleurid protaspid is a phacopid. *J. Paleont.* **30** (1), 104–109, pl. 24.

(36) 1956 Type and other species of Odontopleuridae (Trilobita). Ibid. (3), 504–520, pls. 57–60.

(37) 1956 Photographing small fossils. Ibid. 756–757.

(38) 1956 The trilobite family Isocolidae. Ibid. (5), 1193–1198, pls. 129, 130.

(39) 1956 Beecher's lichid protaspid and *Acanthopyge consanguinea* (Trilobita). Ibid. 1200–1204, pl. 131.

(40) 1956 Proposed use of the plenary powers to suppress certain "nomina dubia" and thus validate the specific name "tuberculatus" as used in the combination "Acidaspis tuberculatus" Hall J. W. in 1859 and, by suppressing the generic name "Acantholoma" Conrad, 1840, to provide an assured basis for the generic name "Leonaspis" Richter (R.) & Richter (E.), 1917 (Class Trilobita). *Bull. zool. Nom.* **12** (1), 22–26.

※ (41) 1956 Silicified Middle Ordovician trilobites; the Odontopleuridae. *Bull. Mus. comp. Zool. Harv.* no. 114 (5), 155–288, 24 pls.

(42) 1956 [with STUBBLEFIELD, C. J.] Proposed use of the plenary powers to validate the generic names "Trinucleus" Murchison, 1839 and "Tretaspis" McCoy, 1849 (class Trilobita). *Bull. zool. Nom.* **12** (2), 49–54.

(43) 1957 Ontogeny of *Elliptocephala, Paradoxides, Sao, Blainia* and *Triarthrus* (Trilobita). *J. Paleont.* **31** (5), 934–936, pls. 115, 116.

(44) 1957 The ontogeny of trilobites. *Biol. Rev.* **32,** 421–469.

(45) 1958 Ontogeny of the trilobite *Peltura scarabaeoides* from Upper Cambrian, Denmark. *Palaeontology,* **1** (3), 200–206, pl. 38.

(46) 1958 [with BOHLIN, B.] New Ordovician Odontopleuridae (Trilobita) from Öland. *Bull. geol. Instn Univ. Upsala,* **38,** 37–45, 3 pls.

(47) 1958 [with KINDLE, C. H.] Stratigraphy of the Cow Head region, western Newfoundland. *Bull. geol. Soc. Amer.* **69,** 315–342, 8 pls.

(48) 1959 Silicified Middle Ordovician trilobites: Remopleurididae, Trinucleidae, Raphiophoridae, Endymionidae. *Bull. Mus. comp. Zool. Harv.* no. 121, 369–496, 36 pls.

(49) 1959 [with KINDLE, C. H.] Some stratigraphic problems of the Cow Head area in western Newfoundland. *Trans. N.Y. Acad. Sci.* (2), **22** (1), 7–18.

(50) 1959 *In* MOORE, R. C. (ed.). *Treatise on Invertebrate Paleontology,* Part **O**, *Trilobita*, 127–144, 365–367, 415–428, 504–509. Geol. Soc. Am. and Univ. Kansas Press.

(51) 1960 *Cordania* and other trilobites from the Lower and Middle Devonian. *J. Paleont.* **34** (3), 405–420, pls. 51–54.

(52) 1960 Unique fossils from Virginia. *Va Miner.* **6** (3), 7 pp.

(53) 1960 Trilobita. In McGraw-Hill, *Encyclopedia of Science and Technology*, 102–106.

(54) 1960 Trilobites. *In* BOUCOT, A. J. *et al.* A late Silurian fauna from the Sutherland River Formation, Devon Island, Canadian Archipelago. *Bull. geol. Surv. Canada,* no. 65, 40.

(55) 1961 Silurian *Hemiarges* (Trilobita) from Cornwallis Island and New York State. *J. Paleont.* **35** (3), 433–444, pls. 55–57.

(56) 1961 Middle Ordovician Pliomeridae (Trilobita) from Nevada, New York, Quebec, Newfoundland. Ibid. (5), 911–922, pls. 99–102.

(57) 1961 A natural history of Trilobites. *Nat. Hist., N.Y.* **70** (7), 8–17; [reprinted with some modifications as] pp. 405–415, 8 pls. *Smithsonian Report* for 1961 (1962).

(58) 1962–1968 A monograph of the Ordovician trilobites of the Bala area, Merioneth. *Monogr. palaeontogr. Soc.* London: Parts 1–4. (Publ. Nos. 497, 504, 512, 520, Parts of volumes **116, 118, 120, 122** for 1962, 1964, 1966, 1968.) 1–138, pls. 1–32.

(59) 1963 [with ROLFE, W. D. I. (eds.)] Phylogeny and evolution of Crustacea. Proceedings of a conference held at Cambridge, Massachusetts, March 6–8, 1962. *Spec. Publs Mus. comp. Zool. Harv.* xi + 192 pp.

(60) 1963 Middle Ordovician trilobites from Lower Head, western Newfoundland. *Bull. Mus. comp. Zool. Harv.* no. 129 (1), 1–119, 36 pls.

(61) 1963 [with KINDLE, C. H.] Middle Ordovician Table Head Formation, western Newfoundland. *Bull. geol. Soc. Amer.* **74,** 745–758, 2 pls.

(62) 1963 *Flexicalymene* Shirley, 1936 (Class Trilobita); proposal to place on the Official List of Generic Names in Zoology. *Bull. zool. Nom.* **20** (2), 157–158.

(63) 1964 *In* NEUMAN, R. B. Fossils in Ordovician tuffs, northwestern Maine, section on Trilobita. *Bull. U.S. geol. Surv.* no. 1181-E, 25–34, pls. 5–7.

(64) 1964 [with WILLIAMS, A.] The Ordovician Period. *In* The Phanerozoic Time-Scale. *Q. Jl geol. Soc. Lond.* **120S,** 241–254.

(65) 1964 Taxonomic basis of Paleoecology. Pp. 19–27, *In* IMBRIE, J. and NEWELL, N. D. (eds.). *Approaches to Paleoecology.* Wiley, New York.

(66) 1965 Photography of small fossils. Pp. 430–433. *In* KUMMEL, B. and RAUP, D. (eds.). *Handbook of paleontological techniques.* W. H. Freeman & Co., San Francisco and London.

(67) 1965 [with COOPER, G. A.] Use of acids in preparation of fossils. Pp. 294–300. *In* KUMMEL, B. and RAUP, D. op. cit. *supra.*

(68) 1965 Trilobites of the Ordovician Table Head Formation, western Newfoundland. *Bull. Mus. comp. Zool. Harv.* no. 132 (4), 275–442, 68 pls.

(69) 1965 *Platycoryphe*, an Ordovician homalonotid trilobite. *J. Paleont.* **39** (3), 487–491, pl. 64.

(70) 1965 [with KINDLE, C. H.] New Cambrian and Ordovician fossil localities in western Newfoundland. *Bull. geol. Soc. Amer.* **76,** 683–688, 2 pls.

(71) 1966 Presidential Address [to Palaeontological Society]. Phylogeny and distribution of Ordovician trilobites. *J. Paleont.* **40** (3), 696–737.

(72) 1966 Trilobites of the Henllan Ash, Arenig Series, Merionethshire. *Bull. Br. mus. (nat. Hist.) Geology*, **11** (10), 389–505, 5 pls.

(73) 1966 [with WILLIAMS, A. and BASSETT, D. A.] The stratigraphy of the Bala district, Merionethshire. *Q. Jl geol. Soc. Lond.* **122**, 219–271, pl. 12 (map).

(74) 1967 [with CAMPBELL, K. S. W.] Silicified Silurian trilobites from Maine. *Bull. Mus. comp. Zool. Harv.* no. 135 (9), 447–483, 19 pls.

(75) 1967 [with ERBEN, H. K.] Scutelluidae Richter & Richter, 1925 (Trilobita): proposed addition to the Official List of Family Group Names. *Bull. zool. Nom.* **24** (4), 230–233.

(76) 1968 [with RICKARDS, R. B.] New tuboid graptolite from the Ordovician of Ontario. *J. Paleont.* **42** (1), 61–69, pl. 14.

(77) 1968 *Cryptolithus* (Trilobita): specific characters and occurrence in Ordovician of eastern North America. Ibid. (3), 702–714, pls. 87–89.

(78) 1968 Zonation and correlation of Canadian and early Mohawkian Series, Pp. 49–60. *In* ZEN, E.-AN, WHITE, W. S., HADLEY, J. B. and THOMPSON, J. B. (eds.). *Studies of Appalachian Geology.* Wiley Interscience, New York.

(79) 1968 Middle Cambrian strata at the Strait of Belle Isle, Newfoundland, Canada. *Spec. Pap. geol. Soc. Amer.* no. 101, 282 [abstract].

(80) 1968 Demonstration: Ordovician faunas from Ny Friesland, north-central Spitsbergen. *Proc. geol. Soc. Lond.* 1968, no. 1648, p. 74 [abstract].

(81) 1968 Trilobitomorpha. *In* MOORE, R. C. *et al.* Developments, trends and outlooks in Paleontology. *J. Paleont.* **42** (6), 1368–1369.

(82) 1969 [with KINDLE, C. H.] Cambrian and Ordovician stratigraphy of western Newfoundland. *Mem. Am. Ass. Petrol. Geol.* **12**, 655–664.

(83) 1969 [with RICKARDS, R. B.] Development of *Glossograptus* and *Skiagraptus*, Ordovician graptoloids from Newfoundland. *J. Paleont.* **43** (3), 800–817, pls. 101, 102.

(84) 1970 J. Barrande's relations with palaeontologists at Cambridge and Harvard Universities. *In* Colloque international sur Joachim Barrande à Prague et à Liblice le 17–20 Mai 1969. *Čas. Miner. Geol.* **15**, 40–41.

(85) 1971 The Burgess Shale: History of research and preservation of fossils. *Symp. Proc. N. Am. pal. Convention*, September 1969, pt. 1, 1170–1201.

(86) 1971 Redescription of *Marrella splendens* [Trilobitoidea] from the Burgess Shale, Middle Cambrian, British Columbia. *Bull. geol. Surv. Canada*, no. 209, 24 pp., 28 pls.

(87) 1971 Silurian calymenid trilobites from the United States, Norway and Sweden. *Palaeontology*, **14** (3), 455–477, pls. 83–89.

(88) 1971 A new calymenid trilobite from the Maquoketa Shale, Iowa. *Smithson. Contrib. Paleobiology*, **3**, 129–136, 2 pls.

(89) 1972 [with RICKARDS, R. B.] The graptolite *Reticulograptus tuberosus sinclairi. J. Paleont.* **46** (1), 154.

(90) 1972 [with HUGHES, C. P.] Ordovician geography and faunal provinces deduced from trilobite distribution. *Phil. Trans. R. Soc. Lond.* B **263** (850), 237–278.

(91) 1972 [with WILLIAMS, A. *et al.*] A correlation of Ordovician rocks in the British Isles. *Spec. Report geol. Soc. Lond.* no. 3, 74 pp.

(92) 1973 Ordovician trilobites. Pp. 13–18. *In* HALLAM, A. (ed.). *Atlas of Palaeobiogeography.* Elsevier, Amsterdam.

(93) 1973 [with HUGHES, C. P.] Organisms and continents through time. Ordovician trilobite distribution and geography. *Spec. Pap. Palaeontology*, **12**, 235–240.

(94) 1974 [with HUGHES, C. P.] Geography and faunal provinces in the Tremadoc epoch. *Spec. Publ. Econ. Min. Paleont.* no. 21, 203–218.

(95) 1974 *Yohoia* Walcott and *Plenocaris* n. gen. Arthropods from the Burgess Shale, Middle Cambrian, British Columbia. *Bull. geol. Surv. Can.* no. 231, 27 pp., 18 pls.

(96) 1974 Ordovician Period. *Encyclopaedia Britannica*, 15th edition, pp. 656–661.

(97) 1975 The enigmatic animal *Opabinia regalis*, Middle Cambrian, Burgess Shale, British Columbia. *Phil. Trans. R. Soc.* B **271** (910), 1–43, pls. 1–16.

(98) 1975 Trilobites with appendages from the Middle Cambrian, Burgess Shale, British Columbia. *Fossils and Strata*, **4**, 409–414, pls. 1–25.

(99)   1977   The Middle Cambrian trilobite *Naraoia*, Burgess Shale, British Columbia. *Phil. Trans. R. Soc.* **B 280** (974), 409–443, pls. 1–16.

(100) 1977   [with ORCHARD, M. J.] Upper Ordovician fauna from northern Iran. *Progr. Abstract internat. Symp. Ordovician System*, Columbus, Ohio, **3**, 32.

(101) 1978   The lobopod animal *Aysheaia pedunculata* Walcott, Middle Cambrian, Burgess Shale, British Columbia. *Phil. Trans. R. Soc.* **B 284** (1000), 165–197, pls. 1–14.

(102) 1979   [with CONWAY MORRIS, S.] The animals of the Burgess Shale. *Scient. Am.* no. 241, 122–133.

(103) 1979   Early Arthropods, their appendages and relationships. *Special Volume Systematics Ass.* no. 12, 253–268.

(104) 1979   [with BRIGGS, D. E. G. and BRUTON, D. L.] Appendages of the arthropod *Aglaspis spinifer* (Upper Cambrian, Wisconsin) and their significance. *Palaeontology*, **22** (1), 167–180, pls. 22–25.

(105) 1980   The significance of the fauna of the Burgess Shale, Middle Cambrian, British Columbia. *Proc. geol. Ass.* **91**, 127–148, 4 pls.

(106) 1980   Exoskeleton, moult stage, appendage morphology, and habits of the Middle Cambrian trilobite *Olenoides serratus*. *Palaeontology*, **23** (1), 171–204, pls. 17–22.

(107) 1981   Memorial to Leif Størmer, 1905–1979. *Memorial geol. Soc. Amer.* no. 11, 4 pp. [and portrait].

(108) 1981   Paedomorphosis and cryptogenesis in trilobites. *Geol. Mag.* **118** (6), 591–602, pls. 1, 2.

(109) 1981   The Sir William Macleay Memorial Lecture 1980. Cambrian animals: their ancestors and descendants. *Proc. Linn. Soc. N.S.W.* **105** (2), 79–87.

(110) 1981   [with BRIGGS, D. E. G.] Relationships of arthropods from the Burgess Shale and other Cambrian sequences. *Open-File Report U.S. geol. Surv.* no. 81-743, 38–41.

(111) 1981   Rare arthropods from the Burgess Shale, Middle Cambrian, British Columbia. *Phil. Trans. R. Soc.* **B 292** (1060), 329–357, pls. 1–14.

(112) 1982   [with BRIGGS, D. E. G.] A new conundrum from the Middle Cambrian Burgess Shale. Abstracts of papers, North American Paleontological Convention III, Montreal, Quebec, Canada. *J. Paleont.* **56** (2, suppl. pt. 2 of 3), 29.

(113) 1982   [with ROSS, R. J. *et al.*] Fission-track dating of British Ordovician and Silurian stratotypes. *Geol. Mag.* **119**, 135–153.

(114) 1982   [with CONWAY MORRIS, S., BRIGGS, D. E. G., HUGHES, C. P. and BRUTON, D. L.] *Atlas of the Burgess Shale* [ed. S. CONWAY MORRIS]. Palaeontological Association, London.

(115) 1983   [with BRUTON, D. L.] *Emeraldella* and *Leancholia*, two arthropods from the Burgess Shale, Middle Cambrian, British Columbia. *Phil. Trans. R. Soc.* **B 300** (1102), 553–582, pls. 1–18.

Acknowledgements. I gladly thank the following colleagues for help in this account: Drs. R. A. Fortey, W. B. Harland, C. P. Hughes, and A. W. A. Rushton.

JAMES STUBBLEFIELD

35 Kent Avenue
Ealing
London W13 8BE

THE TREMADOC ROCKS OF SOUTH AMERICA WITH SPECIAL REFERENCE TO THOSE OF BOLIVIA

by C. P. HUGHES

ABSTRACT. Details are given of the Tremadoc clastic sequences developed in central and southern Bolivia and a review given of other known occurrences of Tremadoc rocks in South America.

ROCKS of Tremadoc age are traditionally referred to the lowest Ordovician throughout South America. Although Tremadoc sequences are known from a variety of regions within South America, the most extensive development occurs in northern Argentina and southern Bolivia. The sequences cropping out in northern Argentina are moderately well-known through the works of Harrington (1938); Harrington and Leanza (1943, 1957); Turner (1959, 1960); Turner and Méndez (1979); Turner and Mon (1979); Aceñolaza (1968, 1973, 1976); and Benedetto (1977, 1978), and it is not the purpose of this short note to attempt to review these works. Rather it is intended to draw attention to the thick, relatively undisturbed sequences in central and southern Bolivia, about which a certain amount has been published, most of which, however, is not widely available outside Bolivia. A brief review of other South American occurrences of Tremadoc rocks is given.

TREMADOC ROCKS OF CENTRAL AND SOUTHERN BOLIVIA

Ordovician rocks were first reported in Bolivia by Kayser (1876), and early important work on these successions and their faunas includes Steinmann and Hoek (1912), Bulman (1931), and Harrington and Leanza (1957). The most recent general account of the geology of Bolivia is that of Ahlfeld and Branisa (1960). Thick sequences of clastic rocks, mainly quartzites, siltstones, and shales, of Tremadoc age are developed in southern Bolivia, particularly in the Tarija region, and a considerable amount of work has been undertaken on these sequences in the last twenty years, but much of it remains unpublished in student theses (Tesis de Grado) of the Universidad Mayor de San Andrés, La Paz, or in unpublished reports of the state oil company, Y.P.F.B. Some specific information has been published in recent years, notably by Botello and Suárez (1973), Suárez (1975, 1976), Rodrigo and Castaños (1978), and Rivas *et al.* (1969). All these, with the exception of Suárez (1975), have been published in Bolivia and are not widely available elsewhere. A very brief account has been published in English by Aceñolaza (1976) which is readily available.

Tremadoc sequences are known from a number of regions in the central and southern parts of the Eastern Cordillera of Bolivia (Brockmann *et al.* 1972; Botello and Suárez 1973; Rivas *et al.* 1969; Justiniano 1972 (unpubl.)). In general the sequences thicken southwards to reach a thickness in excess of 3,000 m near Tarija. They consist of a variety of lithologies, but quartzites, shales, and siltstones predominate. The sequences have been divided into a number of rather thick 'formations' based on a combination of lithological and faunal characteristics, with in some cases the faunal content being taken as the diagnostic feature. A summary of the formational names in use is given in Table 1 and the locations shown on text-fig. 2.

The relationship between the Cambrian Sama Formation and the overlying Iscayachi Formation is a little uncertain. Botello and Suárez (1973) made no indication of any break in the Culpina

[Special Papers in Palaeontology No. 30, pp. 9–14]

TEXT-FIG. 1. Map showing regions in which Tremadoc rocks are
known; outcrops in solid black are diagrammatic.

district and commented that the upper Cambrian Sama Formation terminates with a quartzite bed
that is followed by a presumably Tremadoc sequence beginning with a thin sandy quartzite
containing an inarticulate brachiopod fauna of *Obolus* and *Broeggeria*. They thus appeared to
consider the sequence continuous, although there is no direct evidence that the Sama Formation
is of Cambrian age, since it has not yielded any body fossils, only some *Scolithus*-like trace fossils.
Rivas *et al.* (1969) described a similar sequence from the Carrizal region, and although they
considered the Iscayachi Formation as of Cambrian age, they showed no indication of a break
between the Sama Formation and the Iscayachi Formation. Suárez (1976) in his extensive review
of the Ordovician of Bolivia indicated a clear unconformity between these two formations.

Chapare Region. The lower part of this sequence described by Brockmann *et al.* (1972) consists
of some 2,000 m of unfossiliferous sediments divided into the Avispas and Putintiri Formations
which are in probable fault contact with younger fossiliferous sediments of Llanvirn age. The
Avispas Formation consists of white sandstones, shales, and conglomerates in the upper two-thirds

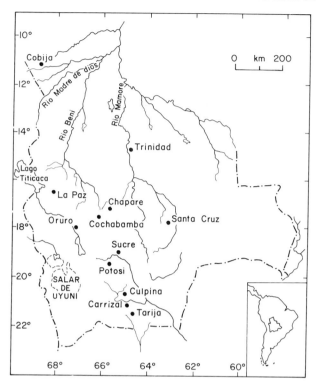

TEXT-FIG. 2. Map of Bolivia showing location of places
mentioned in the text (modified from Rodrigo and Castaños
1978).

TABLE I. Formational names used in Lower Ordovician sequences of the Eastern Cordillera of Bolivia. See text for discussion of the ages of the formations. Chapare region after Brockmann *et al.* (1972); Carrizal/Culpina region after Rivas *et al.* (1969), and Botello and Suárez (1973); Tarija region after Justiniano (1972) as cited by Suárez (1976).

		CENTRAL	SOUTH	
		Chapare	Carrizal/Culpina	Tarija
TREMADOC	ARENIG	AVISPAS Fm.	OBISPO Fm.	CARACHIMAYO Fm.
			CIENEGUILLAS Fm.	
		PUTINTIRI Fm. (Base not seen)	GUANACUNO Fm.	TUCUMILLA Fm.
			ISCAYACHI Fm.	
			SAMA Fm.	SAMA Fm.

resting unconformably on mudstones. The underlying Putintiri Formation consists of limestones and dolomites in the upper part and evaporites with intercalated sandy horizons towards the base. The age is uncertain but the unconformity within the Avispas Formation has been tentatively taken as the local base of the Ordovician. Suárez (1976), on the other hand, considered most of the Putintiri Formation as of Tremadoc age with the Avispas Formation as essentially Arenig in age (see Table 1 herein). Until fossils are found within this sequence its age must remain uncertain.

Culpina Region. This sequence was described by Botello and Suárez (1973) and as indicated above they considered the basal Tremadoc formation, the Iscayachi–Guanacuno Formation, to rest conformably on the underlying Cambrian Sama Formation. Both the lower and upper parts consist of alternating grey-green siltstones and white sandstones, the base of the upper part being marked by a prominent sandstone some 10 m thick. These upper beds have yielded the trilobites *Asaphellus riojanus* and *Angelina* aff. *A. punctolineata*, indicating a lower Tremadoc age. The total thickness of this formation is between 500 and 600 m. The Cieneguillas Formation rests conformably on the Iscayachi–Guanacuno Formation and consists of a lower shale member and an upper member consisting of alternating shales and sandstones; the total thickness of the formation being up to 2,800 m. The lower member has yielded graptolites identified by Botello and Suárez as *Clonograptus*, *Bryograptus* cf. *B. kjerulfi*, and *Dictyonema* aff. *D. yaconense*, indicating a Tremadoc age. The upper member has yielded a didymograptid fauna indicating that it is, at least in part, of Arenig age.

Carrizal Region. The sequence of this region was documented fairly fully by Rivas *et al.* (1969) who give a detailed account of the lithological succession. Unlike Botello and Suárez they considered the Iscayachi and Guanacuno units as two separate formations. The Iscayachi Formation, here 635 m thick, consists of sandstones and shales, and has yielded asaphids, agnostids, and *Parabolina* sp. from near the middle of the formation. The overlying Guanacuno Formation consists of 310 metres of shales and micaceous sandstones and has yielded abundant '*D.*' *flabelliforme*, *Triarthrus tetragonalis*, *Kainella meridionalis*, and *Parabolinopsis* sp.

The Cieneguillas Formation, considered by Rivas *et al.* to be entirely of upper Tremadoc age, consists of some 2,500 metres of alternating shales and sandstones with '*D.*' *flabelliforme* being recorded from shale horizons near the base of the formation and the articulate brachiopod *Dinorthis* sp. from concretions near the top.

Tarija Region. The work on this region by Justiniano (1972) remains unpublished, but reference is made to it by Suárez (1976). The sequence here consists of at least 3,000 m and is believed to be of Tremadoc age, consisting of grey and green shales with intercalated quartzites, the lower levels yielding *Parabolina argentina* and *Jujuyaspis keideli*, with *Kainella meridionalis*, *Parabolinella argentinensis*, and '*Dictyonema*' *flabelliforme* reported from a little higher in the sequence. Justiniano used the term Tucumilla Formation for the whole sequence; this name was first formally published by Suárez (1976).

From the above account of the Tremadoc sequences of Bolivia it may be seen that while thick sequences of Tremadoc age are undoubtedly present in central and southern Bolivia, very little detailed biostratigraphical work has been done. The fossils that are known are often poorly localized and very few are known from precisely collected stratigraphical sequences. Some idea of the fauna present may be obtained from the descriptions of the faunas that have been published (Branisa 1965; Přibyl and Vaněk 1980; Havlíček and Branisa 1980).

TREMADOC ROCKS ELSEWHERE IN SOUTH AMERICA

Venezuela. The Mirales Formation cropping out in the El Baúl region of northern Venezuela, described by Rod (1955), yielded trilobites from somewhere within the lower third of the formation (Martin-Bellizia 1961). Frederickson (1958) assigned all the known specimens to *Parabolina argentina* (Kayser), thus considering the rocks to be of lower Tremadoc age. Although the figured material has not been traced in the collections of the Dirección de Geología in Caracas, the illustrations indicate that all the specimens are distorted and lack cranidial details. The identification and consequent age determination must therefore be regarded as only tentative. To date no further specimens have been collected and micropalaeontological preparations failed to yield any identifiable material.

Colombia. Kay (*in* Trumpy 1943) first reported faunas of Tremadoc age from the Eastern Cordillera of Colombia. Harrington and Kay (1951) described one fauna consisting essentially of inarticulate brachiopods and trilobites from Caño Guapayito which they considered to be of lower Tremadoc age, and another, essentially a trilobite fauna with brachiopods, cystid plates, bellerophontids, and an orthoconic nautiloid from 'Caño 60 Km' which they believed may be of upper Tremadoc age. Tremadoc rocks are also known to exist subsurface in the Llanos. Well-preserved trilobites and graptolites including *J. keideli*, *P. argentina*, and *Dictyonema* sp. were recovered from a depth of about 8,000 ft from Mobil's La Heliera-1 well. Details of the well were given by Stockley (*in* Clark 1960), but no details of the fauna have been published, and the whereabouts of the specimens is not known, although the present author has seen unpublished photographs of the specimens.

Chile. Rocks of Tremadoc age have only recently been recognized in Chile, Cecioni (1979) having recorded '*D.*' *flabelliforme* from near Sotoca in northern Chile. The published illustration, however, is too poor to allow confirmation of Cecioni's identification.

Argentina: Western Cordillera. A carbonate sequence consisting of fossiliferous limestones and dolomites up to 400 m in thickness is known from the Western Cordillera in Mendoza, San Juan, and La Rioja Provinces (Borello 1969; Baldis *et al.* 1981). In some areas the sequence appears to be continuous downwards into rocks of upper Cambrian age (Baldis *et al.* 1981).

CONCLUSIONS

It is only in southern Bolivia that there appears to be any possibility of establishing a continuous section spanning the upper Cambrian to Tremadoc boundary in a non-carbonate sequence. However, until such time as faunas are discovered in the Sama Formation currently considered as Cambrian in age, the position of such a boundary cannot be determined with any precision.

REFERENCES

ACEÑOLAZA, F. G. 1968. Geología estratigráfica de la región de la Sierra de Cajas, Dpto. Humahuaca (provincia de Jujuy). *Revta. Asoc. Geol. Argent.* **23**, 207–222.

—— 1973. El Ordovícico de la Puna salto-catamarqueña. Consideraciones sobre su importancia en la interpretación del desarrollo de la cuenca eo-paleozoica del noroeste Argentino. *Actas Quinto Congr. Geol. Argent.* **4**, 3–18.

—— 1976. The Ordovician System in Argentina and Bolivia, pp. 479–487. *In* BASSET, M. G. (ed.). The Ordovician System: proceedings of a Palaeontological Association symposium, Birmingham, September 1974. 696 pp. University of Wales Press and National Museum of Wales, Cardiff.

AHLFELD, F. and BRANISA, L. 1960. *Geología de Bolivia.* 245 pp. Instituto Boliviana del Petróleo, La Paz.

BALDIS, B. A., BERESI, M. S., BORDONARO, O. and ULIARTE, E. 1981. Estromatolitos, trombolitos y formas afines en el límite Cámbrico-Ordovícico del oeste argentino. *Octavo. Congr. Geol. Arg. Actas* II, 19–30.

BENEDETTO, J. A. 1977. Algunas consideraciones acerca de la posición del límite Cambro-Ordovícico en América del Sur. *Geos.* **23**, 3–11.

—— 1978. Una nueva fauna de trilobites Tremadocianos de la Provincia de Jujuy (Sierra de Cajas), Argentina. *Ameghiniana*, **14** (for 1977), 186–214.

BORELLO, A. V. 1969. Los geosinclinales de la Argentina. *An. Dir. Nac. Geol. Min.* XIV, Buenos Aires.

BOTELLO, R. and SUÁREZ, R. 1973. Estratigrafía de la región de Culpina. Introducción a la zonación bioestratigráfica del Ordovícico inferior en base a su graptofauna. *Bol. Soc. Geol. Boliviana*, no. 20, 97–108.

BRANISA, L. 1965. Los fosiles guías de Bolivia. I. Paleozoico. *Bol. GEOBOL* **6**, 1–282.

BROCKMANN, C., CASTAÑOS, A., SUÁREZ, R. and TOMASI, P. 1972. Estudio geológico de la Cordillera Oriental de los Andes en la zona central de Bolivia (Región del Chapare). *Bol. Soc. Geol. Boliviana*, no. 18, 3–36.

BULMAN, O. M. B. 1931. South American graptolites with special reference to the Nordenskiöld collection. *Ark. Zool.* **22A**, 1–111, pls. 1–12.

CECIONI, A. 1979. El Tremadociano de Sotoca, I Región Norte de Chile. *Actas Segundo Cong. Geol. Chile,* **3**, H159–H164.

CLARK, E. W. 1960. Petroleum developments in South America and Caribbean area in 1959. *Bull. Am. Assoc. Petrol. Geol.* **44**, 1014–1057.

FREDERICKSON, E. A. 1958. Lower Tremadocian trilobite from Venezuela. *J. Paleont.* **32**, 541–543.

HARRINGTON, H. J. 1938. Sobre las faunas del Ordoviciano inferior del norte Argentino. *Revta. Mus. La Plata*, N.S. **1**, sec. paleont. No. 4, 109–289.

—— and KAY, M. 1951. Cambrian and Ordovician faunas of eastern Colombia. *J. Paleont.* **25**, 655–668.

—— and LEANZA, A. F. 1943. La fáunula del Tremadociano inferior de Salitre (Bolivia). *Revta. Mus. La Plata*, N.S. sec. palaeont. vol. 2, pp. 343–356.

—— —— 1957. Ordovician trilobites of Argentina. *Spec. Publ. Univ. Kansas*, **1**, i–x, 1–276.

HAVLÍČEK, V. and BRANISA, L. 1980. Ordovician brachiopods of Bolivia. *Rozpr. česk. Akad. Věd.* **90** (A), 1–54.

JUSTINIANO, E. 1972. Estudio geológico regional del area de Tarija. *Tésis de Grado* (unpubl.). Univ. Mayor San Andrés, La Paz.

KAYSER, E. 1876. Ueber Primordiale und Untersilurische Fossilien aus der Argentinischen Republik: ein Beiträge zur Geologie und Palaeontologie der Argentinischen Republik. *Palaeontographica Supp.* **3** (2), 1–33, pls. 1–5.

MARTIN-BELLIZIA, C. 1961. Geología del Macizo de El Baúl, Estado Cojedas. *Mem. III. Cong. Geol. Venez.* **4**, 1453–1530.

PŘIBYL, A. and VANĚK, J. 1980. Ordovician trilobites of Bolivia. *Rozpr. česk. Akad. Věd.* **90** (2), 1–90.

RIVAS, S., FERNÁNDEZ, A. and ÁLVAREZ, R. 1969. Estratigrafía de los sistemas Ordovícico-Cámbrico y Precámbrico en Tarija, Sud de Bolivia. *Bol. Soc. Geol. Boliviana*, no. 9, 27–49.

ROD, E. 1955. Trilobites in 'metamorphic' rocks of El Baúl, Venezuela. *Bull. Am. Assoc. Petrol. Geol.* **39**, 1865–1869.

RODRIGO, L. A. and CASTAÑOS, A. 1978. *Sinopsis estratigráfica de Bolivia. I Parte; Paleozóico.* 146 pp. Academia Nacional de Ciencias de Bolivia, La Paz.

STEINMANN, G. and HOEK, H. 1912. Das Silur und Cambrian des Hochlandes von Bolivia und ihre Fauna. *Neues Jb. Miner. Geol. Paläont.* **34**, 176–252, pls. 7–14.

SUÁREZ, R. 1975. Zonas graptolitiferas de Bolivia. *Actas Primer. Congr. Argent. Paleont-Bioestr.* Tucumán 1974, **1**, 133–148.

—— 1976. El Sistema Ordovícico en Bolivia. *Revista Téc. Y.P.F.B.* **5**, 111–223.

TRUMPY, D. 1943. Pre-Cretaceous of Colombia. *Bull. Geol. Soc. Am.* **54**, 1281–1304.

TURNER, J. C. M. 1959. Faunas graptolíticas de América del Sur. *Revta. Asoc. Geol. Argent.* **14**, 5–180.

—— 1960. Estratigrafía de la Sierra de Santa Victoria y adyacencias. *Bol. Acad. Nac. Cienc.* Córdoba, **41** (2), 163–196.

—— and MÉNDEZ, V. 1979. Puna. *In* Segundo Simposio de Geología Regional Argentina, vol. I, pp. 13–56. *Bol. Acad. Nac. Cienc.* Córdoba.

—— and MON, R. 1979. Cordillera Oriental. *In* Segundo Simposio de Geología Regional Argentina, vol. I, pp. 57–94. *Bol. Acad. Nac. Cienc.* Córdoba.

C. P. HUGHES

Department of Earth Sciences
University of Cambridge
Cambridge CB2 3EQ
U.K.

A REVIEW OF PERMIAN TRILOBITE GENERA

by R. M. OWENS

ABSTRACT. All trilobite genera based on Permian species as well as some other genera occurring in the Permian are redefined and refigured, and their phylogeny and distribution are discussed. A new proetid subfamily, Weaniinae, is proposed for certain genera formerly included in the Cyrtosymbolinae.

DURING the preparation of a monograph on British Carboniferous trilobites, an assessment of both their generic and suprageneric classification and phylogenetic relationships was attempted. It was soon discovered that this could not be properly carried out without reference to Permian genera. Unfortunately there are few good photographic illustrations of these available, and some of the illustrations most widely referred to (Weller *in* Moore 1959) are line drawings which are for the most part stylized and inaccurate. The aim of this paper is to refigure the type material of the type species of each genus; I have attempted where possible to examine the type specimens, but in some cases it has only proved possible to obtain casts or photographs.

Eighteen genera (sixteen proetids and two brachymetopids) are based upon Permian species; sixteen of these are restricted to the Permian. Additionally, the ranges of four genera whose type species are of Carboniferous age extend into the Permian. Herein the proetids are assigned to five subfamilies (Proetinae, Cyrtosymbolinae, Weaniinae, Cummingellinae, and Ditomopyginae) and the brachymetopids to one (Brachymetopinae), and these are redefined where necessary. The family Phillipsiidae, used in the *Treatise* and redefined by Hahn, Hahn and Brauckmann (1980), is not used. It will be discussed fully elsewhere. Some eighty Permian trilobite species have been described, and most of these can be assigned to the genera listed herein. Some, however, are based upon fragmentary or poorly preserved material, and it is often not possible to tell from existing illustrations to which genus they belong. Terminology used herein follows Harrington *et al. in* Moore (1959, p. O117) and Owens (1973), except that lateral glabellar lobes are labelled L1, L2, etc., and lateral glabellar furrows S1, S2, etc., from posterior forward.

CRITERIA FOR CLASSIFICATION

The classification of proetacean trilobites presents problems because homoeomorphy is rife, making inferences on phyletic relationships hazardous and clear-cut diagnoses difficult to compose. Authors have placed emphasis upon different exoskeletal parts, and a natural phyletic classification is dependent upon selecting those characters which are least affected by the vagaries of homoeomorphy, and which consequently lend clues to the relationships of genera. Adaptive structures which have evolved more than once in response to similar habitats may result in homoeomorphy; these include those involved in feeding and in sensory functions, which would directly affect the shape of the glabella, the size and position of the eye, and the concomitant path of the facial suture. Such cephalic characters should therefore be treated with caution when attempting to infer phyletic relationships.

It has been suggested by a number of authors (e.g. Stubblefield 1936; Fortey and Owens 1979) that cephalic axial characters—particularly the pattern of muscle insertion areas, or lateral glabellar furrows where these are developed—form a derived character (synapomorphic) complex of primary importance in the discrimination of major phyletic lineages. This pattern can be observed to remain broadly constant even within the varied glabellar outline of, for example, *Linguaphillipsia* (pyriform, forward-tapering glabella), *Cummingella* (fiddle-shaped glabella), and *Particeps* (pyriform, forward-expanding glabella); emphasis on muscle insertion areas here produces a rather different classification than one based primarily upon glabellar shape. Osmólska (1970*b*, p. 14) and Owens (1973, pp. 5, 7) also stressed the importance of the structure of the pygidial pleural ribs in the discrimination of proetid genera and their interrelationships. Like muscle insertion area patterns, pygidial structure remains comparatively stable over a wide range of genera; moreover, it also suggests very similar

[*Special Papers in Palaeontology*, No. 30, pp. 15–41, pls. 1–5]

relationships to those indicated by the cephalic axial characters. Thus, in suggesting phyletic relationships I have lent weight to a combination of these characters, regarding others as of secondary importance.

Repositories. The following abbreviations are used: BM—British Museum (Natural History); CGM—Central Geological Museum, VSEGEI, Leningrad; DUT—Delft University of Technology; GIA—Geological Institute, University of Amsterdam; GSI—Geological Survey of India, Calcutta; IGPS—Institute of Geology and Palaeontology, Tohoku University, Sendai; ISGS—Illinois Geological Survey, Urbana; MGP—Museum of the Institute of Geology, University of Palermo; NIGP—Nanking Institute of Geology and Palaeontology; SMF—Senckenberg Museum, Frankfurt am Main; SUP—Department of Geology and Geophysics, University of Sydney; USNM—United States National Museum, Washington; UT—University of Tasmania; UTIEA— University of Tokyo, Institute of Earth Sciences and Astronomy; YPM—Peabody Museum of Natural History, Yale University. The casts figured are in the collection of the National Museum of Wales, Cardiff (NMW).

SYSTEMATIC PALAEONTOLOGY

Family PROETIDAE Salter, 1864
Subfamily PROETINAE Salter, 1864

Diagnosis. Glabellar furrows commonly represented by non-incised muscle insertion areas, but incised in some (particularly later) genera; occipital ring with or without well-defined lobes; preglabellar field commonly short (sag.) or absent; genal spines may or may not be present; thorax of 8(?)–10 segments with preannulus; pygidium with or without border, axis with 4–12 rings, pleural areas with 3–8 pairs of ribs, which are scalloped or have the posterior pleural band elevated above the anterior in exsagittal section; no distinct postaxial ridge.

Range. Ordovician (Caradoc) to Permian (Dzhulfian).

Remarks. In defining the Proetinae, Owens (1973, p. 8) included only *Proetus* (*Pudoproetus*) Hessler from post-Devonian rocks; Osmólska (1970b, p. 33), however, included the Carboniferous genera *Bollandia* Reed, 1943 and *Reediella* Osmólska, 1970b in the subfamily. Owens (1973, p. 8) excluded them because he thought that both their cephalic and pygidial morphology more closely resembled genera of the Schizoproetinae Yolkin, 1968. Subsequent examination of extensive material of both these genera has led to a reconsideration of their affinities, and I am now in agreement with Osmólska in placing them in the Proetinae because their glabellas (although inflated and superficially like those of Ditomopyginae) have a lateral glabellar furrow pattern consistent with those of Proetinae, and their pygidial pleural rib structure is like that of contemporaneous *Proetus* (*Pudoproetus*). Hahn and Hahn (1971) considered both to be subgenera of *Griffithides* Portlock, and although the cephala are broadly similar to that genus, the pygidia are markedly different. Using the criteria discussed above I consider them proetines which are likely to have evolved from *Proetus* (*Pudoproetus*).

The Permian genera *Neoproetus* Tesch, *Kathwaia* Grant and *Neogriffithides* Tumanskaya are all believed to be proetines which have evolved either from *Bollandia* or from *Proetus* (*Pudoproetus*), and the possible phyletic lineages are discussed more fully below. Additionally, the early Silesian genus *Wagnerispina* Gandl, 1977 has pygidial and axial cephalic characters which link it with this group of proetines rather than with cyrtosymbolines in which it was included by its author (1977, p. 132), and it may be associated with the ancestry of *Neogriffithides* and *Kathwaia*.

Genus NEOPROETUS Tesch, 1923

Plate 1, figs. 1–13

Type species (original designation). *Proetus* (*Neoproetus*) *indicus* Tesch, 1923.

Lectotype (here selected). Complete specimen, DUT KA12800 (Pl. 1, figs. 1, 2), Kazanian, Bitauni, Timor, Indonesia.

Species. Neoproetus verrucosus (Gemmellaro), Kazanian, Sosio Valley, Sicily.

Diagnosis. Glabella with bulbous, strongly inflated frontal lobe, overhanging and merged with anterior margin; L1 small, depressed; occipital furrow broad, deep; occipital ring without lobes; genal spine short; thorax of 9 segments; pygidial axis with 8–9 rings defined by deep ring furrows, pleural areas with 7–8 pairs of ribs, broad (exsag.) posterior pleural band elevated well above narrow (exsag.) anterior pleural band, giving interpleural furrows a steep anterior slope; border vertical or reclined; sculpture of terrace ridges and granules.

Remarks. *Neoproetus* shows closest resemblance to the Dinantian proetine *Bollandia* Reed, and I agree with Hahn and Hahn (1975, p. 61) that it is probably derived from this genus. Some *Pudoproetus* Hessler, *Bollandia*, and the closely related genus *Reediella* Osmólska shows the tendency of the posterior pleural band on the pygidium to be elevated above the anterior, and this feature is more accentuated in *Noeproetus*. On the cephalon, the main differences between *Neoproetus* and *Bollandia* are that in the former L1 and the posterior part of the glabella and anterior part of the occipital ring are depressed.

Kobayashi and Hamada (1979) proposed two subgenera of *Neoproetus*, *N. (Triproetus)* and *N. (Paraproetus)*. *N. (Triproetus)* (type species *N. (T.) subovalis* Kobayashi and Hamada, 1979, pl. 1, figs. 4–7, from the Sakmarian Rat Buri Limestone, Changwat Loei, northern Thailand) differs from *Neoproetus* in having glabellar furrows, an inflated L1, and the pygidial pleural ribs of scalloped profile with the interpleural furrows hardly impressed. In cranidial characters *Triproetus* resembles *Paladin* Weller and *Griffithides* more than it does *Neoproetus*, although the pygidium is shorter with fewer axial rings and pleural ribs. It is certainly generically distinct from *Neoproetus*, although its taxonomic position is unclear. Kobayashi and Hamada (1980a) later substituted the name *Siciliproetus* for *Paraproetus* since the latter is preoccupied; its type species is *Phillipsia sicula* Gemmellaro, which is here included in *Neogriffithides*, of which *Siciliproetus* becomes a synonym.

Genus KATHWAIA Grant, 1966

Plate 2, figs. 1–4

Type species (original designation). *Kathwaia capitorosa* Grant, 1966.

Holotype. Enrolled specimen, USNM 145320 (Pl. 2, figs. 1–4), upper part of Middle Productus Limestone, late Guadalupian–early Dzhulfian, Salt Range, Pakistan.

Species. *K. girtyi* (Tumanskaya, 1935), Artinskian, Crimea, U.S.S.R. The specimens from the North Caucasus figured by Weber (1944, pl. 1, figs. 15a, b; pl. 2, figs. 16, 17) as '*Griffithides (Neogriffithides) cf. almensis*' and '*Proetus(?) girtyi*' respectively also belong to *Kathwaia*.

Diagnosis. Glabella with inflated frontal lobe, L1 inflated; eye small, close to glabella; pygidial axis with 8 rings, pleural areas with 6 pairs of ribs which are scalloped in exsagittal section; no border; sculpture granular.

Remarks. Grant (1966, p. 69) considered *Neoproetus* to be the closest contemporary genus to *Kathwaia*, but he pointed out that the latter differs in its more inflated L1, smaller eye, granular sculpture, and other smaller details. Hahn and Hahn (1975, p. 61) also considered *Kathwaia* close to *Neoproetus*, and included both in the Griffithidinae. It is here included in the Proetinae, but is unlikely to be closely related to *Neoproetus*; in addition to the characters noted above, it has a different type of pygidial pleural rib structure (see diagnoses). In this feature, and in the well-developed L1, *Kathwaia* compares quite closely with *Neogriffithides*, and the cephalon figured by Weber (1944, pl. 1, fig. 15a, b), here regarded as belonging to *Kathwaia*, shows closer glabellar resemblance to *Neogriffithides* than does the type species. Both *Kathwaia* and *Neogriffithides* may have a common origin with *Wagnerispina* (see above).

Genus NEOGRIFFITHIDES Tumanskaya, 1935

Plate 1, figs. 14–22

Synonym. Siciliproetus Kobayashi and Hamada, 1980*a.*

Type species (original designation). *Neogriffithides gemmellaroi* Tumanskaya, 1935.

Lectotype (here selected). Cranidium, CGM 2/9733 (Pl. 1, fig. 14), Upper Artinskian, Marta River, Crimea, U.S.S.R.

Species. N. almensis Tumanskaya, Upper Artinskian, Marta River, Crimea, U.S.S.R.; *N. ismailensis* Tumanskaya, Upper Artinskian, Mount Kichkhi-Burnu, Crimea, U.S.S.R.; *N. siculus* (Gemmellaro, 1892), Kazanian, Sosio Valley, Sicily.

Diagnosis. Glabella weakly tapering forwards, distinctly constricted at about mid-length, and not overhanging or coalesced with anterior border; L1 prominent, defined by distinct S1; S2 and S3 short, backwardly oblique, a short distance in front of S1; genal angle rounded; (number of thoracic segments unknown); pygidium short (sag.), without border; axis bluntly terminating, with 8 rings; pleural areas with 5–6 pairs of ribs of scalloped profile; sculpture finely granular.

Remarks. Ruggieri (1959, p. 4) noted that whilst Tumanskaya (1935) had associated a long, multisegmented pygidium (of *Pseudophillipsia* type) with *Neogriffithides* cranidia, Gemmellaro (1892), had associated a short pygidium (of proetine type) with them. Subsequent reconstructions (e.g. Weller *in* Moore 1959, p. O402, fig. 306:6; Hahn and Hahn 1975, pl. 12, fig. 7) follow Tumanskaya. The glabellar furrow pattern of *Neogriffithides* is quite different from that of *Pseudophillipsia* Gemmellaro and its allies (cf. Pl. 1, figs. 14, 17, 18, 20 and Pl. 4, fig. 3), and the cephalon is also quite different from the superficially similar *Ameura* Weller (see Pabian and Fagerstrom 1972, pl. 1, figs. 1–13). Instead, *Neogriffithides* cranidia show a close resemblance to *Proetus* (*Pudoproetus*) (especially *P.* (*P.*) *hahni* Chamberlain, 1977, pl. 3, figs. 1–4, 6), particularly in

---

EXPLANATION OF PLATE 1

Figs. 1–6. *Neoproetus indicus* Tesch, 1923. 1, 2, lectotype, complete exoskeleton with dorsal surface preserved (DUT KA12800), Bitauni, Timor; 1, dorsal and 2, lateral views, × 1·5. (Orig. Tesch 1923, pl. 128, figs. 1a–c.) 3–5, paralectotype, pygidium with exoskeleton preserved (DUT KA12803), Basleo, Timor; 3, dorsal, 4, posterior, and 5, lateral views, × 2. (Orig. Tesch 1923, pl. 128, fig. 4.) 6, paralectotype, pygidium, thorax, and part of cephalon with exoskeleton preserved (DUT KA12802), Noil Fatoe, near Niki-Niki, Timor; dorsal view, × 2. (Orig. Tesch 1923, pl. 128, fig. 3.)

Figs. 7–13. *Neoproetus verrucosus* (Gemmellaro, 1892). Sosio Valley, south of Palermo, Sicily. 7, 8, cranidium with exoskeleton preserved (BM It1416); 7, dorsal and 8, lateral views, × 5. 9–11, pygidium with exoskeleton preserved (BM It1421); 9, lateral, 10, dorsal, and 11, posterior views, × 4. 12, free cheek with exoskeleton preserved (BM It1422), dorsal view, × 5. 13, pygidium with exoskeleton preserved (BM It1420), dorsal view, × 5.

Figs. 14–17, 20. *Neogriffithides gemmellaroi* Tumanskaya 1935. Marta River, Crimea, U.S.S.R. 14, lectotype cranidium (CGM 2/9733), internal mould (plaster cast, NMW 82.19G.1), dorsal view, × 4. (Orig. Tumanskaya 1935, pl. 7, figs. 1, 2.) 15, paralectotype free cheek (CGM 22/9733), mostly exfoliated, dorsal view, × 3·5. (Orig. Tumanskaya 1935, pl. 8, fig. 15.) (Photograph supplied by Dr. L. Popov.) 16, paralectotype cranidium (internal mould), free cheek (external mould), and pygidium (internal mould) (CGM 13/9733), dorsal view, × 4. (Orig. Tumanskaya 1935, pl. 8, fig. 4.) (Photograph supplied by Dr. L. Popov.) 17, paralectotype cranidium with incomplete thoracic segment (CGM 3/9733), internal mould (plaster cast, NMW 82.19G.2), dorsal view, × 4. (Orig. Tumanskaya 1935, pl. 7, figs. 3, 4.) 20, paralectotype cranidium (CGM 4/9733), partially exfoliated (plaster cast, NMW 82.19G.3), dorsal view, × 4. (Orig. Tumanskaya 1935, pl. 7, figs. 5, 6.)

Figs. 18, 19, 21, 22. *Neogriffithides siculus* (Gemmellaro, 1892). Sosio Valley, south of Palermo, Sicily. 18, 19, cranidium with exoskeleton preserved (BM In27392); 18, dorsal and 19, lateral views, × 6. 21, 22, pygidium with exoskeleton preserved (BM It 1424); 21, lateral and 22, dorsal views, × 5.

PLATE 1

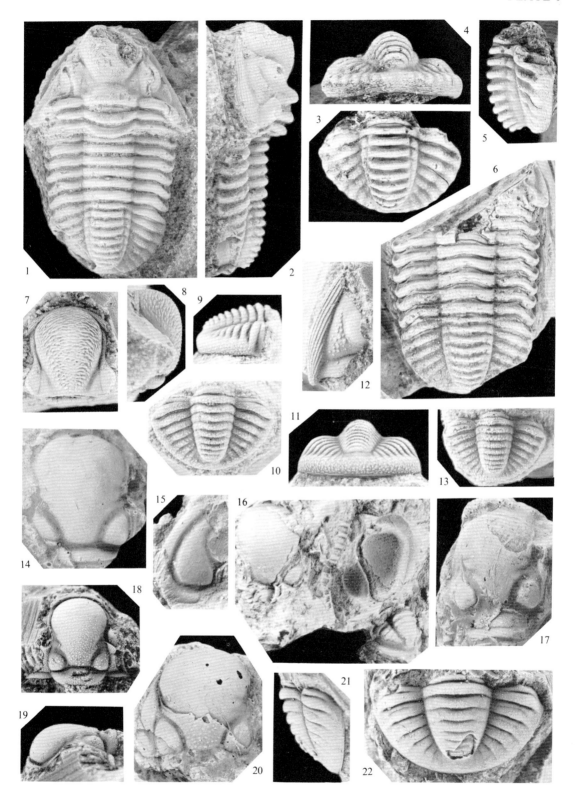

OWENS, *Neoproetus, Neogriffithides*

the disposition of the lateral glabellar furrows. Proetine pygidia occur along with *Neogriffithides* cranidia both in the Crimea and Sicily; these also resemble *P. (P.) hahni* closely. In both areas *Pseudophillipsia* also occurs. Because of the close comparison of *Neogriffithides* cranidia with those of proetines, and because they occur in association with proetine pygidia, I believe that Tumanskaya's association of *Neogriffithides* cranidia with a *Pseudophillipsia* type pygidium was incorrect, and here interpret *Neogriffithides* as a proetine.

In general aspect *Neogriffithides* closely resembles the early Silesian proetine *Wagnerispina wagneri* (Gandl, 1977, pl. 1, fig. 1), to which both it and *Kathwaia* (see above) are probably related.

Subfamily CYRTOSYMBOLINAE Hupé, 1953

Diagnosis. Glabella typically with forward taper, in some genera almost parallel-sided; lateral glabellar furrows incised or represented by smooth muscle insertion areas; lateral occipital lobes absent in most genera; preglabellar field short (sag.) or absent; eyes well-developed in some genera, but in many are reduced or absent; thorax of 9–10 segments; pygidium typically without border, axis with 6–13 rings; 4–10 pleural ribs of scalloped profile in exsagittal section; dorsal exoskeleton normally smooth.

Remarks. A large number of Upper Palaeozoic genera was placed in this subfamily by Richter, Richter and Struve *in* Moore (1959, p. O389), who noted that it contained homoeomorphic genera of perhaps different origin. Subsequently many of these have been revised and transferred elsewhere, and I follow the classification of Hahn and Hahn (1975, p. 34), except to exclude genera herein placed in the Weaniinae (see below), and to include *Hildaphillipsia*.

Range. Upper Devonian (Frasnian) to Permian (Kazanian).

Genus HILDAPHILLIPSIA Hahn and Hahn, 1972

Plate 2, figs. 5–7

Type species (original designation). *Phillipsia hildae* Gheyselinck, 1937.

Type specimens. Gheyselinck (1937, p. 63) based this species upon two enrolled specimens and a pygidium. Of these syntypes only one is now extant (Hahn and Hahn 1972, p. 368), and this specimen, GIA B1379 (Pl. 2, figs. 5–7), from Kazanian, Basleo Beds, south central Timor, Indonesia, is here selected as lectotype.

Species. Hildaphillipsia? salomonensis Gemmellaro, 1892, Kazanian, Pietra di Salomone, Sosio Valley, Sicily.

Diagnosis. Glabella weakly inflated, impinging slightly on anterior border, weakly constricted opposite γ; eye large, close to glabella; cephalic border weakly defined; thorax of 9 segments; pygidium with axis weakly convex in transverse profile with 12 ill-defined axial rings; pleural areas almost smooth, with 8 pairs of ill-defined ribs; no border.

Remarks. Hahn and Hahn (1967, 1972, 1975) believed *Hildaphillipsia* to be derived from *Linguaphillipsia* Stubblefield. Gandl (1980, pp. 309, 310) considered its cranidial characters to closely resemble those of the cyrtosymboline *Archegonus* (*Merebolina*) Gandl from Namurian B, Cantabrian Mountains, northern Spain. He indicated differences in the proportion of the glabella, path of facial suture, eye size, and size of field of the free cheek, and suggested that *Hildaphillipsia* was a cyrtosymboline. Its cranidial and pygidial characters also resemble another cyrtosymboline, *Cyrtoproetus* (here taken to include *Archegonus* (*Angustibole*) Hahn), and I agree with Gandl that it belongs to this subfamily. *Hildaphillipsia* shows tendencies towards effacement, and furrows on the cephalon and pygidium are all very shallow compared with those of *Archegonus* (*Merebolina*) and *Cyrtoproetus*.

Subfamily WEANIINAE subfam. nov.

Diagnosis. Glabella pyriform or fiddle-shaped, L1 defined by incised furrows; S2, S3 shallow, inconspicuous; thorax of 9–10 segments; pygidium normally without border, axis with 7–16 rings, pleural areas with 6–11 pairs of ribs; pleural and interpleural furrows deep, each posterior pleural band distinctly elevated above succeeding anterior pleural band, giving interpleural furrow a steep anterior slope.

Range. Carboniferous (Dinantian) to Permian (Kazanian).

Genera and subgenera. Weania Campbell *in* Campbell and Engel 1963, *Belgibole* Hahn, 1963, *Gitarra* Gandl, 1968, *Nipponaspis* Koizumi, 1972, *Schizophillipsia* Kobayashi and Hamada, 1980*b, Endops* Koizumi, 1972, *Griffithidella* Hessler, 1965, *Carbonocoryphe (Carbonocoryphe)* Richter and Richter, 1950, *C. (Aprathia)* Hahn and Brauckmann, 1975*b, C. (Winterbergia)* Hahn and Brauckmann, 1975*b, Doublatia* Wass and Banks, 1971, ?*Microphillipsia* Ruggieri, 1959.

Remarks. Genera here included in the Weaniinae have been included in various subfamilies. *Weania* was placed by Campbell and Engel (1963, p. 107) in the Cyrtosymbolinae, and Hahn (1965) placed both it and *Belgibole* in *Archegonus* as subgenera, and most subsequent authors have followed that classification. Osmólska (1970*a*, pp. 118, 123) pointed out the close similarities between *Weania, Gitarra,* and *Griffithidella,* and implied that all three might be closely related, possibly with origins in *Pseudowaribole octofer* or a similar species. The pygidial pleural rib structure (see above) of these three genera and the others listed above is basically different from that of *Archegonus* which is of scalloped profile (Owens 1973, p. 5), and it is only in broad glabellar features that these genera and *Belgibole* resemble *Archegonus.*

Intergeneric relationships within the Weaniinae are far from clear, and the most likely lines of derivation are suggested in text-fig. 1. However, they are united in sharing features listed in the diagnosis which serve to distinguish them from the Cyrtosymbolinae and from other proetid subfamilies.

Genus NIPPONASPIS Koizumi, 1972

Plate 2, fig. 8

Type species (original designation). *Nipponaspis takaizumii* Koizumi, 1972.

Holotype. Almost complete internal mould, UTIEA F9004, Kazanian, *Yabeina* Zone, Takakura-yama, north-west of Tamayama Mineral Spring, Yotsukura, Iwaki City, Fukushima Prefecture, Japan.

Species. N. leonensis (Romano, 1971), Carboniferous, Silesian, Westphalian A, northern León, north-west Spain; *N. miosilus* Chamberlain, 1977, Silesian, Ellesmere Island, Canada.

Diagnosis. Glabella fiddle-shaped, constricted across γ–γ, broader posteriorly than anteriorly; 3 pairs of narrow, oblique glabellar furrows of which S1 is longest and defines ovate L1; thorax of 9 segments; pygidium subparabolic without border, 10–11 axial rings defined by shallow ring furrows, pleural areas with 7–8 ribs.

Remarks. Koizumi (1972) compared *Nipponaspis* with *Linguaphillipsia* species, and Hahn and Hahn (1975, p. 53) placed it in the Linguaphillipsiinae. Although the glabella of *Linguaphillipsia* is broadly similar in shape to that of *Nipponaspis,* that of the former is pear-shaped with a parallel-sided frontal lobe and the lateral glabellar furrows tend to be broader and deeper; its pygidium has a border and is proportionately much longer with many more axial rings and pleural ribs, whose structure is quite different, they have a scalloped profile in contrast to the posterior pleural band elevated above the anterior in *Nipponaspis* (not always evident in internal moulds such as the one illustrated (Pl. 2, fig. 8)).

In addition to the type species, Hahn and Hahn (1975, p. 53) included *Gitarra leonensis* Romano, 1971 in *Nipponaspis,* which is similar to *N. takaizumii* in all aspects. The type species of *Gitarra, G. pupuloides* (Leyh, 1897) (see Gandl 1968, pl. 8, figs. 1–7) and *Schizophillipsia, S. yukisawensis* (Kobayashi and Hamada, 1980*b*, pl. 11, figs. 4, 5; pl. 12, figs. 1–14; pl. 13, figs. 3–13), like *Nipponaspis* have fiddle-shaped glabellas with corresponding patterns of lateral glabellar furrows, similar shaped, but broader (trans.) palpebral lobes, and a comparable pygidial outline. They differ in having deeper pygidial pleural and interpleural furrows, with the posterior edge of the posterior pleural band strongly elevated above the succeeding anterior pleural band. In pygidial characters *Gitarra* and *Schizophillipsia* are like the contemporaneous *Carbonocoryphe*, and common ancestry for these genera is probable. It is likely that *Nipponaspis* is derived from *Gitarra* or *Schizophillipsia*, and the broad similarity of these genera to *Linguaphillipsia* is likely to be due to cephalic homoeomorphy.

Genus ENDOPS Koizumi, 1972

Plate 2, fig. 13

Type species (original designation). *Paladin yanagisawai* Endo and Matsumoto, 1962.

Holotype. Complete specimen, IGPS 86672 A/B (Pl. 2, fig. 13), Kazanian, *Yabeina* Zone, Takakura-yama, north-west Tamayama Mineral Spring, Yotsukura, Iwaki City, Fukushima Prefecture, Japan.

Diagnosis. Glabella forward-expanding with small, depressed, isolated L1; frontal lobe of glabella separated from anterior margin by deep, narrow border furrow, thorax of 10 segments; pygidium with 8–9 axial rings; 6 pairs of pleural ribs.

Remarks. Koizumi (1972) compared *Endops* with *Thaiaspis* Kobayashi which has a similar shaped glabella, as well as with *Paladin* Weller and *Ditomopyge* Newell. Hahn and Hahn (1975, p. 65) placed *Endops* in the Ditomopyginae, but did not compare it closely with other genera. The pygidial rib structure is similar in many respects to *Griffithidella* Hessler, 1965, especially to *G. krasnopolskii* (Weber) from the Dinantian of Kazakhstan, U.S.S.R. (see Osmólska 1970a, pl. 1, figs. 5, 11). This and other *Griffithidella* species have similar glabellar shape, with small L1 and deep preglabellar furrow, to *Endops*. *Griffithidella*, like *Endops*, also has 10 thoracic segments. It is likely, therefore, that *Endops* has a common origin with *Griffithidella*.

Genus DOUBLATIA Wass and Banks, 1971

Plate 2, figs. 9–12

Type species (original designation), *Doublatia inflata* Wass and Banks, 1971.

Holotype. Complete specimen, (thorax disarticulated), SUP 12929A, B (Pl. 2, figs. 9, 10), Mid-upper Artinskian, Branxton Formation, *Fenestella* Shale, Mulbing, 16 miles west of Newcastle, New South Wales, Australia.

EXPLANATION OF PLATE 2

Figs. 1–4. *Kathwaia capitorosa* Grant, 1966, Salt Range, Pakistan. Holotype, enrolled specimen with dorsal exoskeleton preserved (USNM 145320). 1, cephalon, 2, lateral view, 3, pygidium, and 4, anterior view, ×4. (Orig. Grant 1966, pl. 13, fig. 1a–d.)

Figs. 5–7. *Hildaphillipsia hildae* (Gheyselinck, 1937), south central Timor. Lectotype (GIA B1379), enrolled specimen with exoskeleton preserved (resin cast, NMW 82.19G.4). 5, cephalon, 6, pygidium, and 7, lateral view. ×5. (Orig. Gheyselinck 1937, pl. 1, fig. 1a, b.)

Fig. 8. *Nipponaspis leonensis* (Romano, 1971), Silesian, Westphalian A, northern León, north-west Spain. Holotype, complete internal mould (BM It8794), ×2·5. (Orig. Romano 1971, pl. 1, fig. 1A; pl. 2, figs. 1, 2.)

Figs. 9, 10. *Doublatia inflata* (Wass and Banks 1971), Mulbing, New South Wales, Australia. Holotype (SUP 12929A, B). 9, dorsal exoskeleton, internal mould, ×2·5 (orig. Wass and Banks 1971, pl. 36, fig. 1). 10, internal mould of free cheek associated with same specimen, ×3·5. (Orig. Wass and Banks 1971, pl. 36, fig. 4.)

Figs. 11, 12. *Doublatia pyriforme* (Wass and Banks 1971), Ray's Hill (fig. 11) and Elephant Pass (fig. 12), Tasmania, Australia. 11, paratype cranidium (UT 90143) (latex cast of external mould), dorsal view, ×4. (Orig. Wass and Banks 1971, pl. 37, fig. 13.) 12, paratype pygidium (UT 55297b) (latex cast of external mould), dorsal view, ×4. (Orig. Wass and Banks 1971, pl. 37, figs. 6, 7.)

Fig. 13. *Endops yanagisawai* (Endo and Matsumoto 1962), Fukushima Prefecture, Japan. Holotype (IGPS 86672 A/B), dorsal view, silicone cast of external mould, NMW 82.19G.5, ×3. (Orig. Endo and Matsumoto 1962, pl. 9, fig. 1a–c; Koizumi 1972, pl. 1, fig. 4a, b.)

PLATE 2

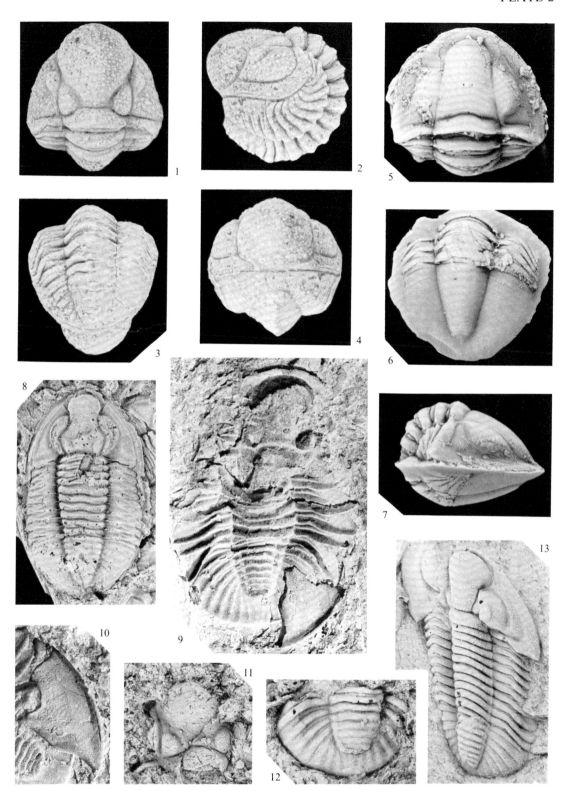

OWENS, Permian trilobites

Species. D. matheri Engel and Laurie, 1978, Lower Permian, Colraine Mudstone, near Wingham, New South Wales, Australia; *D. pyriforme* Wass and Banks, 1971, Sakmarian/Artinskian, basal beds of Enstone Park Limestone, Tasmania, Australia.

Diagnosis. Glabella parallel-sided or with weak forward taper; L1 prominent, ovate or subquadrate; S2, S3 shallow, transverse; thorax with 9 segments; pygidium without border; axis about two-thirds pygidial length with 7–8 rings; pleural areas with 6–7 pairs of pleural ribs.

Remarks. Wass and Banks discussed the possible affinities of *Doublatia* at some length. They concluded that the cephalon most closely resembled *Ditomopyge* and that the pygidium was similar to, among others, *Weania goldringi* Campbell *in* Campbell and Engel 1963 from the early Dinantian of eastern Australia. They postulated that *Doublatia* might have had a long, comparatively isolated history in eastern Australia. In overall aspect *Doublatia* does closely resemble *Weania* and all its characters suggest that it is a weaniine, although it more closely resembles Carboniferous members of this subfamily than it does the contemporaneous *Endops* and *Nipponaspis*.

?Subfamily WEANIINAE subfam. nov.
Genus MICROPHILLIPSIA Ruggieri, 1959

Plate 3, figs. 1–5

Type species (original designation), *Microphillipsia tetraptera* Ruggieri, 1959.

Holotype. Cephalon, MGP P/50051, Kazanian, Pietra di Salomone, Sosio Valley, Sicily.

Species. M.? parvula (Beyrich, 1865), Kazanian, Timor, Indonesia.

Diagnosis. Glabella flask-shaped, weakly constricted at mid-length; S1 partially isolates subquadrate L1; S2, S3 represented by smooth muscle insertion areas; occipital furrow arched strongly forwards, so that occipital ring is twice as long sagittally as laterally; fixed cheek broad between axial furrow and palpebral lobe; eye long, narrow, banana-shaped, well out from axial furrow; cephalic margin angular, with outer part of cephalic border reclined and not visible in dorsal view; short genal spine; pygidium without border furrow, semicircular; axis with 9 rings; pleural areas with 7 pairs of ribs with broader, elevated posterior pleural band, and narrower anterior pleural band; sculpture granular.

Remarks. Ruggieri (1959) questionably associated a small pygidium with cephala of *M. tetraptera*. Another pygidium of a similar type (Pl. 2, figs. 4, 5) from Sosio shows sculpture like that of the cephalon, and since it is associated with no other trilobite, it seems reasonable to assume that it belongs to *Microphillipsia*.

The relationships of *Microphillipsia* are not clear. Hahn and Hahn (1975) placed it in the Ditomopyginae, but apart from the superficial resemblance of the glabella to *Paladin*, there are no further points of close resemblance to ditomopygines. The structure of the cephalic border recalls that of *Weania*, although the eye and glabellar shape are rather different. The pygidia questionably assigned to *Microphillipsia* have a pleural rib structure like that of *Weania* and other weaniines. On the basis of these comparisons, it is possible that *Microphillipsia* is a weaniine, and on present evidence this subfamily seems the most likely for inclusion of the genus.

Subfamily CUMMINGELLINAE Hahn and Hahn, 1967

Diagnosis. Glabella broadly hourglass-shaped, anterior and posterior width (tr.) usually subequal, but either may be wider; lateral glabellar furrows incised or represented by smooth muscle insertion areas; free cheek narrow; thorax of 9 segments; pygidium with or without distinct border, and broad axis with 7–15 rings and 5–8 pleural ribs of scalloped profile.

Range. Carboniferous (Dinantian) to Permian (Dzhulfian).

Permian genera. Paraphillipsia Tumanskaya, 1935.

Remarks. Hahn and Hahn (1967, 1972, 1975) included *Ameura* Weller, 1936 which has Carboniferous and Permian species in the Cummingellinae on account of its glabellar outline. However, the pygidium is typically ditomopygine, and it is likely that the 'cummingelline' glabellar outline is an expression of homoeomorphy.

Genus PARAPHILLIPSIA Tumanskaya, 1935

Plate 3, figs. 6–13, 16, 17

Type species (original designation). *Paraphillipsia karpinskyi* Tumanskaya, 1935.

Holotype. Enrolled specimen, CGM 59/9733 (Pl. 3, fig. 6), Upper Artinskian, Salghir River, near Simpheropol, Crimea, U.S.S.R.

Species. P. inflata Kobayashi and Hamada, 1979, Upper Sakmarian, Rat Buri Limestone, Changwat Loei, northern Thailand; *P. kussicum* Tumanskaya, 1935, Upper Artinskian, Marta River, Crimea; *P. levigata* Kobayashi and Hamada, 1980c, 'Middle Permian', Shimoyama Limestone, Shikoku Island, west Japan; *P. middlemissi* (Diener, 1897), Dzhulfian, Himalayas; *P. netschaevi* Tumanskaya, 1935, Upper Artinskian, Crimea; *P. pahara* Weller, 1935, 'Permian', Tibet; *P. tauricum* Tumanskaya, 1935, Upper Artinskian, Crimea; *P. tschernyshevi* (Netschaev *in* Weber, 1932), Asselian, Turkestan, U.S.S.R.; *P. v-n-weberi* Tumanskaya, 1935, Upper Artinskian, Crimea; *P.* sp. ('*Neoproetus* sp.' of Hahn, Hahn and Ramovš, 1970, pl. 1, fig. 3), Artinskian, Slovenia, Yugoslavia.

Diagnosis. Glabella weakly constricted at mid-length, as wide or a little wider anteriorly than posteriorly; glabellar furrows weak or non-incised, S1 defining trapezoidal L1; eye small, opposite narrowest point of glabella; anterior glabellar lobe fused with anterior border; pygidium with smooth border area; axis with 7–8 rings; pleural areas with 5–6 pairs of ribs; dorsal surface smooth or granulose.

Remarks. Paraphillipsia bears a general resemblance to *Cummingella* Reed, a typically Dinantian genus, whose stratigraphical range extends to Westphalian C (Kobayashi and Hamada 1980b), both in cephalic and pygidial morphology. It differs principally in having a distinct 'waist' at mid-length of the glabella, a subquadrate L1, and a smaller number of pygidial axial rings and pleural ribs. Examination of casts of the type species of *Humilogriffithides, H. divinopleurus* Inai 1936, from the late Silesian of Manchuria, China has shown that it is also similar to *Paraphillipsia* and *Cummingella* species; and *Particeps productus* (Weber) (Osmólska 1970b, pl. 15, figs. 1, 4, 6, 12) and *P. kargini* (Weber) (Osmólska 1970b, pl. 15, figs. 3, 7) from the Namurian of the Urals and Donetz Basin respectively, show particularly close cranidial similarity to *Paraphillipsia. Particeps* Reed probably has its origins in *Cummingella*, and it is presumed that *Paraphillipsia* has its origins in the *Cummingella–Humilogriffithides–Particeps* complex.

Subfamily DITOMOPYGINAE Hupé, 1953

Diagnosis. Glabella forward expanding, frontal lobe coalesced with anterior border or separated from it by anterior border furrow; L1 normally prominent and defined by deep S1; S2, S3 short, inconspicuous; preoccipital lobe commonly present; thorax of 9 segments with preannulus; pygidium typically elongated with well-defined border; axis normally narrow and elongated with 8–27 rings; pleural areas with deep, incised pleural furrows, interpleural furrows normally very weak and running along highest point of each rib, their position sometimes indicated only by a row of fine granules. Pygidium commonly mucronate.

Permian genera. Acropyge Qian, 1977; *Anisopyge* Girty, 1908; *Delaria* Weller, 1944; *Iranaspidion* Kobayashi and Hamada, 1978; *Pseudophillipsia* Gemmellaro, 1892; *Timoraspis* Hahn and Hahn, 1967; *Vidria* Weller, 1944. The type species of *Ditomopyge* Newell, 1931, *Ameura* Weller, 1936, and *Paladin* Weller, 1936 are of Carboniferous age. All three genera have Permian representatives. The range of *Pseudophillipsia* extends back into the Silesian.

Genus DITOMOPYGE Newell, 1931

Plate 3, figs. 14, 15, 18–20; Plate 4, figs. 1, 2

Synonyms. Cyphinium Weber, 1933; *Permoproetus* Tumanskaya, 1935; *Neophillipsia* Gheyselinck, 1937.

Type species. Phillipsia (Griffithides) scitulus Meek and Worthen, 1865, Carboniferous (late Silesian), Springfield, Illinois, U.S.A. (see Pabian and Fagerstrom 1972, p. 808).

Permian species. Ditomopyge amorni Kobayashi and Hamada, 1979, upper Sakmarian, Changwat Loei, northern Thailand; *D. artinskiensis* (Weber, 1933), Artinskian, north Urals, U.S.S.R.; *D. bjornensis* Ormiston, 1973, probably Lower Artinskian, Ellesmere Island; *D. decurtata* (Gheyselinck, 1937), Wolfcampian, Kansas, U.S.A.; *D. fatmii* Grant, 1966, late Guadalupian–early Dzhulfian, Salt Range, Pakistan; *D. gortanii* (Tumanskaya, 1935), Upper Artinskian, Crimea, U.S.S.R.; *D. meridionalis* Teichert, 1944, Artinskian, west Australia; *D. netchaevi* (Weber, 1932), lower Artinskian, Pamir, Turkestan, U.S.S.R.; *D. postcarbonarius* (Gemmellaro, 1892), Upper Artinskian, Sosio Valley, Sicily; *D. proetoides* (Mansuy, 1913), 'Lower Permian', Kham-Kheut, Vietnam; *D. sylvensis* (Weber, 1944), Artinskian, west Urals, U.S.S.R.; *D. teschi* (Tumanskaya, 1935), Upper Artinskian, Crimea, U.S.S.R.; *D. whitei* (Pabian and Fagerstrom, 1972), Lower Permian, Big Blue Series, Nebraska, U.S.A.; *D. yungchangensis* Wang, 1937, probably Lower Permian, Kansu Province, China.

EXPLANATION OF PLATE 3

Figs. 1–5. *Microphillipsia tetraptera* Ruggieri, 1959. Sosio Valley, south of Palermo, Sicily. 1–3, cephalon with dorsal exoskeleton preserved (BM It1415); 1, dorsal, 2, anterior, and 3, anterior oblique views, × 7. 4, 5, pygidium with dorsal exoskeleton preserved, probably belonging to this species (BM It1414); 4, dorsal and 5, lateral views, × 9.

Figs. 6, 7, 11, 12, 17. *Paraphillipsia karpinskyi* Tumanskaya, 1935. 6, holotype (CGM 59/9733), enrolled specimen with dorsal exoskeleton preserved, dorsal view of cephalon, × 3. (Orig. Tumanskaya 1935, pl. 2, figs. 1–7.) Salghir River, Crimea, U.S.S.R. (Photograph supplied by Dr. L. Popov.) 7, incomplete cephalon, mostly exfoliated (CGM 62/5217), dorsal view, × 4·5. (Orig. Weber 1944, pl. 1, fig. 22.) East of mouth of Urushten River, north Caucasus, U.S.S.R. (Photograph supplied by Dr. L. Popov.) 11, 12, pygidium with dorsal exoskeleton preserved (CGM 63/5217); 11, dorsal and 12, lateral views, × 4 (plaster cast, NMW 82.19G.6). (Orig. Weber 1944, pl. 1, fig. 19.) Malaia Laba River basin, north Caucasus, U.S.S.R. 17, cranidium with dorsal exoskeleton preserved (CGM 61/5217), dorsal view, × 3 (plaster cast, NMW 82.19G.7). (Orig. Weber 1944, pl. 1, fig. 18.) Urushten, north Caucasus, U.S.S.R.

Figs. 8, 9. *Paraphillipsia tschernyshewi* (Netschaev *in* Weber, 1932). 8, complete specimen, with dorsal exoskeleton preserved (CGM 259/349), dorsal view, × 3·5 (plaster cast, NMW 82.19G.8). (Orig. Netschaev *in* Weber 1932, pl. 3, fig. 31a–e.) Darvaz, Tangigor Gorge, near Safed-Daron, Turkestan, U.S.S.R. 9, cephalon partially exfoliated (CGM 60/5217), dorsal view, × 4·5 (plaster cast, NMW 82.19G.9). (Orig. Weber 1944, pl. 1, fig. 17.) Ferghana, Turkestan, U.S.S.R.

Figs. 10, 16. *Paraphillipsia v-n-weberi* Tumanskaya, 1935, Mount Kichkhi-Burnu, Crimea, U.S.S.R. 10, lectotype cephalon partially exfoliated (CGM 60/9733), dorsal view, × 3 (plaster cast, NMW 82.19G.10). Orig. Tumanskaya 1935, pl. 3, figs. 1–3.) 16, paralectotype pygidium with dorsal exoskeleton preserved (CGM 61/9733), dorsal view, × 3 (plaster cast, NMW 82.19G.11). (Orig. Tumanskaya 1935, pl. 3, figs. 4, 5.)

Fig. 13. *Paraphillipsia pahara* Weller 1935, Tibet. Cranidium, partially exfoliated (lectotype, here selected; YPM 15027A), dorsal view, × 4. (Orig. Weller 1935, fig. 1.)

Figs. 14, 15, 18–20. *Ditomopyge teschi* (Tumanskaya, 1935), Marta River, Crimea, U.S.S.R. 14, pygidium, internal mould (CGM 122/9733), dorsal view, × 3·5. (Orig. Tumanskaya 1935, pl. 10, fig. 13: syntype of *Permoproetus teschi*.) (Photograph supplied by Dr. L. Popov.) 15, pygidium, internal mould CGM 118/9733), dorsal view, × 4 (plaster cast, NMW 82.19G.12). (Orig. Tumanskaya 1935, pl. 10, fig. 9: syntype of *Permoproetus teschi*.) 18, pygidium and part of thorax, internal mould (CGM 117/9733), dorsal view, × 5 (plaster cast, NMW 82.19G.13). (Orig. Tumanskaya 1935, pl. 10, figs. 6, 7: syntype of *Permoproetus teschi*.) 19, 20, cranidium, internal mould (CGM 115/9733); 19, dorsal and 20, lateral views, × 4·5 (plaster cast, NMW 82.19G.14). (Orig. Tumanskaya 1935, pl. 10, figs. 3, 4: syntype of *Permoproetus teschi*.)

Fig. 21. *Ditomopyge gortanii* Tumanskaya, 1935, Marta River, Crimea, U.S.S.R. Cranidium, partially exfoliated (CGM 125/9733), dorsal view, × 4 (plaster cast, NMW 82.19G.15). (Orig. Tumanskaya 1935, pl. 11, figs. 1, 2: syntype of *Permoproetus gortanii*.)

PLATE 3

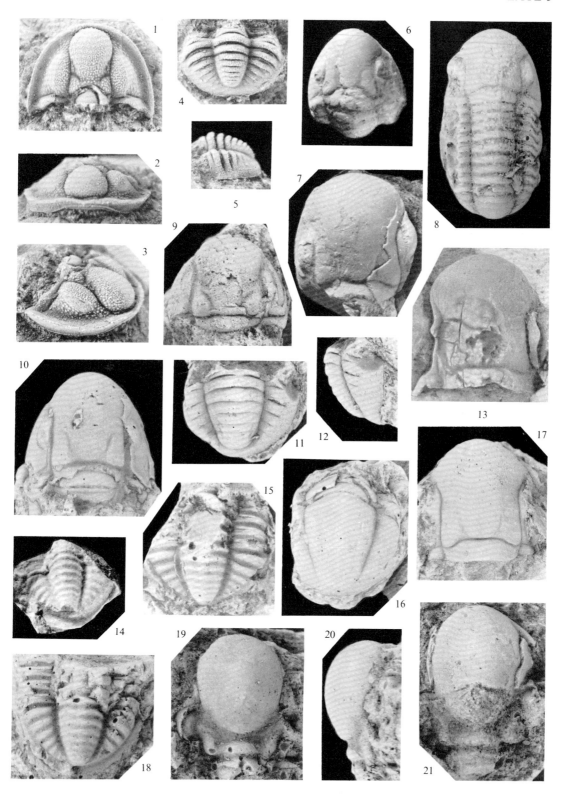

OWENS, Permian trilobites

Diagnosis. Ditomopygine with preoccipital lobe, pygidium with well-defined border, non mucronate, but with paired spines in immature specimens, with 8–18 axial rings and 7–12 pleural ribs.

Remarks. *Permoproetus* Tumanskaya, 1935, type species *P. teschi* (Pl. 3, figs. 14, 15, 18–21), from the Upper Artinskian of the Crimea, U.S.S.R., closely resembles *Ditomopyge*, differing only in having smaller eyes and palpebral lobes and rather fewer pygidial axial rings (8–9) and pleural ribs (6) than in most *Ditomopyge* species. There is, however, a range in eye size and in numbers of axial rings and pleural ribs (see diagnosis) which merge with those of *Permoproetus*, and there is no clear separation at generic level. *Permoproetus* is therefore regarded as being congeneric with *Ditomopyge*.

<h2 style="text-align:center">Genus PSEUDOPHILLIPSIA Gemmellaro, 1892</h2>

Plate 4, figs. 3–6, 10

Type species (original designation). *Phillipsia sumatrensis* Roemer, 1880.

Type specimens. Roemer's syntypes were housed in the Geological Institute of the University of Wrocław (formerly Breslau), Poland. It is not known whether this material still survives (Hahn *et al.* 1970, p. 324). Additional material, possibly syntypes, is preserved in the Senckenberg Museum, Frankfurt am Main. All from Kazanian, west coast of Sumatra, Indonesia.

Permian species. *P. acuminata* Mansuy, 1912, 'Middle Permian', Luang-Prabang area, Laos; *P. anshunensis* Qian, 1977, Upper Permian (Longtan), Anshun, Guizhou Province, China; *P. armenica* Weber, 1944, 'Permian', Armenia, U.S.S.R.; *P. borissiaki* Tumanskaya, 1935, Upper Artinskian, Crimea, U.S.S.R.; *P. chongqingensis* Lu, 1974, southwestern China; *P. elegans* Gemmellaro, 1892, Upper Artinskian, Sosio Valley, Sicily; *P. gemmellaroi* Canavari *in* Greco 1935, Upper Artinskian, Sosio Valley, Sicily; *P. heshanensis* Qian, 1977, Upper Permian, Guizhou Province, China; *P. hungarica* (Schréter, 1948), 'Middle Permian', Nagyvisnyó, Bükk Mountains, Hungary; *P.? imbricatus* (Kobayashi and Hamada 1980c), 'Middle Permian', Shiga Prefecture, Honshu, Japan; *P. intermedia* Kobayashi and Hamada, 1980c, 'Middle Permian', Gifu Prefecture, Honshu, Japan; *P. lipara* Goldring, 1957, probably early Permian, Haushi, Oman; *P. mustafensis* Tumanskaya, 1935, Upper Artinskian, Crimea, U.S.S.R.; *P. obtusicauda* (Kayser, 1883), Tatarian, Kiangsi Province, South China; *P. œhlerti* (Gemmellaro, 1892), Upper Artinskian, Sosio Valley, Sicily; *P. paffenholzi* (Weber, 1944), 'Upper Permian', Armenia, U.S.S.R.; *P. pyriformis* Qian, 1977, Upper Permian (Dalong), Qinglong, Guizhou Province, China; *P. qinglongensis* Qian, 1977, Upper Permian (Dalong), Qinglong, Guizhou Province, China; *P. solida* Weber, 1944, Upper(?) Permian, North Caucasus, U.S.S.R.; *P. sosiensis* (Gemmellaro, 1892), Upper Artinskian, Sosio Valley, Sicily; *P. spatulifera* Kobayashi and Hamada, 1980d, 'late Middle Permian', Miyagi Prefecture, Honshu, Japan; *P. steatopyga* Goldring, 1957, probable early Permian, Wadi Lusaba, Oman; *P. subcircularis* Qian, 1977, Upper Permian (Longtan), Anshun, Guizhou Province, China; *P. timorensis* (Gheyselinck, 1937), Kazanian, Besleo, Timor.

Diagnosis. Ditomopygine with glabella separated from anterior border; preoccipital lobe and L1 isolated from remainder of glabella by deep, broad, furrows; L2–L4 commonly represented by ovate, raised areas; pygidium with narrow border; axis with 17–27 rings; pleural areas with 9–17 pairs of narrow (exsag.) ribs; terminal ventral sagittal septum present on pygidial axis.

Remarks. *Pseudophillipsia* has been discussed in detail by Goldring (1957), and more recently by Hahn and Brauckmann (1975a), who divided it into two subgenera, the nominate one, and *Carniphillipsia* (type species *P. ogivalis* Gauri, 1965). *Carniphillipsia* differs in lacking the raised, ovate L2–L4, in having shallower, narrower furrows between L1 and the median preoccipital lobe, a less forwardly expanded glabella, fewer pygidial axial rings (17–21), and pleural ribs (9–13). In it Hahn and Brauckmann included all the Carboniferous species of *Pseudophillipsia* as well as *P. lipara* and *P. steatopyga* of probable early Permian age. They also included *P. oehlerti* and *P. sosiensis* with doubt. Other species with a similar morphology include *P. chongqingensis*, *P. heshanensis*, and *P. qinglongensis*. Hahn and Brauckmann (1975a, p. 118) suggested a *Ditomopyge-Carniphillipsia-Pseudophillipsia* lineage; *Carniphillipsia* has characters of both genera—in the cephalon like *Ditomopyge* and in the pygidium like *Pseudophillipsia*, and could equally well be included as a subgenus of either genus, or possibly regarded as a genus in its own right. For the present it is included in *Pseudophillipsia*, using the acquisition of a very elongated, parabolic pygidium as an (arbitrary) dividing point between *Pseudophillipsia* and *Ditomopyge*.

In glabellar characteristics and in the number of pygidial pleural ribs and axial rings *P.* (*Carniphillipsia*) is like *Iranaspidion* Kobayashi and Hamada, but the latter differs in having a bilobate L1 and a weak median sulcus on the glabella immediately in front of the median preoccipital lobe. It is likely that *Iranspidian* has its origins in *P.* (*Carniphillipsia*). *Iranaspidian* and members of both subgenera of *Pseudophillipsia* have a median slit at the posterior end of the pygidial axis (see Pl. 4, figs. 4, 5, 11; Kobayashi and Hamada 1978, fig. 4c; Goldring 1957, pl. 1, fig. 7c); this is evident on the dorsal surface in the former and on internal moulds of the latter. Ventrally the slit is expressed as a median septum in *P.* (*C.*) *lipara* (Goldring 1957, p. 199). Its function is unknown.

Genus IRANASPIDION Kobayashi and Hamada, 1978

Plate 4, figs. 11–14

Type species (original designation). *Iranaspidion sagittalis* Kobayashi and Hamada, 1978.

Holotype. Enrolled specimen (Pl. 4, figs. 12, 13), Upper Guadalupian, southwestern part of Kuh-e-hambast Range, Abadeh region, central Iran.

Diagnosis. Ditomopygine with L1 divided into two nodes; L2–L4 not isolated and upstanding; weak sulcus on glabella anterior to median preoccipital lobe; pygidium with 20 axial rings and 11 pairs of pleural ribs; small median slit at extreme posterior end of pygidial axis.

Remarks. Kobayashi and Hamada (1978, p. 157) noted that this genus is allied to *Pseudophillipsia* and *Ditomopyge*. Its close similarity to *P.* (*Carniphillipsia*) is discussed above. Whether the distinguishing features are really of generic significance is doubtful, and there is a strong case for its inclusion in *P.* (*Carniphillipsia*).

Genus ACROPYGE Qian, 1977

Plate 4, fig. 15

Type species (original designation). *Acropyge multisegmenta* Qian, 1977.

Holotype. Pygidium (NIGP) (Pl. 4, fig. 15), late Permian (Dalong), Qinglong, Guizhou Province, People's Republic of China.

Species. *A. brevica* Yin, 1978, Permian, Guizhou Province, People's Republic of China; *A. encrinuroides* (Weber, 1944), Permian, Armenia, U.S.S.R.; *A. lanceolata* Kobayashi and Hamada, 1978, Upper Guadalupian, southwestern part of Kuh-e-hambast Range, Abadeh region, central Iran; *A. weggeni* Hahn and Hahn, 1981, highest Upper Permian, central Elburz Mountains, northern Iran.

Diagnosis. Cephalon like *Pseudophillipsia*; pygidium elongate triangular, with narrow border; axis with 20–28 rings, pleural areas with 11–14 pairs of narrow (exsag.) ribs; posterior margin extended into point.

Remarks. Yin (1978, pl. 141, fig. 12) figured a cephalon which he ascribed to *A. brevica*; it is, according to Hahn and Hahn (1981, p. 218), like that of *Pseudophillipsia*. The pygidium broadly resembles that of *Anisopyge*, but the structure of the pleural ribs and axis is different; apart from the pointed posterior margin, it is not unlike *Pseudophillipsia*, to which it is evidently most closely related.

Genus TIMORASPIS Hahn and Hahn, 1967

Plate 4, figs. 7–9; Plate 5, fig. 1

Type species (original designation). *Griffithides breviceps* Gheyselinck, 1937.

Holotype. Enrolled specimen, GIA B1388 (Pl. 4, figs. 7–9), Basleo Beds, Kazanian, south central Timor, Indonesia.

Species. A pygidium from the Upper Artinskian, Sosio Valley, Sicily (Pl. 5, fig. 1) compares closely with that of *Timoraspis breviceps* and represents a second species of *Timoraspis*.

Diagnosis. Cephalon like *Ditomopyge*, but with L1 and preoccipital lobe depressed and more or less coalesced; L1 with weak transverse depression; pygidium with 13–14 axial rings, pleural areas with 10 pairs of weakly defined ribs, pleural and interpleural furrows both shallow.

Remarks. Timoraspis is obviously closely related to *Ditomopyge*, differing only in being somewhat effaced and with the preoccipital lobe and L1 depressed. The reconstructions of Gheyselinck (1937, p. 88) and Hahn and Hahn (1975, pl. 12, figs. 8*a*, *b*) exaggerate the very shallow transverse depression on L1, and show incorrectly the posterior margin of the pygidium to be acuminate rather than rounded.

Genus ANISOPYGE Girty, 1908

Plate 5, figs. 2–5, 10, 14

Type species (original designation). *Phillipsia perannulata* Shumard, 1858.

Neotype (Cisne 1971, p. 525). Pygidium, USNM 118168 (Pl. 5, fig. 10), Guadalupian, Capitan Limestone, Guadalupe Mountains, Texas, U.S.A.

Species. Anisopyge inornata Girty, 1909, Yeso Formation (earlier Leonardian), Mesa del Yeso, New Mexico, U.S.A.; *A. hyperbola* Chamberlain, 1972, Chocal Limestone (Leonardian), Huehuetenango, Guatemala.

Diagnosis. Glabella with preoccipital lobe; anterior border furrow narrow and clearly defined; S1–S3 broad, rather shallow, defining small, ovoid slightly raised L2 and L3 of independent convexity from remainder of glabella; pygidial axis with 19–24 narrow (sag., exsag.) rings; pleural areas with 7–9 pairs of ribs on which interpleural furrows are hardly apparent; pleural furrows deep and broad; pygidial border vertical or declined.

Remarks. The type species of *Anisopyge*, *A. inornata* and *A. hyperbola*, are distinctive in having pygidia with a large number of axial rings and comparatively small number of pleural ribs. Other

---

Figs. 1, 2. *Ditomopyge decurtata* (Gheyselinck, 1935), near Wichita, Kansas, U.S.A. Enrolled specimen, with dorsal exoskeleton preserved (USNM 145322). 1, cephalon and 2, pygidium, × 3. (Orig. Grant 1966, pl. 13, fig. 4a–c.)

Figs. 3–6. *Pseudophillipsia sumatrensis* (Roemer, 1880), Sumatra, Indonesia. 3, cephalon, partially exfoliated (SMF 24224), dorsal view, × 2 (plaster cast, NMW 82.19G.16). (Orig. Hahn *et al.* 1970, pl. 1, fig. 1.) 4, pygidium, internal mould (SMF X738a), dorsal view, × 2 (plaster cast, NMW 82.19G.17). 5, 6, pygidium, partially exfoliated (SMF X2078); 5, dorsal and 6, lateral views, × 2·5 (plaster cast, NMW 82.19G.18). (Orig. Hahn *et al.* 1970, pl. 1, fig. 2.)

Figs. 7–9. *Timoraspis breviceps* (Gheyselinck, 1937), south central Timor. Holotype enrolled specimen, with dorsal exoskeleton preserved (GIA B1388). 7, cephalon, 8, lateral view, and 9, pygidium, × 3 (resin cast, NMW 82.19G.19). (Orig. Gheyselinck 1937, pl. 3, fig. 1.)

Fig. 10 *Pseudophillipsia* sp. Toenioen Eno, Timor. Pygidium, with dorsal exoskeleton preserved (BM It1239), dorsal view, × 2·5.

Figs. 11–14. *Iranaspidion sagittalis* Kobayashi and Hamada 1978, Kuh-e-hambast Range, central Iran. 11, 14, paratype pygidium, with dorsal exoskeleton preserved; 11, dorsal and 14, lateral views, × 3. (Orig. Kobayashi and Hamada 1978, fig. 4.) (Photographs supplied by Professor T. Kobayashi.) 12, 13, holotype, enrolled specimen, with dorsal exoskeleton preserved; 12, cephalon and 13, pygidium × 2·5. (Orig. Kobayashi and Hamada 1978, fig. 3.) (Photographs supplied by Professor T. Kobayashi.)

Fig. 15. *Acropyge multisegmenta* Qian, 1977, Guizhou Province, People's Republic of China. Holotype pygidium, internal mould, dorsal view, × 4. (Orig. Qian 1977, pl. 1, fig. 9.) (Photograph supplied by Y. Qian.)

PLATE 4

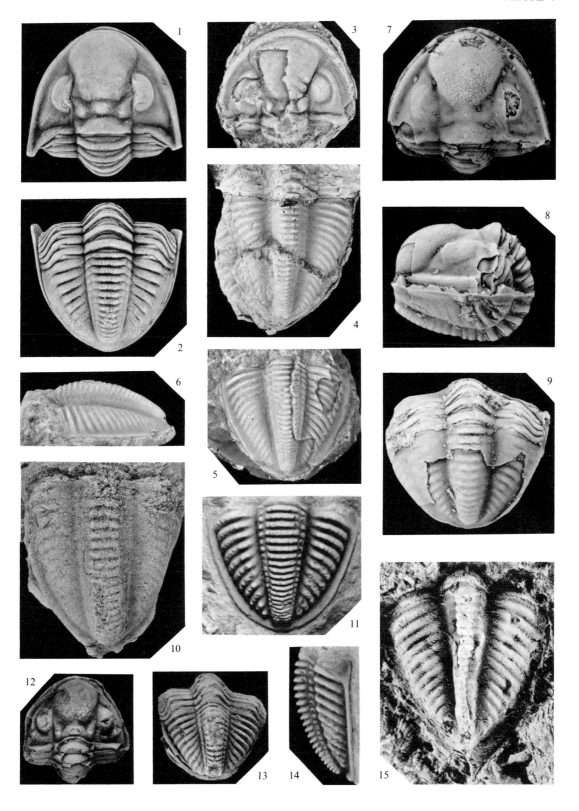

OWENS, Permian trilobites

species recently included in *Anisopyge* are *A. mckeei* Cisne, 1971, *A. whitei* Pabian and Fagerstrom, 1972, and *A. sevilloidia* Chamberlain, 1970. None of these has the same characteristic pygidium as the other three species. *A. mckeei* has both cephalic and pygidial characters more similar to those of *Ameura* Weller, while *Anisopyge sevilloidia* more closely resembles *Delaria* in these aspects. *A. whitei*, on the other hand, shows close affinities with *Ditomopyge* species. Chamberlain considered *Anisopyge* to have evolved from *Sevillia* via *A. sevilloidia*. Hahn and Hahn (1967) and Pabian and Fagerstrom (1972) believed that *Anisopyge* evolved from *Ditomopyge* and the latter (1972, p. 813,4) discussed a possible *D. scitula–A. whitei–A. perannulata* lineage. In the light of present knowledge, I consider this lineage the most likely. Since *A. whitei* is here considered to retain more *Ditomopyge* characters than it has acquired *Anisopyge* ones, I place it in the former genus.

Genus DELARIA Weller, 1944

Plate 5, figs. 6, 7, 11

Type species (original designation). *Anisopyge? antiqua* Girty 1908.

Lectotype (Cisne 1971, p. 526). Pygidium, USNM 118169 (Pl. 5, figs. 6, 7), Upper Leonardian, Bone Spring Limestone, Guadalupe Mountains, Texas, U.S.A.

Species. Delaria macclintocki Cisne, 1971, Kaibab Limestone (Upper Leonardian), west central U.S.A.; *D. snowi* Cisne, 1971, Kaibab Limestone, west central U.S.A.; *D. sevilloidia* (Chamberlain, 1970), Lower Leonardian–Lower Guadalupian, Wyoming, U.S.A.

Diagnosis. Glabella with preoccipital lobe; L1 small, ovate; S1, S2, S3 broad, very shallow, defining ovate, slightly raised L2 and L3; preglabellar furrow shallow; glabella impinges slightly on to anterior border; pygidium approximately semicircular, axis with 13–14 rings, 7–8 pleural ribs; border broad, downturned.

EXPLANATION OF PLATE 5

Fig. 1. *Timoraspis* sp., Sosio Valley, south of Palermo, Sicily. Pygidium partially exfoliated (BM It1428), dorsal view, ×4·5.

Figs. 2, 3, 4, 5, 10, 14. *Anisopyge perannulata* (Shumard, 1858), Guadalupe Mountains, Texas, U.S.A. 2, 3, pygidium, partially exfoliated (USNM 118168); 2, dorsal and 3, lateral views, ×3. 4, cranidium, with exoskeleton preserved (USNM 118168), dorsal view, ×2·5. (Orig. Cisne 1971, pl. 68, figs. 1, 2.) 5, incomplete free cheek, with exoskeleton preserved (USNM 118168), dorsal view, ×3. 10, neotype pygidium, with exoskeleton preserved (USNM 118168), dorsal view, ×4. (Orig. Weller 1944, pl. 49, fig. 6; Cisne 1971, pl. 68, figs. 1, 2.) 14, pygidium and part of thorax, with exoskeleton preserved (USNM 118168), dorsal view, ×4.

Figs. 6, 7, 11. *Delaria antiqua* Girty, 1908, Guadalupe Mountains, Texas, U.S.A. 6, 7, lectotype pygidium, with exoskeleton preserved (USNM 118169); 6, dorsal and 7, lateral views, ×4·5. (Orig. Weller 1944, pl. 49, fig. 2; Cisne 1971, pl. 68, figs. 24, 29.) 11, cranidium, partially exfoliated (USNM 118169), dorsal view, ×4. (Orig. Weller 1944, pl. 49, fig. 1; Cisne 1971, pl. 68, fig. 33.)

Figs. 8, 9. *Vidria vespa* (Weller, 1944, Glass Mountains, Texas, U.S.A. 8 holotype pygidium, silicified (USNM 109003), dorsal view, ×2. (Orig. Weller 1944, pl. 49, fig. 4.) (Photograph supplied by Mr. F. C. Collier.) 9, paratype pygidium, silicified (USNM 109003), dorsal view, ×3. (Orig. Weller 1944, pl. 49, fig. 5.) (Photograph supplied by Mr. F. C. Collier.)

Figs. 12, 13. *Paladin brevicauda* (Gheyselinck 1937), south central Timor. Enrolled specimen with exoskeleton preserved (DUT KA15623). 12, cephalon and 13, pygidium, ×3. (Orig. Gheyselinck 1937, pl. 4, fig. 1.)

Fig. 15. *Cheiropyge himalayensis* Diener, 1897, Chitichun no. 1, Himalayas, India. Pygidium with exoskeleton preserved (GSI 6068), dorsal view, ×3 (plaster cast, NMW 82.19G.20). (Orig. Diener 1897, pl. 1, fig. 2a–c.)

Fig. 16. *Paladin baungensis* (Gheyselinck, 1937), Baung, Timor. Cephalon with part of thorax, with exoskeleton preserved (DUT KA12804), dorsal view, ×1·75. (Orig. Tesch 1923, pl. 178, fig. 5; Gheyselinck 1937, pl. 2, fig. 1.)

Figs. 17, 18. *Brachymetopus kansasensis* (Weller, 1944), Lawrence, Kansas, U.S.A. Complete specimen (ISGS 3497). 17, cephalon and part of thorax (internal mould). 18, pygidium and part of thorax (external mould), ×6·5. (Orig. Weller 1944, pl. 49, fig. 9.)

PLATE 5

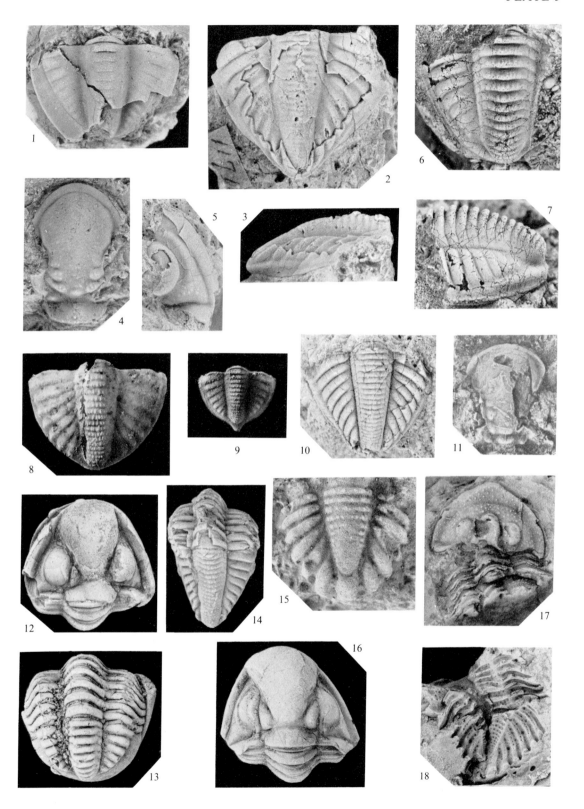

OWENS, Permian trilobites

Remarks. Because it has isolated L1–L3 and preoccipital lobe and a more or less semicircular pygidium, Chamberlain's *A. sevilloidia* is transferred from *Anisopyge* to *Delaria*.

Genus VIDRIA Weller, 1944

Plate 5, figs. 8, 9

Type species (original designation). *Vidria vespa* Weller, 1944.

Holotype. Pygidium, USNM 109003 (Pl. 5, fig. 8), Guadalupian, Word Formation, Glass Mountains, Texas, U.S.A.

Species. V.? *contendens* Cisne, 1971, Upper Leonardian, Kaibab Limestone, Arizona, U.S.A.

Diagnosis. Pygidium subtriangular; axis with 17–18 rings, overhanging posterior margin; pleural areas with 7 pairs of ribs, which broaden slightly abaxially.

Remarks. Weller (1944, p. 323) based the type species of *Vidria* on two pygidia, the smaller of which has a short mucro. Until the cephalon is discovered, the relationships of *Vidria* with other genera will remain uncertain, but on present evidence I agree with Cisne that a close relationship with *Delaria* is likely. Cisne (1971) included his species *contendens* with doubt in *Vidria*, pointing out a similar pleural rib pattern and outline to *V. vespa*. He also noted the similarity between *V.?* *contendens* and *Delaria macclintocki*.

Family BRACHYMETOPIDAE Prantl and Přibyl, 1951
Subfamily BRACHYMETOPINAE Prantl and Přibyl, 1951

Diagnosis. See Owens and Thomas 1975, p. 810.

Range. Silurian (Wenlock) to Upper Permian.

Genus BRACHYMETOPUS M'Coy, 1847

Plate 5, figs. 17, 18

Type species. Phillipsia maccoyi Portlock, 1843, Dinantian; Courceyan or Chadian Stage, Kildare, Ireland.

Permian species. Brachymetopus caucasicus Licharev *in* Weber, 1944, 'Upper Permian', Nikitina Ravine, North Caucasus, U.S.S.R.; *B. kansasensis* (Weller, 1944), Lower Permian, Haskell Limestone, Kansas, U.S.A.; *B.* sp., Wolfcampian, Alaska (described and figured by Chamberlain (1977, p. 763, pl. 1, figs. 9, 10) as *Cheiropyge himalayensis*).

Remarks. There are few brachymetopid species known from the Permian, and these have generally been included in *Cheiropyge* Diener. However, the three listed above have a different structure of the pygidial pleural ribs and postaxial region from *Cheiropyge*, but are similar to late Silesian species of *Brachymetopus*, e.g. *B. pseudometopina* Gauri and Ramovš, 1964, and are therefore included here in that genus.

Genus CHEIROPYGE Diener, 1897

Plate 5, fig. 15

Type species (by monotypy). *Cheiropyge himalayensis* Diener, 1897.

Holotype. Pygidium, GSI 6068 (Pl. 5, fig. 15), Tatarian, Chitichun Limestone, Chitichun no. 1, Himalayas, India.

Species. C. maueri Weber, 1944, Artinskian, Sylva River, western Urals, U.S.S.R.

Diagnosis. Pygidial axis with 12–14 rings defined by shallow ring furrows; postaxial area inflated, steeply declined; 6 pairs of pleural ribs defined by deep, broad pleural furrows; interpleural furrows shallow; margin denticulate.

Remarks. Cheiropyge broadly resembles *Brachymetopus* species with a denticulate or spinose pygidial margin. It differs from them in having more swollen pleural ribs in which the interpleural furrows are very much reduced, and in the steeply declined, inflated postaxial area. Until the cephalon of the genus is known, further comparison with *Brachymetopus* is not possible.

Genus LOEIPYGE Kobayashi and Hamada, 1979

Type species (original designation). *Loeipyge spinifer* Kobayashi and Hamada (1979, pl. 1, figs. 1a–d).

Holotype. Incomplete pygidium, Sakmarian, Rat Buri Limestone, Changwat Loei, northern Thailand.

Remarks. This genus is based upon one incomplete pygidium and therefore cannot be satisfactorily diagnosed. The specimen has a large number of narrow axial rings with short, median thorn-like spines at irregular intervals. There are at least 7 broad, granulose pleural ribs separated by broad, shallow pleural furrows; interpleural furrows are not evident, and the margin is entire. Kobayashi and Hamada (1979) compared it with various contemporaneous genera, and concluded that it was an 'aberrant off-shoot of the Griffithidinae'; they also noted a resemblance to *Brachymetopus*. The structure and ornamentation of the pleural ribs and axis suggest that *Loeipyge* is a brachymetopid. The pleural ribs with effaced interpleural furrows are like those of *Cheiropyge*, but *Loeipyge* differs in having an entire margin and a much less steeply delined postaxial area. These differences are possibly not of generic significance, but further assessment must await more complete material of both genera.

PHYLOGENY AND DISTRIBUTION

Phylogeny. Some remarks about inferred relationships are given under each genus, and the phylogeny proposed is summarized in text-fig. 1. Five proetid subfamilies (Proetinae, Cyrtosymbolinae, Weaniinae, Cummingellinae, and Ditomopyginae) and one brachymetopid (Brachymetopinae) are represented in the Permian. Of these the Ditomopyginae is the most important, in terms of numbers of both genera and species. Some eighty species of Permian trilobites have been described, of which about 65% are ditomopygines, 11% cummingellines, 9% proetines, 6% weaniines, 6% brachymetopids, and 2% cyrtosymbolines; in addition, there are some species whose affinities are unknown. This composition is similar to that of the later Carboniferous (Silesian). Following an early Dinantian radiation, many genera became extinct by the late Dinantian or early Silesian. There followed a radiation of ditomopygines which became the dominant trilobite subfamily during the latter part of the Carboniferous and in the Permian. All other ditomopygines appear to have their origins in *Paladin*, which appeared in mid-Dinantian times; its origins probably lie in *Linguaphillipsia* which in turn may have a common origin with a dechenelline such as *Schizoproetus*. The range of *Paladin* extends through the Silesian, and two species from the Kazanian of Timor, *P. baungensis* (Gheyselinck) and *P. brevicauda* (Gheyselinck) (Pl. 5, figs. 12, 13, 16) are the last known representatives of the genus. *Ditomopyge*, which is likely to have arisen from *Paladin* (the essential difference being the acquisition of a prominent median preoccipital lobe), is important in the Silesian and ranges to the late Permian. *Ditomopyge* gave rise to two similar-looking genera, *Pseudophillipsia* and *Anisopyge* in Old World Tethys and West U.S.A., respectively. Both have L2–L4 represented by raised, ovate areas and a multisegmented pygidium; they appear to have developed independently from one another. Genera closely related to *Pseudophillipsia* are discussed above; occurring with *Anisopyge* in western U.S.A. are *Delaria* and *Vidria*, both apparently associated with the *Anisopyge* lineage. *Ameura*, another genus characteristic of this area has been included in the Cummingellinae (see above), but the structure and disposition of the lateral glabellar furrows rather suggest a relationship with *Paladin* (see Whittington 1954).

The Cummingellinae are represented by one genus, *Paraphillipsia*, in the Permian. Its possible relationships with other cummingellines are outlined above. The Cyrtosymbolinae and Weaniinae are represented by one and three (four?) genera respectively in the Permian; relationships with Carboniferous genera are uncertain.

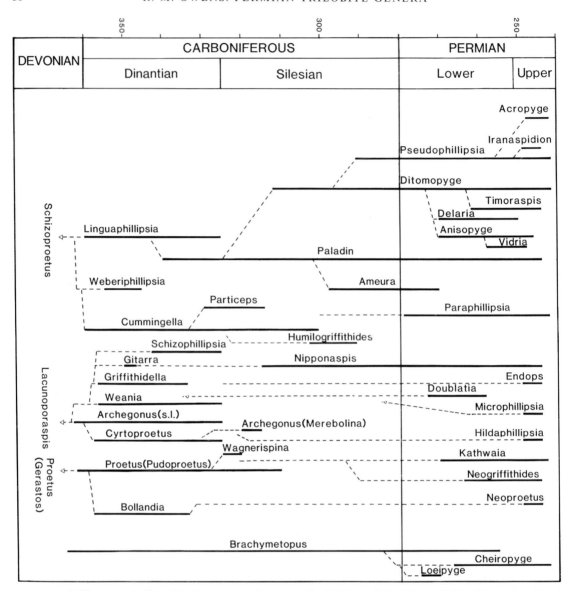

TEXT-FIG. 1. Known stratigraphical ranges and suggested phyletic relationships of Permian and selected Carboniferous trilobite genera. Solid lines represent ranges, broken lines inferred derivations.

There are three proetine genera in the Permian which have their origins in two different branches of the Proetinae. *Kathwaia* and *Neogriffithides* are presumed to be derived ultimately from *Proetus* (*Pudoproetus*), with which they share many cephalic and pygidial characteristics (see above). *Neoproetus* is apparently related to *Bollandia*, a Dinantian genus; there is, however, a long gap in the stratigraphical record between the last *Bollandia* and first *Neoproetus* which spans the Silesian and Lower Permian. The morphology of these genera is so similar, however, that there can be little doubt as to their close relationship. The origin of *Bollandia* is likely to be in *Proetus* (*Pudoproetus*) (see above). Brachymetopidae, represented by *Brachymetopus*, *Cheiropyge*, and *Loeipyge* are discussed above.

In conclusion, it may be surmised that there was a last radiation of the Trilobita in the Permian;

greatest diversity seems to have been achieved early in Upper Permian times, presumably concomitant with the widespread development of 'reefs'. Only six genera—*Pseudophillipsia*, *Ditomopyge*, *Paraphillipsia*, *Kathwaia*, *Acropyge*, and *Cheiropyge*—appear to have survived the earlier part of the Upper Permian (Kazanian), and are found in the latest Permian of the Himalayas, Iran, and China.

Distribution. Permian trilobites are known from Eurasia, Australasia, and North and Central America; they are rarely a common element of Permian faunas. Chamberlain (1977, pp. 760, 763)

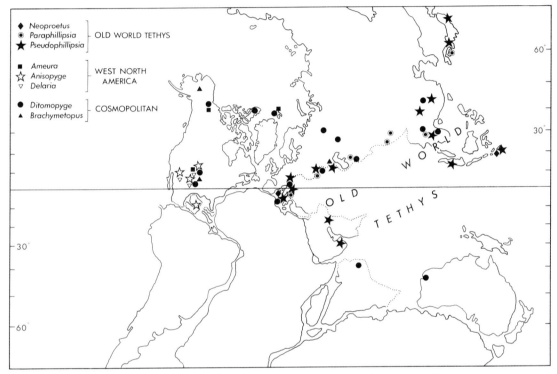

TEXT-FIG. 2. Distribution of selected trilobite genera on a 'Permian' World Map (after Smith, Briden and Drewry 1973, text-fig. 10).

recognized two faunal realms in the Permian: a Eurasian-Arctic one typified by *Ditomopyge* of the 'D. artinskiensis group' and a Midcontinent (i.e. North America)-Andean one typified by *Ditomopyge* of the 'D. scitula group'. Having surveyed Permian genera on a world-wide basis it is now possible to present a more detailed picture. The only cosmopolitan genera appear to be *Ditomopyge* and *Brachymetopus*, the former being the presumed ancestor of the homoeomorphic genera *Anisopyge* and *Pseudophillipsia* (see above) which are apparently of mutually exclusive distribution, the former restricted to western U.S.A. and the latter to Eurasia (Old World Tethys). Species from outside western U.S.A. assigned to *Anisopyge* have been found not to belong to that genus. Ditomopygines closely related to *Anisopyge* (*Delaria* and *Vidria*) are similarly restricted in their distribution, whilst *Ameura* has been found in western U.S.A., Yukon, and Spitsbergen. *Timoraspis*, *Acropyge*, and *Iranaspidion* are ditomopygines which are restricted to Old World Tethys. The four other proetid subfamilies (Proetinae, Cyrtosymbolinae, Weaniinae, and Cummingellinae) and the brachymetopids *Cheiropyge* and *Loeipyge* are found only in Old World Tethys. The record of the cummingelline *Paraphillipsia* from Alaska (Chamberlain 1977) is considered doubtful: one pygidium described as *P. aglypta* Chamberlain probably belongs to a different proetid.

To date, Permian trilobites have not been used as zonal indices, although some species, such as *Ameura trigonopyge* (Osmólska), which occurs in the early Permian of Spitsbergen and the Yukon might well be important in this respect. The same applies to genera such as *Pseudophillipsia*, *Paraphillipsia*, and *Ditomopyge* which are widely distributed and with large numbers of species. At present little is known of the stratigraphical and geographical distribution of individual species.

Most, if not all, Permian trilobites occur in shallow water or inshore deposits, particularly in 'reefs'; a study by Cisne (1971) on species from a small area in western North America has shown that there is a marked decrease in diversity in an offshore direction. Unlike the Carboniferous and earlier periods there therefore appear to have been no or very few trilobites adapted to deeper-water conditions. This could be a contributory factor to their final extinction. The great adaptive radiations which repeatedly took place in 'reef' and other shallow water environments throughout the Palaeozoic probably had their origins in a comparatively small number of forms which invaded shallower waters from deeper water, offshore environments. At the close of the Permian widespread marine regression meant that many shallow-water habitats disappeared, and with them the highly adapted trilobites which occupied them. Since there were no longer any deeper-water trilobites remaining, these elegant arthropods finally disappeared from the oceans which they had graced for so long.

Acknowledgements. It would have been impossible to assemble the material required for this paper without the generous assistance of a large number of people throughout the world, and it is a particular pleasure to thank the following for loan of specimens or for supplying casts or photographs: Dr. C. Blotwijk, Delft; Mr. F. C. Collier, Washington; Dr. R. A. Fortey, London; Mrs. P. L. Green, Tasmania; Dr. L. S. Kent, Urbana; Professor T. Kobayashi, M.J.A., Tokyo; Mrs. J. S. Lawless, Yale; Professor K. Mori, Sendai; Dr. L. Popov, Leningrad; Qian Yi-yuan, Nanking; Dr. M. V. A. Sastry, Calcutta; Dr. W. Struve, Frankfurt am Main; Dr. R. E. Wass, Sydney; Dr. J. H. Werner, Amsterdam. Miss Sian Bassett translated Italian texts, and Sir James Stubblefield, F.R.S., and Professor H. B. Whittington, F.R.S., kindly read a draft of the manuscript and offered valuable suggestions for its improvement.

REFERENCES

BEYRICH, E. Über eine Kohlenkalk-Fauna von Timor. *Phys. Math. Abh. K. Akad. Wiss. Berlin*, 1863–1865, 61–96, pls. 1–3.

CAMPBELL, K. S. W. and ENGEL, B. A. 1963. The faunas of the Tournaisian Tulcumba Sandstone and its members in the Werrie and Belvue synclines, New South Wales. *J. geol. Soc. Aust.* **10**, 55–122, pls. 1–39.

CHAMBERLAIN, C. K. 1970. Permian trilobite species from central Wyoming and west Texas. *J. Paleont.* **44**, 1049–1054, pl. 140.

—— 1972. Evolution of the Permian trilobite *Anisopyge*. Ibid. **46**, 503–508.

—— 1977. Carboniferous and Permian trilobites from Ellesmere Island and Alaska. Ibid. **51**, 758–771, 3 pls.

CISNE, J. L. 1971. Palaeoecology of trilobites of the Kaibab Limestone (Permian) in Arizona, Utah and Nevada. Ibid. **45**, 525–533, pl. 68.

DIENER, C. 1897. The Permocarboniferous fauna of Chitichun No. 1. *Mem. geol. Surv. India Palaeont. indica*, Ser. 15, Himalayan Fossils, **1**, 1–105, pls. 1–13.

ENDO, R. and MATSUMOTO, E. 1962. Permo-Carboniferous trilobites froom Japan. *Sci. Rep. Saitama Univ.* Ser. B, **4**, 149–172, pls. 8–10.

ENGEL, B. A. and LAURIE, J. R. 1978. A new species of the Permian trilobite *Doublatia* from the Manning District, New South Wales. *Alcheringa*, **2**, 49–54.

FORTEY, R. A. and OWENS, R. M. 1979. Enrollment in the classification of trilobites. *Lethaia*, **12**, 219–226.

GANDL, J. 1968. Stratigraphische Untersuchungen im Unterkarbon des Frankenwaldes unter besonderer Berücksichtigung der Trilobiten, 1: Die Trilobiten im Unterkarbon des Frankenwaldes. *Senckenberg. leth.* **49**, 39–117, 9 pls.

—— 1977. Die Karbon-Trilobiten des Kantabrischen Gebirges (NW Spanien), 2: Die Trilobiten der Alba-Schichten (Unter-Visé bis Namur A). Ibid. **58**, 113–217, 7 pls.

—— 1980. Die Trilobiten des Kantabrischen Gebirges (NW Spanien), 3: Trilobiten mis 'Kulm-Charakter' aus dem Namur B. Ibid. **60**, 291–351, 4 pls.

GAURI, K. L. 1965. Uralian stratigraphy. Trilobites and brachiopods of the western Carnic Alps (Austria). *Jb. geol. Bundesanst. Wien*, A, Sonder-Band, **11**, 1–94, 17 pls.

——and RAMOVŠ, A. 1964. *Eolyttonia* (Brach.) and *Brachymetopus* (Tril.) from the Upper Carboniferous (Orenburgian) of Karawanken, Yugoslavia. *N. Jb. Geol. Paläont.* Abh. **119**, 103–112, pl. 14.

GEMMELLARO, G. G. 1892. I crostacei de calcari con *Fusulina* della valle del Fiume Sosio nella provencia di Palermo in Sicilia. *Mem. Soc. ital. Sci. XL*, Ser. 3, **8**, 1–40, 5 pls.

GHEYSELINCK, R. F. C. R. 1937. *Permian trilobites from Timor and Sicily with a revision of their nomenclature and classification.* Scheltema and Holkema's Bockhandel en Uitgeversmaatschappij N.V., Amsterdam. xvi + 108 pp., 3 pls.

GIRTY, G. H. 1908. The Guadalupian fauna. *Prof. Pap. U.S. geol. Surv.* no. 58, 1–651, 31 pls.

——1909. *In* GIRTY, G. H. and LEE, W. T. The manzano Group of the Rio Grande Valley, New Mexico. *Bull. U.S. geol. Surv.* no. 389, 41–141, pls. 6–12.

GOLDRING, R. 1957. *Pseudophillipsia* (Tril.) from the Permian (or Uralian) of Oman, Arabia. *Senckenberg. leth.* **38**, 195–210, 1 pl.

GRABAU, A. W. 1936. Early Permian fossils of China, part II. Fauna of the Maping Limestone of Kwangsi and Kweichow. *Palaeont. sin.* Ser. B, **8**, 1–441, pls. 1–31.

GRANT, R. E. 1966. Late Permian trilobites from the Salt Range, West Pakistan. *Palaeontology*, **9**, 64–73, pl. 13.

GRECO, B. 1935. La fauna permiana del Sosio conservata nei musei di Pisa, di Firenze e di Padova. *Palaeontogr. ital.* **35** (N.S. 5), 101–190, pls. 12–15.

HAHN, G. 1963. Trilobiten der unteren *Pericyclus*-Stufe (Unterkarbon) aus dem Kohlenkalk Belgiens. Teil 1: Morphologie, Variabilität und postlarvale Ontogenie von *Cyrtosymbole* (*Belgibole*) *belgica* n.sg., n.sp. *Senckenberg. leth.* **44**, 209–249, pls. 37–38.

——1965. Revision der Gattung *Archegonus* Burmeister 1843 (Trilobita). Ibid. **46**, 229–262.

——and BRAUCKMANN, C. 1975a. Revision zweier Trilobiten-Arten aus dem Perm Asiens. *Geologica Palaeont.* **9**, 117–124.

————1975b. Zur Evolution von *Carbonocoryphe* (Trilobita; Unter-Karbon). *Senckenberg. leth.* **56**, 305–333, 1 pl.

——and HAHN, R. 1967. Zur Phylogenie der Proetidae (Trilobita) des Karbons und Perms. *Zool. Beitr.* N.F. **13**, 303–349.

————1969. Trilobitae carbonici et permici I. (Brachymetopidae, Otarionidae, Proetidae: Proetinae, Dechenellinae, Drevermanninae. Cyrtosymbolinae.) *Fossilium Cat.* I. Animalia, **118**, 1–160.

————1970. Trilobitae carbonici et permici II. (Proetidae: Griffithidinae). Ibid. **119**, 162–331.

————1971. Revision von *Griffithides* (*Bollandia*) (Tril.; Unter-Karbon) *Palaeontographica*, A137, 109–154, pls. 25–27.

————1972. Trilobitae carbonici et permici III. *Fossilium Cat.* I. Animalia, **120**, 332–531.

————1975. Die Trilobiten des Ober-Devon, Karbon und Perm. *Leitfossilien*, no. 1, viii + 127 pp., 12 pls.

————1981. Über *Acropyge* (Trilobitae; Ober-Perm). *Senckenberg. leth.* **61**, 217–225.

——and BRAUCKMANN, C. 1980. Die Trilobiten des belgischen Kohlenkalkes (Unter-Karbon). 1. Proetinae, Cyrtosymbolinae und Aulacopleuridae. *Geologica Palaeont.* **14**, 165–188, pls. 1, 2.

——and RAMOVŠ, A. 1970. Perm-Trilobiten aus Slowenien, NW-Jugoslavien. *Senckenberg. leth.* **51**, 311–333, 1 pl.

HESSLER, R. R. 1963. Lower Mississippian trilobites of the family Proetidae in the United States. Part I. *J. Paleont.* **37**, 543–563, pls. 59–62.

——1965. Lower Mississippian trilobites of the family Proetidae in the United States. Part II. Ibid. **39**, 248–264, pls. 37–40.

HUPÉ, P. 1953. Classe de trilobites. *In* PIVETEAU, J. (ed.). *Traité de paléontologie*, **3**, 44–246. Paris.

INAI, Y. 1936. *Humilogriffithides*, a new ally of *Griffithides*. *Proc. imp. Acad. Japan*, **12**, 299–302, 1 pl.

KAYSER, E. 1883. Oberkarbonische Fauna von Lo-Ping. *In* RICHTHOFEN, F. VON. China. Ergebnisse einiger Reisen und darauf gegründeter Studien. 4. Palaeontologischer Theil, 106–208, pls. 19–29. Berlin.

KOBAYASHI, T. and HAMADA, T. 1978. Two new late Upper Permian trilobites from central Iran. *Proc. Japan Acad.* **54** (B), 157–162.

————1979. Permo-Carboniferous trilobites from Thailand and Malaysia. *Geol. Palaeont. SE Asia*, **20**, 1–21, pls. 1–3.

————1980a. A nomenclatural note on *Neoproetus* (*Paraproetus*). *Trans. Proc. palaeont. Soc. Japan*, N.S. **117**, 254.

————1980b. Carboniferous trilobites of Japan etc. *Spec. Pap. palaeont. Soc. Japan*, **23**, 1–132, 22 pls.

————1980c. Three new species of Permian trilobites from west Japan. *Proc. Japan Acad.* **56** (B), 120–124.

———— 1980d. A new Permian species of *Pseudophillipsia* (Trilobita) with spatulate genal spines. Ibid. **56** (B), 195–199.

KOIZUMI, H. 1972. New genera of the trilobite family Phillipsiidae from the Takakura-yama Group (Permian), Abukuma Massif in Japan. *Chikyū Kagaku* [Earth Science], **26** (1), 19–25, 2 pls. [In Japanese; English summary.]

LEYH, C. F. 1897. Beiträge zur Kenntnis des Paläozoikum der Umgebung von Hof a. Saale. *Z. deutsch. geol. Ges.* **49**, 504–560, pls. 17, 18.

LU, Y.-H. 1974. In: *Handbook of stratigraphy and palaeontology of SW China.* Nanking Institute of Geology and Palaeontology, Academia Sinica. iii + 454 pp., 202 pls.

MANSUY, H. 1912. Mission du Laos. I. Geologie des environs de Luang-Prabang. *Mém. Serv. géol. Indoch.* **1** (4), 1–32, pls. 1–4.

———— 1913. Faunes des calcaires a Productus de l'Indochine. Premiere serie. Ibid. **2** (4), 1–133, pls. 1–13.

M'COY, F. 1847. On the fossil botany and zoology of the rocks associated with the coal of Australia. *Ann. Mag. nat. Hist.* **20**, 145–157, 226–236, 298–312, pls. 9–17.

MEEK, F. B. and WORTHEN, A. H. 1865. Contributions to the palaeontology of Illinois and other western states. *Proc. Acad. nat. Sci. Philad.* **17**, 245–273.

MOORE, R. C. (ed.). 1959. *Treatise on Invertebrate Paleontology, Part 0, Arthropoda* 1. Geological Society of America and University of Kansas Press (Lawrence). xix + 560 pp., 415 figs.

NEWELL, N. D. 1931. New Schizophoriidae and a trilobite from the Kansas Pennsylvanian. *J. Paleont.* **5**, 260–269, pl. 31.

ORMISTON, A. R. 1973. Lower Permian trilobites from northern Yukon Territory and Ellesmere Island, District of Franklin. *Bull. geol. Surv. Can.* no. 222, 129–139, pl. 16.

OSMÓLSKA, 1970a. On some rare genera of the Carboniferous Cyrtosymbolinae Hupé, 1953 (Trilobita). *Acta palaeont. pol.* **15**, 115–136, pls. 1, 2.

———— 1970b. Revision of non-cyrtosymbolinid trilobites from the Tournaisian-Namurian of Eurasia. *Palaeont. pol.* **23**, 1–165, 22 pls.

OWENS, R. M. 1973. British Ordovician and Silurian Proetidae (Trilobita). *Palaeontogr. Soc.* [*Monogr.*], 1–98, 15 pls.

———— and THOMAS, A. T. 1975. *Radnoria,* a new Silurian proetacean trilobite, and the origins of the Brachymetopidae. *Palaeontology,* **18**, 809–822, pls. 95, 96.

PABIAN, R. K. and FAGERSTROM, J. A. Late Paleozoic trilobites from southeastern Nebraska. *J. Paleont.* **46**, 789–816, 1 pl.

PORTLOCK, J. E. 1843. *Report on the geology of Londonderry and parts of Tyrone and Fermanagh.* Dublin and London. i–xxxi, 784 pp., pls. 1–38, A–I, map.

PRANTL, F. and PŘIBYL, A. 1951. A revision of the Bohemian representatives of the family Otarionidae R. and E. Richter (Trilobitae). *Stat. geol. Úst. Česk. Rep.* **17** for 1950, 353–512, 5 pls. [Czech and English text; Russian summary.]

QIAN, YI-YUAN. 1977. Upper Permian trilobites from Qinglong and Anshun of Guizhou. *Acta palaeont. sin.* **16** (2), 279–286, 1 pl. [In Chinese; English summary.]

RICHTER, R. and RICHTER, E. 1950. Tropidocoryphinae im Karbon (Tril.). *Senckenbergiana,* **31**, 277–284, 1 pl.

ROEMER, F. 1880. Über eine Kohlenkalk fauna der Westküste von Sumatra. *Palaeontographica,* **A27**, 1–11, pls. 1–3.

ROMANO, M. 1971. A new proetid trilobite from the Lower Westphalian of north-west Spain. *Trab. Geol. Fac. Cienc. Univ. Oviedo,* **4**, 379–383, 2 pls.

RUGGIERI, G. 1959. Una nuova trilobite del Permiano del Sosio (Sicilia). *Memorie Ist. geo. miner. Univ. Padova.* **21**, 1–10, 1 pl.

SALTER, J. W. 1864. A monograph of British trilobites from the Cambrian, Silurian and Devonian formations (1). *Palaeontogr. Soc.* [*Monogr.*], 1–80, pls. 1–6.

SCHRÉTER, Z. 1948. Trilobiták a Bükk Hegységbol. *Földt. Közl.* **78**, 25–39.

SHUMARD, B. F. 1858. Notice of new fossils from the Permian strata of New Mexico and Texas, collected by Dr. George G. Shumard, geologist to the United States Government expedition for obtaining water by means of Artesian wells along the 32nd parallel, under the direction of Capt. John Pope, U.S. Corps Top. Eng. *Trans. Acad. Sci. St. Louis,* **1** (2), 290–297.

SMITH, A. G., BRIDEN, J. C. and DREWRY, G. E. 1973. Phanerozoic world maps. *In* HUGHES, N. F. (ed.). Organisms and continents through time. *Spec. Pap. Palaeont.* no. 12, 1–42.

STUBBLEFIELD, C. J. 1936. Cephalic sutures and their bearing on current classifications of trilobites. *Biol. Rev.* **11**, 407–440.

TEICHERT, C. 1944. Permian trilobites from western Australia. *J. Paleont.* **18**, 455–463, pl. 77.

TESCH, P. 1923. Trilobiten aus der Dyas von Timor und Letti. *In* WANNER, J. (ed.). Paläontologie von Timor nebst Kleineren Beiträgen zur Paläontologie einiger anderen Inseln des Ostindischen Archipels. Part II: 123–132, pl. 178.

TUMANSKAYA [TOUMANSKY], O. G. 1935. Permo-Karbonovye otlozheniya Kryma. Chast' II. Permo-karbonovye trilobity Kryma [The Permo-Carboniferous Beds of the Crimea. Part II. The Permo-Carboniferous trilobites of the Crimea.] *Glavnoe geologo-gidro-geodesicheskoe upravlenie.* Leningrad and Moscow, 63 pp., 12 pls. [In Russian; English summary.]

WANG, Y. 1937. On a new Permian trilobite from Kansu. *Bull. geol. Soc. China,* **16**, 357–370, 1 pl.

WASS, R. E. and BANKS, M. R. 1971. Some Permian trilobites from eastern Australia. *Palaeontology,* **14**, 222–241, pls. 36, 37.

WEBER, V. N. 1932. Trilobity Turkestana [Trilobites of Turkestan]. *Trudy vses. geol.-razv. Ob"ed. NKTP.* iv + 157 pp., 4 pls. [In Russian; English summary.]

—— 1933. Trilobity Donetskogo basseyna [Trilobites of the Donetz Basin]. Ibid. 95 pp., 4 pls. [In Russian; English summary.]

—— 1944. Trilobity kamennougol'nych i permskich otlozheniy SSSR. II. Permskie trilobity [Trilobites of the Carboniferous and Permian systems of the U.S.S.R. II. Permian trilobites]. *Monografii Paleont. SSSR.* **71**, 30 pp., 2 pls. [In Russian; English summary.]

WELLER, J. M. 1935. Permian trilobites from the central Himalayas. *Mem. Conn. Acad. Arts Sci.* **9**, 31–35.

—— 1936. Carboniferous trilobite genera. *J. Paleont.* **10**, 704–714, pls. 95.

—— 1944. Permian trilobite genera. Ibid. **18**, 320–327, pl. 49.

WHITTINGTON, H. B. 1954. Two silicified Carboniferous trilobites from west Texas. *Smithson. misc. Collns,* **122**, 1–16, pls. 1–3.

YIN, K. 1978. Trilobites, pp. 440–445, pl. 121 in *Atlas of Southwest Chinese fossils in Kueichou District.* **2**: *Carboniferous–Quaternary.* Peking. [In Chinese.]

R. M. OWENS

Department of Geology
National Museum of Wales
Cardiff CF1 3NP

A NEW SILURIAN ARTHROPOD FROM LESMAHAGOW, SCOTLAND

by P. A. SELDEN *and* D. E. WHITE

ABSTRACT. A single incomplete specimen of an arthropod of uncertain affinities, *Pseudarthron whittingtoni* gen. et sp. nov., from the Ludlow Series of the Lesmahagow Silurian Inlier, Scotland, is described. It has an oval shape and is characterized by a small cephalic region and a thorax of at least seven tergites, each traversed by a pronounced ridge and shallow furrow. The expansive dorsal covering is considered to have served to conceal and protect the appendages. *Pseudarthron* was probably a member of a fresh- or brackish-water, vagrant epifaunal benthos. Certain morphological features indicate a possible relationship to the Upper Ordovician *Triopus draboviensis* Barrande, 1872 and suggest similarities to the Lower Devonian *Cheloniellon calmani* Broili, 1932. *P. whittingtoni* also has features in common with the Upper Devonian *Oxyuropoda ligioides* Carpenter and Swain, 1908 and with *Camptophyllia eltringhami* Gill, 1924 and *C. fallax* Gill, 1924 from the Coal Measures (Upper Carboniferous). Of these, *Cheloniellon* is chelicerate-like but the remainder, like *Pseudarthron*, are of doubtful affinities.

IN conjunction with a recent revision of the Hamilton (23SW) Sheet by the South Lowlands Unit, Institute of Geological Sciences (I.G.S.), Edinburgh, the Palaeontology Unit investigated the Silurian biostratigraphy of the Lesmahagow Inlier. During the course of this work, a single example of an unusual arthropod was found, which is described below as *Pseudarthron whittingtoni* gen. et sp. nov.

It occurred in greenish-grey, slightly silty, shaly mudstone at the top of the south face of the easternmost (C) of three disused quarries to the east of South Hill farm (text-fig. 1), in the northern part of the Inlier.

From this mudstone several rod-like structures (GSE 13873; 3E 4549–4557, 4559, 4563 of the IGS Scottish collections) were also collected, with dimensions of the order of 10 mm long and 2–3 mm wide (text-fig. 2A). Most have an imbricate structure and well-preserved examples are sinistrally striated, both of which are features exhibited by some coprolites (Häntzschel *et al.* 1968, p. 2, fig. 1; pl. 3, figs. 8, 20, 21). Their chemical composition is not known but they appear to consist of greenish-grey clay. If they are coprolites, the animal from which they originated might be fish or arthropod, but, on the basis of size, not *P. whittingtoni*.

No other fossils were found at this locality during the present survey. However, the I.G.S. Scottish collections contain several fish specimens including *Ateleaspis tessellata* Traquair, *Birkenia elegans* Traquair, *Lanarkia horrida* Traquair, and *L. spinosa* Traquair from the 'third quarry and third field east of South Hill farm' which is almost certainly quarry C. The collection also includes *Pachytheca*?, *Dictyocaris* sp., and several examples of *Taitia catena* Crookall, all problematical fossils of possible algal affinities, together with possible coprolites identical with those found with *Pseudarthron whittingtoni*. These fish and problematical fossils are preserved in a finely laminated siltstone and were collected by D. Tait in 1900. This fish bed probably immediately underlies the bed from which *P. whittingtoni* was obtained and is evidently no longer exposed.

Approximately along strike from quarry C, in the middle quarry (B of text-fig. 1), near the top of the section, '*Glauconome*' is fairly common in a maroon, finely micaceous, argillaceous siltstone (GSE 3E 4538–4548). Specimens of this problematical fossil from the '*Glauconome*' Band of the Fish Bed Formation in the Hagshaw Hills Inlier have been described and figured by Rolfe (1961, p. 260, pl. 15A). He concluded that it is not a bryozoan, but if 'a peculiarly "jointed" plant, it is of doubtful affinities'.

Stratigraphy. Walton (1965, pp. 195–198) summarized the lithostratigraphy of the Lesmahagow Inlier, based on unpublished work by Jennings (1961). The Silurian succession consists of three major divisions—the Priesthill (oldest), Waterhead and Dungavel (youngest) Groups. Within the Waterhead Group, the Dippal Burn Formation and, approximately 150 m higher in the sequence, the Slot Burn Formation each contain fish beds. Although the fish-bearing horizons of the two formations are lithologically and faunally indistinguishable (Ritchie 1968, p. 320), the fish bed formerly exposed near South Hill farm has been mapped as part of the Slot Burn Formation.

[Special Papers in Palaeontology No. 30, pp. 43–49]

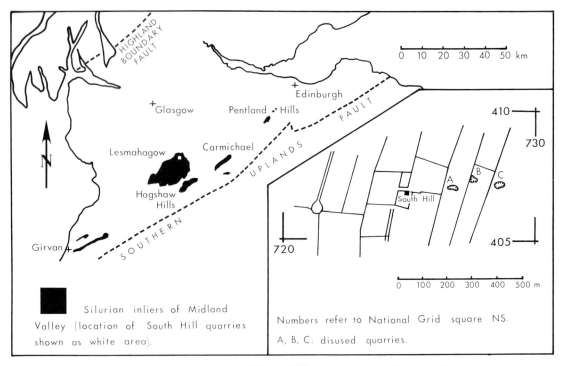

TEXT-FIG. 1. Location map of South Hill quarries, Lesmahagow Inlier.

On the basis of the degree of development of certain skeletal structures, Westoll (1945, p. 354) considered fish from the Waterhead Group to be younger than those from the K_1 dolomitic limestones of Oesel, Estonia, which are within the Rootsiküla Stage. This stage ranges from late Wenlock to early Ludlow (Kaljo 1970). Similarly, Westoll considered the fish from the Waterhead Group to be slightly older than those of the Rudstangen fish bed, Ringerike, south Norway, of late Ludlow age. On these bases, the fish beds of the Waterhead Group are believed to be of Ludlow age. This is supported by independent records, by Dr. Jancis Ford and Dr. J. B. Richardson, of spores no older than late Ludlow (Ludfordian) from the Logan Formation, which immediately overlies the Slot Burn Formation (Dr. Richardson, British Museum (Natural History), pers. comm.).

Preservation. The part (GSE 13871) and counterpart (GSE 13872) appear to be preserved as ventral and dorsal moulds respectively of the dorsal shield. The surfaces of the fossil appear slightly darker and finer grained than the matrix, possibly indicating the carbonized remains of parts of the cuticle. The finely granulated appearance of the fossil may be original but could be due to impression of the coarser grained matrix. On both ventral and dorsal moulds, but especially on the latter, some iron-stained material (which does not react with dilute HCl) lies in pockets and furrows. This may represent post-diagenetic mineralization rather than the remains of cuticular material. Minute clusters of Fe and Mn? oxides are also thickly distributed throughout the matrix.

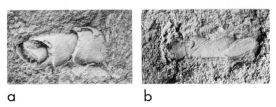

a   b

TEXT-FIG. 2. Possible coprolites associated with *Pseudarthron whittingtoni*; *a*, GSE 13873; *b*, GSE 13874; both × 3.

SYSTEMATIC PALAEONTOLOGY

ARTHROPODA
Phylum UNCERTAIN
Genus PSEUDARTHRON gen. nov.

Type and only known species. Pseudarthron whittingtoni sp. nov.

Derivation of name. Greek: *Pseudes* (false) and *Arthron* (joint) referring to the transverse tergal ridges which give the impression of supernumerary tergites.

Diagnosis. Dorsoventrally flattened, trilobed arthropod of oval shape. Cephalic region small in relation to thoracic region. Thoracic region of at least seven tergites with broad pleurae: first pleura and anterior margin of second curved anteriorly, fourth to possible eighth pleura showing successively greater posterior curvature. Pronounced ridge and shallow furrow traversing axis and pleurae parallel to anterior margin of each tergite at a point just anterior of its mid-length, becoming less distinct towards narrow doublure of lateral margin. (Appendages and details of cephalic and caudal regions not known.)

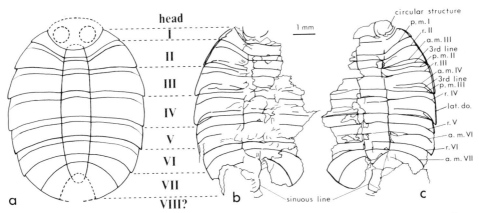

TEXT-FIG. 3. *Pseudarthron whittingtoni* gen. et sp. nov.; *a*, reconstruction of dorsal surface, tergites numbered I–VIII?. *b*, *c*, camera-lucida drawings of ventral mould, GSE 13871, and dorsal mould, GSE 13872, respectively. a. m., anterior margin of tergite; lat. do., lateral doublure; p. m., posterior margin of tergite; r., transverse ridge; Roman numerals indicate tergite numbers.

Pseudarthron whittingtoni sp. nov.

Text-figs. 3, 4

Holotype. GSE 13871, 13872, part and counterpart of only known specimen; Institute of Geological Sciences, Edinburgh.

Horizon and locality. Slot Burn Formation, Waterhead Group, Ludlow Series; old quarry, 270 m ENE of South Hill farm [NS 72854070] (Quarry C, text-fig. 1).

Derivation of name. In honour of Professor H. B. Whittington, F.R.S.

Diagnosis. As for the genus.

Description. The overall shape of *Pseudarthron* is oval. The head region is incompletely preserved. It is suboval in outline (the major axis at right angles to that of body), and small in relation to the rest of the body. Under low-angle light, a circular structure is apparent laterally. The preserved tergites number seven, but there is

TEXT-FIG. 4. *Pseudarthron whittingtoni* gen. et sp. nov.; *a*, GSE 13871, ventral mould; *b*, GSE 13872, dorsal mould; both ×6·5.

the possibility that an eighth existed posteriorly. Measured sagittally, the length of the tergites increases from the first to the fourth, whilst the fifth to seventh are shorter again. The axis accounts for 0.23 of the total width of the animal at tergite V.

At first sight the number of transverse lines on the axial region appears to indicate, particularly on GSE 13872, successive long and short tergites. The bolder lines, however, when followed across the pleurae, fade slightly at their lateral extremities and do not terminate at the posterior angles of the tergites. These transverse ridges are in the form of escarpments in cross-section, with the dip slope to the anterior, and a slight furrow at the base of the scarp slope. The posterior pleural angles can be seen clearly on both part and counterpart. These angles represent the junction between the lateral margin of the exoskeleton, and the less prominent lines crossing it, and indicate that the latter represent tergal margins. Both anterior and posterior tergal margins can be identified, for example at the tergite IV–tergite V junction. At the junctions of tergites II–III and III–IV, however, a third faint line can be seen, which may represent an anterior or posterior doublure. A narrow lateral doublure is present on each tergite.

The posterior part of the specimen is incompletely preserved. The pleurae of tergite VII expand posterolaterally. A sinuous line at the posterior extremity of the fossil may indicate an overlap in the midline of posteriorly expanded pleurae of an eighth tergite. However, the sinuous line disappears under alcohol, whilst other pleural margins remain, and therefore it may be an artefact.

Approximate dimensions, in mm, are as follows: total length *c*. 16, maximum width of axis 2·5, length of head *c*. 2·6, width of head 5, axial length (sagittally) of tergite/maximum width of tergite: I 1·1/7·2, II 1·2/9·9, III 1·8/11, IV 2·1/11, V 1·7/10·6, VI 1·6/9·8, VII 1·6/8·5, ?VIII *c*. 2·3?

AFFINITIES

From its general appearance, and without preserved appendages, *Pseudarthron* cannot be readily assigned to one of the recognized groups of arthropods. However, it bears certain resemblances to some previously described arthropods of uncertain affinities, especially the only known specimen of the Upper Ordovician *Triopus draboviensis* Barrande, 1872, now apparently lost (Chlupáč 1965). This resembles *Pseudarthron* in its oval outline and in details of its tergal characters: axial width about 0·25 of maximum width, seven tergites (possible eighth not preserved), pleurae expanded laterally, anterior borders of anterior tergites curved forwards, posterior borders of posterior tergites curved backwards, and the presence of prominent transverse ridges which fade laterally. In both genera the head and tail regions are incompletely preserved. *Triopus* differs from *Pseudarthron* in possessing a large cephalic shield in relation to the rest of the body (as evidenced by the posterior border which is preserved), an axial region with longitudinal ridges, impersistence of the pleural ridge and furrow system across the axis, and slightly greater expansion of the pleurae laterally. We consider that the first three of these characters are so distinct and recognizable as to merit the separation of *Pseudarthron* and *Triopus* at the generic level. At 35 mm in length the *Triopus* specimen is about twice the size of that of *Pseudarthron* (16 mm long).

Barrande (1872) referred *Triopus*, with reservations, to the trilobites. Novák (MS, *in* Chlupáč 1965, p. 22) and Neumayr (1887, p. 97) assigned it to the xiphosurans, and Jahn (1893) suggested that it could be a chiton. Chlupáč (1965) considered the affinity of *Triopus* with aglaspidids as unquestionable, comparing it with *Neostrabops* Caster and Macke, 1952. He also suggested that *Triopus* may represent the opisthosoma of one of the merostomes, such as *Zonozoe* or *Drabovaspis*, which are found in the same beds as *Triopus* but are known only from carapaces. Bergström (1968) took the last suggestion further, and attempted a restoration of *Triopus* as a xiphosurid, with a *D. complexa* carapace and a hypothetical telson. In our opinion *Triopus* (whether or not bearing the carapace of *Drabovaspis* or another supposed merostome) defies classification due to the lack of preserved appendages. Even where appendages are preserved in Palaeozoic xiphosuran-like arthropods, relationships may remain obscure: restudy of *Aglaspis spinifer* by Briggs *et al.* (1979) revealed that this animal is not a chelicerate, as previously supposed, and the relationship of the aglaspidids is problematic (see also Bergström 1980). Three xiphosuran specimens are known from the Lesmahagow Silurian Inlier, all from the Priesthill Group: *Neolimulus falcatus* Woodward, 1868, *Pseudoniscus* sp. (Eldredge 1974), and *Cyamocephalus loganensis* Currie, 1927 (Eldredge and Plotnick 1974). None of these are well preserved, and none particularly resembles *Pseudarthron*.

Pseudarthron shows some striking similarities to the early Devonian *Cheloniellon calmani* Broili (Broili 1932, 1933; Stürmer and Bergström 1978) in the overall shape of the animal, the tagmosis and trilobation, the expansions to the anterior and posterior pleurae, the transverse lines, and lateral doublures of the tergites. *Cheloniellon* differs from *Pseudarthron* in its greater size (64 mm for the smallest known *Cheloniellon versus* 16 mm) and apparently greater number of tergites (9 *versus* 8? in *Pseudarthron*).

The remarkable pyrite preservation in the Hunsrück Slate enabled Stürmer and Bergström (1978) to use an X-ray technique to redescribe *Cheloniellon*, and particularly the appendages, in fine detail. *Cheloniellon* was thus shown to have many features in common with the Xiphosura, as it also has with *Sidneyia* (Bruton 1981). Thus there is good evidence to place *Cheloniellon* close to the Chelicerata, but separate from that group as one of the many arthropod lines which arose from unknown, possibly Precambrian, ancestors (see Whittington 1979).

Two other arthropods of uncertain affinities which bear some resemblance to *Pseudarthron* are *Oxyuropoda ligioides* Carpenter and Swain, from the Upper Devonian of Kilkenny, Eire (Carpenter and Swain 1908; Rolfe 1969*b*), and *Camptophyllia eltringhami* Gill (also *C. fallax*), from the Coal Measures of Co. Durham, England (Gill 1924). The head of *Pseudarthron* may be similar in shape to that of *Oxyuropoda*, and circular structures are present in both (Rolfe 1969*b*, fig. 394). Also, each tergite of *Oxyuropoda* is traversed by a line at about 0·33 of its length (sagittally) from its anterior margin. *Oxyuropoda*, however, has a more slender outline than *Pseudarthron*, fewer thoracic tergites, and an abdominal region (not present in *Pseudarthron*). *Camptophyllia* is also more elongate in shape than *Pseudarthron* and its paratergal lobes are not directly comparable with the pleurae of *Pseudarthron*. *Camptophyllia* was referred with doubt to the Arthropleurida (Rolfe 1969*a*), and *Oxyuropoda* is regarded as Arthropoda *incertae sedis* (Rolfe 1969*b*). We consider that the similarities between *Pseudarthron*, *Oxyuropoda*, and *Camptophyllia* are more likely to be due to convergence than phylogenetic relationship.

Camptophyllia, *Cheloniellon*, and *Oxyuropoda* have each at some time been compared to Crustacea Isopoda and oniscomorph diplopods. All three differ from trilobites and other 'early armoured arthropods' (Manton 1977) principally in the lack of a large cephalic shield. The hemicylindrical shape of oniscomorph diplopods and many isopods evolved partly as an enrolment device (Manton 1979). Consequently, most fossils of these forms are preserved lying on one side on the bedding plane, in a curled attitude. *Pseudarthron* was dorsoventrally flattened in life and was probably incapable of complete enrolment. The earliest recorded isopod is from the Upper Carboniferous of Mazon Creek, Illinois, U.S.A. (Schram 1970) and it is considered by Schram (1981) that the Peracarida (which includes the isopods) probably originated no earlier than the late Devonian. Against this background, and in view of the lack of diagnostic myriapod or crustacean characters in *Pseudarthron*, there appears to be no possibility of an affinity with these groups.

We prefer to regard *Pseudarthron* as an arthropod of uncertain affinities until specimens bearing appendages are discovered, leading to a more precise assignation.

PALAEOECOLOGY

In contrast to the underlying Priesthill Group, in which brachiopods, trilobites, ostracods, and molluscs are well represented, the Waterhead Group contains no fossils indicative of a marine environment. Fossils are confined to the fish beds and associated strata of the Dippal Burn and Slot Burn Formations, and consist of fish, eurypterids (especially *Lanarkopterus dolichoschelus* (Størmer)), problematical ?algal forms and *P. whittingtoni*. We agree with the unpublished conclusions of Jennings (1961), quoted by Ritchie (1968, p. 321), that the sediments of the Waterhead Group above and below the fish beds indicate a deltaic environment; the finely laminated lithology of the fish beds themselves and their lack of fossils of undoubted marine habitat are considered to indicate the temporary establishment of a quiet, possibly fresh- or brackish-water, lagoonal environment in which the ?algae, eurypterids, and fish lived. The mudstone in which *P. whittingtoni* occurred is associated with a fish bed, and probably accumulated in a comparable environment.

We consider that the expansive dorsal covering of *Pseudarthron*, like that of *Triopus*, *Cheloniellon*, some trilobites, and other Palaeozoic arthropods, was an adaptation for concealment and protection of the appendages during walking on the substrate.

We conclude that *Pseudarthron* was probably a member of a fresh- or brackish-water, vagrant epifaunal benthos; other aspects of its mode of life are not known.

Acknowledgements. We are grateful to Dr. A. W. A. Rushton, I.G.S., London, whose helpful suggestions improved the manuscript. We thank Dr. M. J. Gallagher and Mr. A. D. McAdam, I.G.S., Edinburgh, for supplying stratigraphical and locality details, Mr. P. J. Brand, I.G.S., Edinburgh, for arranging the loan of specimens in his care and providing curatorial information, and Dr. M. J. Benton, University of Oxford, for useful discussion. Photographs were prepared by Mr. R. Aspin, I.G.S., London and Miss S. Maher, University of Manchester. D. E. W. publishes by permission of the Director, Institute of Geological Sciences (N.E.R.C.).

REFERENCES

BARRANDE, J. 1872. *Système Silurien du centre de la Bohême. 1ère partie. Recherches paléontologiques. Supplément au Vol. 1. Trilobites, Crustacés divers et Poissons.* Prague.

BERGSTRÖM, J. 1968. Eolimulus, a lower Cambrian xiphosurid from Sweden. *Geol. Fören. Stockholm Förh.* **90**, 489–503.

—— 1980. Morphology and systematics of early arthropods. *Abh. naturwiss. Ver. Hamburg*, **23**, 7–42.

BRIGGS, D. E. G., BRUTON, D. L. and WHITTINGTON, H. B. 1979. Appendages of the arthropod *Aglaspis spinifer* (Upper Cambrian, Wisconsin) and their significance. *Palaeontology*, **22**, 167–180.

BROILI, F. 1932. Ein neuer Crustacee aus dem rheinischen Unterdevon. *Sber. bayer. Akad. Wiss.* for 1932, 27–38.

—— 1933. Ein zweites Exemplar von *Cheloniellon. Sber. bayer. Akad. Wiss.* for 1933, 11–32.

BRUTON, D. L. 1981. The arthropod *Sidneyia inexpectans*, Middle Cambrian, Burgess Shale, British Columbia. *Phil. Trans. R. Soc.* **B 295**, 619–653.

CARPENTER, G. H. and SWAIN, J. 1908. A new Devonian isopod from Kiltorcan, County Kilkenny. *Proc. R. Irish Acad.* **27 B**, 61–67.

CHLUPÁČ, I. 1965. Xiphosuran merostomes from the Bohemian Ordovician. *Sborn. geol. Věd. Rada P*, **5**, 7–38.

ELDREDGE, N. 1974. Revision of the suborder Syniphosurina (Chelicerata, Merostomata), with remarks on merostome phylogeny. *Am. Mus. Novitates*, no. 2543, 1–41.

—— and PLOTNICK, R. E. 1974. Revision of the pseudoniscine merostome genus *Cyamocephalus* Currie. Ibid., no. 2557, 1–10.

GILL, E. L. 1924. Fossil arthropods from the Tyne Coalfield. *Geol. Mag.* **61**, 455–471.

HÄNTZSCHEL, W., EL-BAZ, F. and AMSTUTZ, G. C. 1968. Coprolites. An annotated bibliography. *Mem. Geol. Soc. Am.* **108**, i–vii, 1–132.

JAHN, J. J. 1893. *Duslia*, eine neue Chitonidengattung aus dem Böhmischen Untersilur, nebst einigen Bemerkungen über die Gattung *Triopus* Barr. *Sber. Akad. Wiss. Wien* **102**, 591–603.

JENNINGS, J. S. 1961. 'The geology of the eastern part of the Lesmahagow Inlier.' Unpublished Ph.D. Thesis, University of Edinburgh.

KALJO, D. L. (ed.). 1970. *Silur Estonii*. Eesti NSV Teaduste Akadeemia Geologia Institut. Valgus, Tallinn. Pp. 1–343. [In Russian with Estonian and extensive English summaries.]

MANTON, S. M. 1977. *The Arthropoda: Habits, Functional Morphology, and Evolution.* Clarendon Press, Oxford. Pp. i–xx, 1–527.

—— 1979. Uniramian evolution with particular reference to the pleuron. *In* CAMATINI, M. (ed.). *Myriapod Biology*. Academic Press, London and New York. Pp. 317–343.

NEUMAYR, M. 1887. *Erdgeschichte*. Bd. II. Bibliographisches Institut. Leipzig and Vienna. Pp. i–xii, 1–880.

RITCHIE, A. 1968. *Lanarkopterus dolichoschelus* (Størmer) gen. nov., a mixopterid eurypterid from the Upper Silurian of the Lesmahagow and Hagshaw Hills inliers, Scotland. *Scott. J. Geol.* **4**, 317–338.

ROLFE, W. D. I. 1961. The geology of the Hagshaw Hills Silurian Inlier, Lanarkshire. *Trans. Edinb. geol. Soc.* **18**, 240–269.

—— 1969a. Arthropleurida. *In* MOORE, R. C. (ed.). *Treatise on Invertebrate Paleontology*. New York and Lawrence, Geol. Soc. Am., pt. **R**. Arthropoda 4, R607–R620.

—— 1969b. Arthropoda *incertae sedis. In* MOORE, R. C. (ed.). Ibid. R620–R625.

SCHRAM, F. R. 1970. Isopod from the Pennsylvanian of Illinois. *Science*, no. 169, 854–855.

—— 1981. On the classification of Eumalacostraca. *J. crustacean Biol.* **1**, 1–10.

STÜRMER, W. and BERGSTRÖM, J. 1978. The arthropod *Cheloniellon* from the Devonian Hunsrück Shale. *Paläont. Z.* **52**, 57–81.

WALTON, E. K. 1965. Lower Palaeozoic Rocks. *In* CRAIG, G. Y. (ed.). *The Geology of Scotland*. Oliver and Boyd, Edinburgh and London. Pp. 161–227.

WESTOLL, T. S. 1945. A new cephalaspid fish from the Downtonian of Scotland, with notes on the structure and classification of ostracoderms. *Trans. R. Soc. Edinb.* **61**, 341–357.

WHITTINGTON, H. B. 1979. Early arthropods, their appendages and relationships. *In* HOUSE, M. R. (ed.). *The Origin of Major Invertebrate Groups*. (Systematics Association Special Volume No. 12.) Academic Press, London and New York. Pp. 253–268.

P. A. SELDEN

Department of Extra-Mural Studies
University of Manchester
Manchester M13 9PL

D. E. WHITE

Institute of Geological Sciences
Exhibition Road
South Kensington
London SW7 2DE

THE ANOMALOUS BATHYURID TRILOBITE
CERATOPELTIS AND ITS HOMOEOMORPHS

by R. A. FORTEY *and* J. S. PEEL

ABSTRACT. The early Ordovician trilobite *Ceratopeltis* Poulsen, 1937, was hitherto known from a single, abraded pygidium. Abundant new material from eastern North Greenland shows that it is a peculiar bathyurid. Its development of marginal pygidial spines is matched by several other unrelated trilobites in the earlier Ordovician, inhabitants of separate former epicontinental seas.

IN 1937 Poulsen described the early Ordovician trilobite faunas from East Greenland. These included a variety of trilobites belonging to the Bathyuridae, a family which Whittington (1963) recognized as typical of the shallow-water, tropical carbonates which accumulated over wide areas of the North American–Greenland continent at that time (Whittington and Hughes 1972; Cocks and Fortey 1982). Poulsen described one new genus—*Ceratopeltis*—from a single example, a pygidium (Pl. 6, figs. 1, 2) which is somewhat weathered but shows the unusual feature of a pair of large marginal spines originating from near the forward margin of the border. Poulsen himself was in doubt about the affinities of this specimen, citing various (mostly Cambrian) trilobites which had a similar pygidial development. Since then *Ceratopeltis* has remained something of an enigma, and eventually was placed among an assortment of 'unrecognisable genera' in the *Treatise on Invertebrate Paleontology* (Moore 1959).

In 1979 Peel collected a number of slabs of thin-bedded limestone from north-eastern Greenland. These slabs are covered with the cranidia, free cheeks, and pygidia of a single species of trilobite. The pygidia show long marginal spines carried laterally, and a second, short postaxial pair. Comparison with the type of *Ceratopeltis* makes it overwhelmingly probable that the new collections represent the same genus and species; this paper is intended to clarify the relationships of this previously obscure trilobite.

MATERIAL

Poulsen (1937) described the holotype of *Ceratopeltis latilimbata* from the Cape Weber Formation of Canadian (early Ordovician age) at Kap Weber, north-eastern Greenland (text-fig. 1). The new material was collected from near the base of the Amdrup Member of the Wandel Valley Formation at Kap Holbæk, Danmark Fjord, north-eastern Greenland (text-fig. 1). The Wandel Valley Formation ranges in age from late Canadian to Chazyan (Arenig–Llanvirn), but the presence of *Ceratopea billingsi* Yochelson, in the Amdrup Member, above the horizon yielding *Ceratopeltis latilimbata*, indicates a late Canadian age (Peel 1982). *Ceratopea billingsi* is also present in east Greenland in the lower part of the Narwhale Sound Formation, which overlies the Cape Weber Formation.

The slabs from Kap Holbæk are covered with more than fifty fragments of the species. All the cranidia, free cheeks, and pygidia are of a single type, and there seems no reason to suppose that more than one species is represented; all are less than $\frac{1}{2}$ cm long. The specimens have weathered out naturally on the surface of the slabs, but many are too abraded to be useful. Taken together they provide a clear picture of the dorsal morphology. They resist further manual preparation. Weathering often serves to obscure or run together the posterior pair of spines; it was probably this which led Poulsen to recognize the anterior pair only. However, it is evident that the holotype owes much of its shape to subsequent preparation. Matrix and fossil are almost indistinguishable and the posterior pair of spines were overlooked in the course of excavation of the posterior margin. Sedimentary infilling between the spines can be recognized (cf. Pl. 6, figs. 1, 2). The convexity of the margin immediately posterior to the anterior pair of spines in the holotype appears to result from a combination of preparation and original convexity. The new material from Kap Holbæk is often slightly crushed (e.g.

[Special Papers in Palaeontology, No. 30, pp. 51–57, pl. 6]

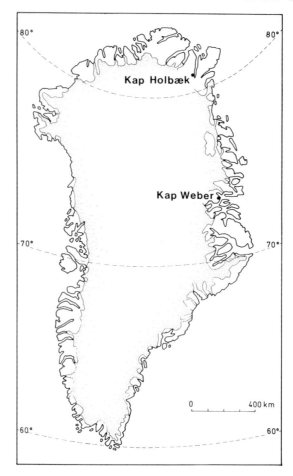

TEXT-FIG. 1. *Ceratopeltis* localities in Greenland.

Pl. 6, fig. 5) which possibly serves to emphasize the concave nature of the margin immediately posterior to large, anterior pair of spines.

SYSTEMATIC PALAEONTOLOGY

Family BATHYURIDAE Walcott, 1886
Subfamily BATHYURINAE Walcott, 1886
Genus CERATOPELTIS Poulsen, 1937

Type species. Ceratopeltis latilimbata Poulsen, 1937, original designation.

Ceratopeltis latilimbata Poulsen, 1937

Plate 6; text-fig. 2

Holotype. Pygidium, Geological Museum, Copenhagen, MMH 3712.

Figured material. Specimens from G.G.U. sample 274925 preserved in the Geological Museum, Copenhagen. Cranidia, MGUH 16.053–16.054; pygidia, MGUH 16.047, 16.049–16.052, 16.055; free cheek, MGUH 16.048.

Additional material. Geological Survey of Greenland (G.G.U.), Copenhagen, sample 274925.

Description. All cranidia and pygidia are less than $\frac{1}{2}$ cm long. The sagittal length of both was evidently about the same, and the whole trilobite is unlikely to have been much more than 2 cm long. Transverse convexity is relatively low; apart from a gentle downward slope on the preglabellar field, and outwards from the eyes, pleural regions are generally flat, and only the axis is convex.

Cranidium with width across posterior margin exceeding sagittal length. The vaulted glabella is widest at the occipital ring, parallel sided, or tapering forwards slightly to broadly rounded front margin. Axial and occipital furrows deep. Most specimens show no signs of glabellar furrows. In view of the fact that the occipital furrow is always clearly shown even on abraded specimens, it seems most unlikely that there were glabellar furrows of any depth. An abraded specimen (Pl. 6, fig. 9) shows evidence of two short pairs of

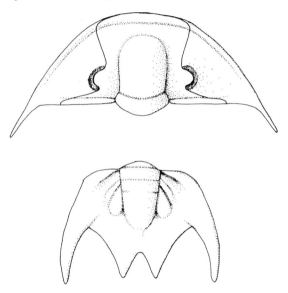

TEXT-FIG. 2. Reconstruction of cephalon and pygidium of *Ceratopeltis*.

furrows, but a better preserved specimen (e.g. Pl. 6, fig. 10) does not; the former is probably an artefact of weathering. Short preglabellar field of about the same (sag.) length as the occipital ring. Curved palpebral lobes about one-third glabellar length, not close to glabella, placed far back such that their hind ends are opposite the occipital furrow. Postocular cheek thus narrow and spine-like laterally, defined by relatively deep furrow. Preocular cheeks broad (tr.) and gently sloping to border. Anterior sections of facial suture diverge at about 30°–40° to the sagittal line taking relatively straight course until curving adaxially on border; posterior sections highly divergent. Cranidial rim well defined, narrow, gently convex.

Triangular free cheek shows continuation of rim as lateral border, fading out towards genal spine. Postocular suture presumably cut margin very near base of spine, which is quite short. Plate 6, fig. 4 clearly shows the moderate-sized curved eye.

Pygidium including spines slightly wider than long. Narrow axis tapers gently, extending to two-thirds sagittal length of pygidium, and carrying one well-defined axial ring, and probably a second; long terminal piece. One partially abraded specimen (Pl. 6, fig. 3) shows detailed pleural structure. Only two segments are defined, and their expression is confined to a small, gently convex adaxial area. Two deep pleural furrows stop at the broad border, the second sloping backwards more steeply than the anterior one. One much weaker pair of interpleural furrows. The regions behind the second pleural furrows are inflated as a pair of drop-like lobes. The well-preserved pygidium on Plate 6, fig. 5 shows that the anterior pleural furrows pass into the proximal parts of the first pair of pygidial spines; these are developed from the first pygidial segment. They are long, and slightly inward-curved distally on most specimens. The second pair, by contrast, are developed from the end of the postaxial field and are extensions of the border. They are short, triangular with an acute V between them. The anterior pair extend back almost on a level with the posterior pair. Little surface sculpture is preserved, apart from fine transverse lines on the postaxial part of the pygidium. Ventral preservation (Pl. 6, fig. 11) shows broad, pygidial doublure.

Discussion. In spite of its remarkable spinose pygidium *Ceratopeltis* is obviously a bathyurid. The long, anteriorly rounded glabella, sutures, and free cheek are typical of the family. Notwithstanding the large pygidium we include it within the Bathyurinae rather than the Bathyurellinae. Relatively few bathyurids have so few segments expressed on the pygidial pleurae, and all these are bathyurines. The genus *Catochia* Fortey, 1979, from the early Ordovician part of the St. George Group, western Newfoundland, has a small pygidium with only two pleural segments; the pygidial pleural development of *C. ornata* is quite similar to that of *Ceratopeltis*. The free cheek of *Ceratopeltis* is not unlike that of the early bathyurid *Peltabellia* Whittington. Bathyurellinae typically have broad, blade-like genal spines.

Only one bathyurid has pygidial morphology which can be compared with *Ceratopeltis*. This is the type and only species of *Lutesvillia* Cullison, 1944, *L. bispinosa*, from the Mid-Canadian Theodosia Formation of Missouri. There is an obvious resemblance between this form and *Ceratopeltis*. The cranidium differs only in that the glabella is somewhat acuminate anteromedially, and the cranidial border furrow is subdued. The pygidium of *Lutesvillia* also has few, short pleural furrows stopping at a broad border; the lateral spines are rather short and stubby. The main difference resides in the absence of the second pair of spines in *Lutesvillia* (Cullison, 1944, pl. 35, fig. 3). It seems likely that *Ceratopeltis* and *Lutesvillia* are members of a single phylogenetic group, the Greenland form developing an extra pair of spines as extensions of the border. This difference may not be of much taxonomic importance. If they were to be accommodated within a single genus *Ceratopeltis* would be the senior name.

There is one other bathyurid with pygidial marginal spines: *Pseudoolenoides* Hintze, 1953. This is from a younger horizon (Whiterock) than *Ceratopeltis*. It differs from the latter in lacking a preglabellar field, in having incised 1P glabellar furrows, and in its narrow pygidium without a distinct border. It is closest to *Acidiphorus* Raymond in most characters, and unrelated to the *Ceratopeltis–Lutesvillia* group.

HOMOEOMORPHS OF CERATOPELTIS

The pygidial structure of *Ceratopeltis* is unusual among trilobites, and it might be thought improbable that a similar morphology could have developed independently in another trilobite group. Remarkably, though, there are two contemporary genera which are almost exactly comparable with regard to the marginal pygidial spines. These are *Hungioides* Kobayashi, 1936, and *Omeipsis*, Kobayashi, 1951. The former is a dikelokephaline with a wide distribution from Bohemia and Thuringia to south-west China. An undescribed Arenig species from the Nora Formation of Queensland, Australia, is under study by Fortey and J. S. Shergold. It is a near contemporary of *Ceratopeltis*. *Omeipsis* is a member of the family Taihungshaniidae; it has been redescribed by Lu (1975). In all three genera the anterior pair of spines originate as pleural extensions of the first pygidial segment, whereas the second pair of spines are shorter and appear to originate from the border, without obvious relationship to the pygidial segmentation. The taihungshaniids may belong to the superfamily Cyclopygacea *sensu* Fortey 1981; the Dikelokephalinidae are probably related to the Ceratopygacea. Hence, the three homoeomorphs are derived from three

---

PLATE 6

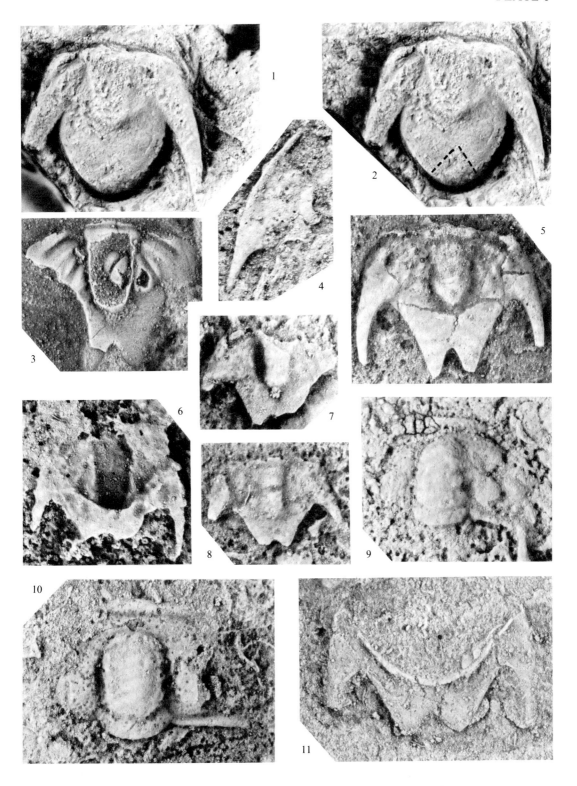

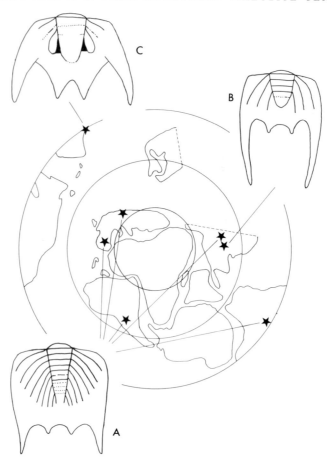

TEXT-FIG. 3. Occurrence of the homoeomorphs *Hungioides* (A)
and *Omeipsis* (B) in the Lower Ordovician of Gondwanaland.
Ceratopeltis (C) is currently known only from Greenland.

different ptychoparioid superfamilies, and have no phylogenetic relationship one to another. This example shows that it is necessary to be cautious before using pygidial characters of the type discussed here in any assessment of phylogeny.

On an early Ordovician continental reconstruction (text-fig. 3) Greenland lay near the palaeo-equator. The distribution of *Hungioides* is quite extensive: to the occurrences above we may add the Iberian Peninsula (I. Rabano, pers. comm.) and Argentina, if the inadequately defined *Argentinops* Přibyl and Vaněk, 1980, is regarded as a synonym of *Hungioides*, as seems likely. These occurrences are all on the earlier Ordovician Gondwanaland, and far removed from Greenland. We have little information on the facies in which *Hungioides* occurs, but in central Australia at least it is common in inshore, carbonate/clastic rocks, a platformal habitat not very different from that in the earlier Ordovician of Greenland. *Omeipsis* is so far known only from southern China (Kweichou). Presumably all three homoeomorphs evolved in parallel at virtually the same time in response to some specific niche in the epeiric habitat. The biogeographically restricted, inshore family Bathyuridae was recruited into the role in Greenland to produce *Ceratopeltis*, the equally restricted Taihungshaniidae and Dikelokephalinidae produced closely similar morphologies in the perimeter of the Gondwanaland continent.

Acknowledgements. Magister Søren Floris (Geological Museum, Copenhagen) kindly arranged the loan of Poulsen's holotype. Peel publishes with permission of the Director, Geological Survey of Greenland.

REFERENCES

COCKS, L. R. M. and FORTEY, R. A. 1982. Faunal evidence for oceanic separations in the Palaeozoic of Britain. *J. Geol. Soc. London*, **139** (4), 465–478.

CULLISON, J. S. 1944. The stratigraphy of some Lower Ordovician formations of the Ozark Uplift. *Bull. Sch. Min. Metall. Univ. Missouri*, **15** (2), 1–112.

FORTEY, R. A. 1979. Early Ordovician trilobites from the Catoche Formation (St. George Group), western Newfoundland. *Bull. geol. Surv. Canada*, no. 321, 61–114.

—— 1981. *Prospectatrix genatenta* (Stubblefield) and the trilobite superfamily Cyclopygacea. *Geol. Mag.* **118** (6), 603–614.

HINTZE, L. F. 1953. Lower Ordovician trilobites from western Utah and eastern Nevada. *Bull. Utah Geol. Miner. Surv.* **48**, 1–249.

KOBAYASHI, T. 1936. Three contributions to the Cambro-Ordovician faunas. *Jap. J. Geol. Geogr.* **13**, 163–184.

—— 1951. On the Ordovician trilobites in central China. *J. Fac. Sci. Univ. Tokyo*, ser. II **8** (1), 1–87.

LU, YEN-HAO 1975. Ordovician trilobite faunas of central and southwestern China. *Palaeontologica Sin.* N.S. B, **11**, 1–463. [In Chinese and English.]

PEEL, J. S. 1982. The Lower Palaeozoic of Greenland. Proc. IIIrd Arctic Symposium Calgary 1981. *Can. Soc. Petrol. Geol. Mem. 8*, 309–330.

POULSEN, C. 1937. On the Lower Ordovician faunas of East Greenland. *Meddr Grønland*, no. 119, 1–72.

PŘIBYL, A. and VANĚK, J. 1980. Ordovician trilobites of Bolivia. *Rozpr. cesl. Akad. Ved.* 90 (2), 1–90.

WHITTINGTON, H. B. 1963. Middle Ordovician trilobites from Lower Head, western Newfoundland. *Bull. Mus. comp. Zool. Harvard*, no. 129, 1–118.

—— and HUGHES, C. P. 1972. Ordovician geography and faunal provinces deduced from trilobite distributions. *Phil. Trans. R. Soc. Lond.* ser **B, 263**, 235–278.

R. A. FORTEY

Department of Palaeontology
British Museum (Natural History)
Cromwell Road,
London SW7 5BD

J. S. PEEL

Geological Survey of Greenland
Øster Voldgade 10
DK-1350 Copenhagen K
Denmark

PROGENESIS IN TRILOBITES

by K. J. MCNAMARA

ABSTRACT. Progenesis in trilobites is considered to have occurred by two different mechanisms. These are herein termed *terminal progenesis* and *sequential progenesis*. When terminal progenesis has operated the descendant paedomorph will have initially followed the same meraspid ontogenetic pathway as its ancestor. Thus, degree of morphological change between moults and the time spent intermoult will have been the same. Precocious maturation by earlier inhibition of the juvenile hormone would have resulted in the descendant paedomorph moulting fewer times than its ancestor and producing fewer thoracic segments. Sequential progenesis occurred by a reduction in the period between moults during the meraspid stage of development, by precocious activation of moulting hormone at each moulting event. Consequently, the degree of morphological change apparent between instars will have been reduced. Although the descendant has passed through the same number of moults as its ancestor before onset of maturity, the reduced period between moults means that on attaining maturity the descendant will be paedomorphic and smaller than its ancestor. However, the same number of thoracic segments will have been generated as in the ancestor. This process is quite different from neoteny in trilobites, whereby the *rate* of morphological development between moults was reduced. In terminal and sequential progenesis the rate of morphological development was the same as in the ancestor. Examples of these two processes in trilobites are given.

ONE of Harry Whittington's most important contributions to the study of trilobites has been his recognition of the importance of studying ontogenetic development in order to assess phylogenetic relationships. In particular he has reiterated, in a recent review of the role of paedomorphosis in trilobites (Whittington 1981), the importance of comparing early growth stages to indicate taxonomic affinities. One of the obvious outcomes of such studies is an analysis of the effect of variations in rate of development in the early part of the trilobite's ontogeny. Whittington (1957), quite correctly, was sceptical of the assertions of Stubblefield (1936) and Hupé (1953*a, b*) that paedomorphosis was an important feature of trilobite evolution. The evidence at that time was, as Whittington pointed out, tenuous to say the least. In his recent review of the topic, Whittington (1981) has outlined how, since his earlier review, some detailed studies have been carried out (Robison 1967; Robison and Campbell 1974; McNamara 1978, 1981*a*; Ludvigsen 1979; Fortey and Rushton 1980). These studies have demonstrated that paedomorphosis can explain the origin of some Cambrian and Ordovician species.

With the publication by Gould (1977) of a detailed review of the relationships between ontogeny and phylogeny, in which great emphasis was placed on the different processes affecting growth and development, the path was clear for an even more detailed appraisal of the role of paedomorphosis in the evolution of trilobites. This could be accomplished by an analysis of different paedomorphic processes. Gould (1977) illustrated how retention of ancestral juvenile characters in a descendant (*paedomorphosis*) could occur by two quite different processes: *neoteny* and *progenesis*. When neoteny operates, rate of morphological development through ontogeny is reduced. If sexual maturity is attained in the descendant at the same time as in the ancestor, the morphological stage reached is less advanced; thus the descendant adult resembles a juvenile of the ancestor. It may be of similar size to the ancestor, or even larger, if onset of maturity is delayed. Examples in trilobites are *Pseudogygites* (Ludvigsen, 1979) and some species of *Xystridura* (McNamara, 1981*a*).

Evolution of ancestral juvenile characters in the descendant adult can also occur by precocious onset of maturity. This was termed progenesis by Giard (1887). As Alberch *et al.* (1979) have illustrated in their classification of heterochronic processes, when progenesis has operated, ontogenetic development is initially the same in the descendant as in the ancestor. However, earlier onset of maturity results in cessation of development, and the retention of ancestral juvenile

[Special Papers in Palaeontology No. 30, pp. 59-68]

characters in the descendant adult. As a consequence of the shorter period spent in the juvenile phase of growth, the progenetic form will, at maturity, be smaller than its progenitor. This size difference between progenetic adults and the adults of their progenitors, allows a relatively simple characterization of a progenetic species. Size is less diagnostic for neotenic species as they may reach the same, or a larger, size than the descendant adult.

In recent times a number of progenetic trilobites have been recognized (McNamara 1978, 1981*a*, *b*, 1982; Whittington 1981). Some of these examples were described prior to Gould's (1977) work, and application of heterochronic terminology was often less than rigorous. For instance, Robison and Campbell (1974) described the paedomorphic *Thoracocare* as a neotenic species, whilst Fortey and Rushton (1980) described species of *Acanthopleurella* as paedomorphic, but distinguished them neither as neotenic, nor as progenetic. Both *Thoracocare* and *Acanthopleurella* are clearly progenetic forms (see below). Other trilobites which also show progenetic characters include species of *Olenellus* (McNamara 1978) and some zacanthoidid genera, such as *Vanuxemella* (McNamara 1981*b*). Progenesis has also been described in the Middle Cambrian Xystridurinae (McNamara 1981*a*). The clear retention by *Galahetes fulcrosus* Öpik of morphological characters present in meraspids of all earlier and contemporaneous xystridurines, and the much smaller adult size attained by *Galahetes*, indicate attainment of sexual maturity at an earlier stage of growth than in other xystridurines. In other words, *Galahetes* fulfils the criteria necessary for characterization as a progenetic trilobite.

However, *Galahetes* has one particular anomalous character: it possesses the same number of thoracic segments as its xystridurine ancestor. If, as is generally accepted, thoracic segments in trilobites were generated at a reasonably even rate throughout the juvenile meraspid phase and, at the attainment of the full complement of thoracic segments, sexual maturity was reached and morphological development was greatly reduced or ceased, how was *Galahetes* able to retain ancestral juvenile characters, and a much smaller size? Why does it not have fewer thoracic segments, like other progenetic trilobites?

The aim of this paper is essentially to re-analyse the heterochronic process which allowed the evolution of *Galahetes*. It will be proposed that because trilobites, like other arthropods, grew by moulting, progenesis was able to occur in two quite different ways. This is because moulting and attainment of maturity are under the control of two different hormonal systems in arthropods. The morphological effects of variations in these hormonal systems, analogues of hormonal systems in living arthropods, will be discussed and their role in trilobite evolution assessed.

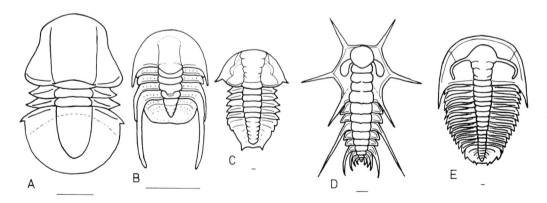

TEXT-FIG. 1. A–D, terminal progenetic trilobites; E, sequential progenetic trilobite. A, *Thoracocare minuta* (Resser) (after Robison and Campbell 1974, fig. 1H). B, *Acanthopleurella grindrodi* Groom (after Fortey and Rushton 1980, fig. 20a). C, *Vanuxemella nortia* Walcott (after Rasetti *in* Moore 1959, fig. 168,3). D, *Olenellus armatus* Peach (after McNamara 1978, text-fig. 1). E, *Galahetes fulcrosus* Öpik (after McNamara 1981*a*, fig. 8B). Bars represent 1 mm.

TERMINAL PROGENESIS IN TRILOBITES

A number of examples of progenetic trilobites, which satisfy the criteria necessary for description as examples of the activity of 'pure' progenesis, have been described (text-fig. 1): *Thoracocare* (Robison and Campbell 1974), species of *Olenellus* (McNamara 1978), species of *Acanthopleurella* (Fortey and Rushton 1980), and the zacanthoidids *Vanuxemella*, *Albertella*, and *Paralbertella* (McNamara 1981*b*). By comparison with assumed ancestral forms, these 'pure' progenetic trilobites are characterized by their retention, as adults, of ancestral juvenile characters; their small size; and a reduced number of thoracic segments. Although not known with certainty, as it has not been possible to compare directly meraspid development of both ancestor and descendant, it is possible to reasonably confidently suggest that both the ancestor and the 'pure' progenetic descendant followed the same ontogenetic trajectory prior to attainment of maturity (Alberch *et al.* 1979, fig. 15b). In other words, size and shape changes were the same and changed at the same rate.

Trilobites, like other arthropods, grew by moulting. Thus the morphology and number of instars will be the same in ancestor and descendant. Prior to maturity ancestor and descendant instars will, therefore, be identical and impossible to separate taxonomically. 'Pure' progenesis occurs in trilobites simply by a variation in onset of maturity. 'Pure' progenesis in arthropods, which express their morphology in a discontinuous manner throughout growth, is herein termed *terminal progenesis*, in order to distinguish it from other forms of arthropod progenesis described below.

TABLE 1. A comparison of the major characteristics of the paedomorphic processes in trilobites (relative to the ancestral state).

	Adult size	Rate of morphological development	No. of moults	Time between moults
Terminal progenesis	Smaller	Same	Fewer	Same
Sequential progenesis	Smaller	Same	Same	Reduced
Neoteny	Same	Reduced	Same	Same

The saltatory mode of morphological expression in arthropods does not necessarily reflect an underlying episodic rate of growth. Cellular division, and thus morphological development, is a nearly continuous process (Phillips *et al.* 1980) and can only be expressed when there is an increase in size at time of moulting because of the rigid nature of the exoskeleton. As Alberch *et al.* (1979) have stressed, growth involves both morphological and size changes. The two, however, need not always progress in tandem. Thus in trilobite growth, and in the growth of other arthropods, morphological development will be largely continuous whereas size increase is a discontinuous process. Consequently, the morphological expression of an instar is dependent upon the stage and rate of moulting.

If, as shown, meraspid growth in terminal progenetic trilobites followed the same morphological pathway as meraspid growth in the ancestor, the period between moults must have been the same. Thus if moulting occurred at the same rate in the descendant as in the ancestor, so long as the rate of morphological development between moults was the same in both, ontogenetic pathways will have been identical. Should rate of morphological change have been reduced, however, but moulting have occurred at the same rate, such that size increase at each moult was identical, then neoteny will have ensued (Table 1). The same number of moults, and thus the same number of thoracic segments, will have been produced through juvenile growth irrespective of the number of thoracic segments generated per moult.

The variation in size of individual meraspids of the same degree led Evitt (1953) and Whittington (1957) to suggest that trilobites may have undergone more than one moult per meraspid degree. It has also been suggested (Whittington 1957) that under certain circumstances more than one segment may have been released into the thorax at one moult. Be that as it may, a reduction in rate of morphological development between moults would have resulted in a neotenic descendant adult as large as the ancestor but possessing juvenile characters (text-fig. 3). An example is the Ordovician trilobite *Pseudogygites*, which evolved from *Isotelus* by such a reduction in rate of development (Ludvigsen 1979).

Where progenesis has occurred, morphological differences between adult and descendant may be appreciable. The significance of this to modern views on saltatory modes of evolution has been discussed elsewhere

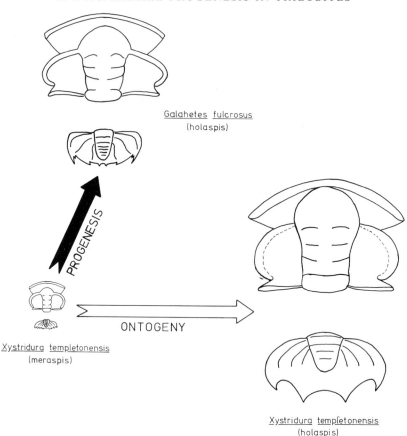

TEXT-FIG. 2. Illustration of xystridurine cranidia and pygidia depicting the suggested evolution of *Galahetes* from *Xystridura* by progenesis. Note the close similarity of *Galahetes* holaspid to the meraspid of *Xystridura*. All species of *Xystridura* attain a larger maximum size than *Galahetes*. The *Xystridura* reconstructions are based on *X. templetonensis*. Holaspid reconstructions × 2; meraspid reconstructions × 4.

(McNamara 1982). In trilobites, where the juvenile stage was probably pelagic even when the adult was benthic, the occurrence of progenesis may have resulted in the selection of forms which spent their entire lives as pelagic species. A consequence of progenesis is the large variation in both morphology and size between ancestor and descendant. Many resultant characters, however, may have been of little or no initial adaptive significance, a common occurrence when alterations to regulation of development provided the morphological variability (Gould and Lewontin 1979). It is possible that, on occasions, selection may have been for the small size alone.

SEQUENTIAL PROGENESIS IN TRILOBITES

In a recent analysis of the evolutionary relationships of the anomalous xystridurine *Galahetes fulcrosus* (McNamara 1981*a*), it was suggested that *G. fulcrosus* evolved by progenesis. This species possesses two of the three progenetic criteria: a paedomorphic adult form and a much smaller maximum adult size than other xystridurine species. The progenetic nature of *Galahetes* has been illustrated by a series of measurements of cranidial parameters which have been compared with the ontogenetic sequence of an ancestral xystridurine (McNamara 1981*a*, figs. 2–5). These measurements clearly demonstrate that whereas attainment of the holaspid state in the ancestral xystridurine occurs at a cranidial length of 5·0–5·5 mm, the holaspid condition in *Galahetes* is attained at a cranidial length of only 2·0–2·5 mm.

The xystridurine cranidium undergoes large morphological change during meraspid growth (Öpik 1975; McNamara 1981a). As *Galahetes* spent a shorter period of time in this rapid growth phase the degree of morphological change was less than in other xystridurines. Consequently, juvenile xystridurine characters are preserved in the adult *Galahetes*. These characters include: retention of a broader cranidium and a preglabellar field; less tapered frontal lobe; more well-defined, but narrower, eye lobes which are set farther from the glabella; second glabellar furrows which join the axial furrow; less anteriorly divergent anterior branches of the facial suture; poorly developed thoracic doublure; and a greater number of axial segments in the pygidium (text-fig. 2). There can be no doubt that *Galahetes* is not just a small form of another xystridurine species, because early holaspids of species such as *Xystridura templetonensis* (Chapman) are morphologically more advanced than holaspids of *Galahetes* three times as large (McNamara 1981a, fig. 7).

The attainment of the adult complement of thoracic segments in *Galahetes* occurs at a cranidial length of 2·0–2·5 mm. This correspondence between attainment of the full adult number of thoracic segments and the reduction in degree of morphological change also occurs in the other xystridurine species. However, unlike other progenetic trilobites, *Galahetes* does not have fewer thoracic segments than its ancestor. All adult xystridurines possess thirteen thoracic segments. Thus a process different from terminal progenesis must be invoked in order to explain the evolution of *Galahetes*.

The simplest explanation which, by analogy with modern arthropods, can be attributed to hormonal mechanisms, involves a variation in the rate of moulting. In order to produce the same number of thoracic segments in the descendant as in the ancestor, the rate of moulting must have been increased. Thus, if the time spent between moults in the descendant was shortened compared with the ancestor, the period during which morphological development occurred was reduced. Thus at each successive moult the progenetic descendant will have fallen farther and farther behind the ancestral form in morphological development. Each successive moulting event will have been precociously initiated. Thus each descendant moult may be considered to have undergone a mini-progenesis. Such a phenomenon is herein termed *sequential progenesis*.

It must be stressed that when sequential progenesis occurred the rate of morphological development was not reduced. A shortening of the period between moults would have meant that, in addition to a reduction in the time available for morphological development, the period during which size increase between moults could have occurred was reduced. If, as in the ancestor, the same number of moults occurred before attainment of maturity, with the operation of sequential progenesis the resultant adult would have produced the same number of thoracic segments as its ancestor, but have increased less in size and undergone less morphological development. This explanation is based upon the assumption that the ratio of number of moults to number of thoracic segments generated was unchanged between ancestor and descendant.

The distinction between terminal progenesis and sequential progenesis in trilobites can be clearly seen by referring to text-fig. 3. For convenience the ancestral form is considered to have passed through thirteen moults in the meraspid period, to correspond with the thirteen thoracic segments present in the Xystridurinae. This is not meant to imply that one thoracic segment was produced per moult. The exact number is unknown. In this diagrammatic graph, which illustrates the morphological changes between moults, instars are given the morphological states A–M. The degree of morphological development is shown to decrease progressively throughout juvenile growth. By reducing the time spent in each intermoult period by, in this case, a half, a descendant trilobite also passing through thirteen moults will, at maturity, have only reached morphological state D. This is sequential progenesis. The form is paedomorphic, smaller than the ancestor, but has the same number of thoracic segments. The increase in rate of moulting resulted in precocious maturation. A terminal progenetic trilobite will follow the same initial ontogenetic pathway as the ancestor. By precocious onset of maturity the derived paedomorph will have passed through fewer moults than the ancestor, and have generated fewer thoracic segments. For comparison, the ontogenetic pathway of a neotenic descendant is also illustrated. The size attained is also the same as the ancestor, because the same number of moults was produced, and at the same rate, but rate of morphological development between moults was reduced. Thus morphological state D is also only reached but by a third quite different process.

An alternative interpretation for the evolution of *Galahetes* is that the trilobite underwent terminal progenesis after passing through an early meraspid period when thoracic segments were generated more frequently than in the ancestor. Suppose, for example, that the ancestral xystridurine moulted three times before release of each successive thoracic segment, whereas the progenetic form released a thoracic segment after only two moults. In this scenario thirteen thoracic segments would be attained at a smaller size. Sexual maturity would, then, have had to have been initiated after fewer moults. In other words, two different changes would have had to have occurred in conjunction: a more rapid release of segments and attainment of maturity after fewer moults than the ancestor, coinciding with production of the ancestral number of thoracic segments. But maturity in trilobites would not have been controlled by the production of a set number of thoracic segments. On the contrary, the

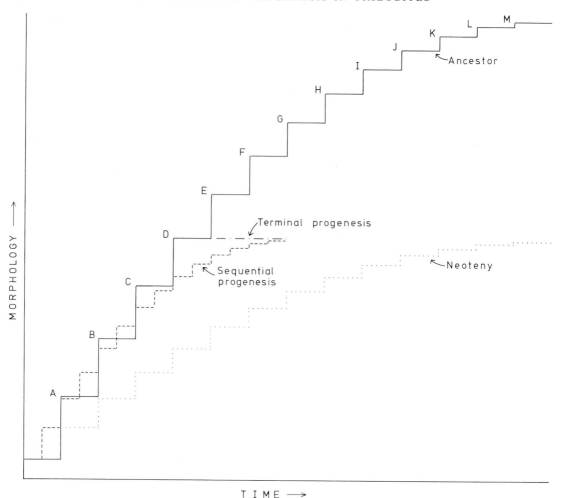

TEXT-FIG. 3. Diagrammatic plot of the meraspid period of growth of an ancestral trilobite (continuous line), whose moults express the morphological states A–M. Morphological development between moults is shown to progressively decrease throughout the meraspid period. Paedomorphic descendants may attain only morphological state D in their adult form in three ways (excluding post-displacement (Alberch *et al.* 1979), which is not relevant to this discussion): by *terminal progenesis* (broken and dotted line), whereby precocious maturation by premature cessation of secretion of juvenile hormone causes moulting to an adult form of state D; by *sequential progenesis* (broken line), whereby a reduction (in this case by a half) of the period between moults, by premature secretion of the moulting hormone, restricts the degree of morphological development between moults; or by *neoteny* (dotted line), whereby rate of morphological development is reduced. The same number of moults and thoracic segments are produced in the neotenic and sequential progenetic descendants as in the ancestor; the terminal progenetic form produces less.

number of thoracic segments would have been under the control of hormonal mechanisms which induce the onset of maturity. In living arthropods the number of moults is critical, because of the known relationship between the activity of moulting and juvenile hormones (see below). Furthermore, the reduction in number of thoracic segments in terminal progenetic trilobites suggests that, in those situations, juvenile growth in ancestors and descendants was identical, with the same number of moults relative to the number of thoracic segments released.

The model of sequential progenesis presented above is therefore considered to be the most parsimonious explanation for the observed phenomena in *Galahetes* and is in accord with observed effects of alterations to moulting hormonal systems in living arthropods. *Galahetes* would have undergone the same number of moults as its ancestor; rate of thoracic segment production per moult would have been the same, as would the number of segments produced (Table 1).

Other trilobites which have been described as paedomorphic forms may be interpreted as having arisen by sequential progenesis. The olenellids *Daguinaspis* and *Choubertella* were considered by Hupé (1953a) to be paedomorphic descendants of *Fallotaspis*. *Daguinaspis* and *Choubertella* are small genera, barely reaching the size of the smallest species of *Fallotaspis*. In addition, the occurrence of the same number of thoracic segments in all three genera suggest that *Daguinaspis* and *Choubertella* might have evolved by sequential progenesis.

Although the progenetic *Olenellus armatus* Peach may be considered to be a terminal progenetic species (McNamara 1978), *O. reticulatus* Peach, another progenetic species, shares the same number of thoracic segments as its presumed progenitor, *O. lapworthi* Peach and Horne. Its smaller maximum size and paedomorphic features suggest that it, too, may have evolved by sequential progenesis.

It is possible, of course, that both terminal and sequential progenesis may operate together. Identifying the effect of these two processes in one paedomorph is a difficult exercise. If progenesis can occur by two different processes in trilobites (and potentially in other arthropods), then it is probable that extension of the descendant ontogenetic trajectory beyond that of the ancestor (hypermorphosis) may also occur by two different processes, *sequential hypermorphosis* and *terminal hypermorphosis*. To date, examples of hypermorphosis in trilobites have not been recognized. A terminal hypermorphic trilobite would be expected to have generated more thoracic segments than its ancestor, whereas the sequential hypermorphic trilobite would have produced the same. Distinguishing between peramorphs of trilobites produced by acceleration and sequential hypermorphosis could prove difficult. An accelerated form, however, is unlikely to be larger than the ancestral form, whereas the sequential hypermorph is.

FACTORS CONTROLLING TRILOBITE DEVELOPMENTAL REGULATION

A copious literature dealing with the effect of endocrine systems on insect and crustacean morphogenesis (e.g. Wigglesworth 1936, 1965; Wyatt 1972; Doane 1973; Gilbert and King 1973; Willis 1974; Gilbert 1974; Ashburner 1980; Aiken 1980) has, apart from unravelling the mechanisms of the various hormonal systems, revealed the close similarity in the control of moulting and maturation in these two groups of arthropods. In particular, the moulting hormone ecdysone has been found to induce moulting in a number of groups of arthropods, including insects, crustaceans, arachnids, and xiphosurans (Wyatt 1972). Similarly, the other major hormone controlling morphogenesis, the juvenile hormone, acts in a similar manner in both insects and crustaceans. As Wyatt (1972, p. 410) has suggested, it is thus likely that ecdysones were used to regulate moulting in arthropods prior to the divergence of the major arthropod groups in the Precambrian.

The probable closer relationship between trilobites and crustaceans than between crustaceans and insects (Eldredge 1977), yet the close similarity in endocrine systems between these latter two groups, further suggests that trilobite morphogenesis is likely to have been controlled by hormonal systems analogous to those found in living arthropods. Certainly, many of the processes which induced morphological changes between ancestors and descendants in trilobites can be explained by reference to experiments which have been carried out on living arthropods.

The moulting sequence and the morphological expression of the moult in arthropods is under endocrine control. Neurosecretory cells in the brain produce a brain hormone which stimulates the synthesis and release of the moulting hormone, ecdysone. This moulting hormone initiates the moulting process. Independent secretion of the juvenile hormone by specific glands modifies the expression of the moult and acts in conjunction with the moulting hormone. It is possible that the juvenile hormone may even act directly on regulatory genes, repressing their activity (Williams and Kafatos 1971; Ashburner 1980). While the juvenile hormone is present at a sufficiently high titre, the organism will display juvenile morphological characteristics. When this hormone ceases to be produced, metamorphosis to the adult phase occurs.

The classic work of Wigglesworth (1936) on the effects of experimentally inducing precocious maturation in the assassin bug *Rhodnius* by artificially excising the source of the juvenile hormone, revealed the great importance of this hormone to the timing of maturation and the morphological character of the adult. Similar experiments on the silkworm *Bombyx* (Bounhiol 1936, 1937) and the phasmid *Dixippus* (Pflugfelder 1937) have shown how paedomorphs could be produced by artificially reducing the number of juvenile moults. It seems likely that terminal progenesis in trilobites occurred by a similar process: precocious onset of maturation caused by premature inactivation of the juvenile hormone. The effect of temperature in controlling terminal progenesis in the mealworm *Tenebrio mollitor* (Singh-Pruthi 1924) and *Rhodnius* (Wigglesworth 1965) have led to the suggestion (McNamara 1978) that temperature variations may have been influential in controlling progenesis in olenellid trilobites.

It is probable that sequential progenesis in trilobites was also under endocrine control, but resulted from the activity of a different hormonal system. The moulting sequence in arthropods is initiated by the secretion of ecdysone. This activates the epidermal cells to secrete a new cuticle. Alteration to the timing of the initiation of deposition of the new cuticle will affect the degree of intermoult morphological development. Thus, if cuticle deposition occurs soon after the previous moult there will be little morphological change between instars. If, however, the onset of deposition of new cuticle is inhibited by delay in ecdysone secretion, then there will be a greater degree of morphological development between instars (Wigglesworth 1965). Similarly, it has been shown (Wigglesworth 1933) that artificially inducing the bed-bug *Cimex* to precociously deposit its cuticle, results in the appearance of certain morphological characters in an intermediate stage of development. Similar effects have been obtained by accelerating onset of moulting in *Rhodnius* (Wigglesworth 1961) and in *Hyalophora* (Staal 1968; Willis 1974).

Thus it is considered that sequential progenesis in trilobites resulted from changes to the regulation of ecdysone secretion. By precocious secretion of ecdysone at each moulting event through meraspid development, the sum of the morphological changes undergone during this period of development was reduced with respect to the ancestor. Cessation of juvenile hormone production would have occurred after the same number of moults had been produced as in the ancestor. This would therefore have resulted in the generation in the descendant of the same number of thoracic segments as in the ancestor, but retention of juvenile ancestral characters in its adult phase.

EVOLUTIONARY SIGNIFICANCE OF CHANGES TO DEVELOPMENTAL REGULATION

The evolutionary significance of alterations to regulatory gene systems, and the effect of these alterations on changes to developmental rates has been stressed by Valentine and Campbell (1975) and Valentine (1977). Adopting Britten and Davidson's (1971) model for gene regulation, whereby small numbers of specific regulatory genes act to modify the timing of expression of a large number of influential structural genes, they have argued that such resultant developmental changes may have played a crucial role during early metazoan evolution. Changes to developmental regulation will, they argue, have allowed the rapid evolution of a wide range of morphotypes capable of exploiting and adapting to a wide range of ecological habitats. If, as in living arthropods, hormones controlling moulting and maturation in trilobites also acted directly upon regulatory genes, then any alteration to the timing of activation of these hormones, perhaps caused by environmental factors, could have had profound morphological effects on the organism. Aiken (1980) has shown that regulation of moulting and maturation in decapod crustaceans may be influenced by many environmental factors, such as food, light, temperature, salinity, parasitism, mutilation, social environment, substrate, and water quality.

As I have argued elsewhere (McNamara 1981b), the higher incidence of progenesis in trilobites in the Cambrian than in later trilobites, suggests that early in the evolution of the Trilobita hormonal mechanisms were more suspectible to changes in regulation, perhaps influenced by fluctuating environmental conditions. This view is supported by examination of the taxonomic status of numbers of thoracic segments in Cambrian and post-Cambrian trilobites. Numbers of thoracic segments may vary interspecifically in some early Cambrian genera (such as *Emuella* and *Olenellus*); vary between genera in many middle and late Cambrian forms (such as in the Zacanthoididae and Oryctocephalidae); but vary only between families in some post-Cambrian forms (such as the Trinucleidae) or even between suborders (such as the Phacopina). Clearly, there are some post-Cambrian trilobite families and genera in which the number of thoracic segments vary, but there are few supragenetic Cambrian taxa in which the number of thoracic segments does not vary between genera or species.

Consequently, the appearance of a wide range of trilobite morphotypes in the early Cambrian and their rapid diversification may, to a large degree, have occurred as the result of strong selection pressure on changes to developmental regulation. The capacity for the initiation of morphological change in trilobites by disruption of two hormonal systems increased the potential for heterochronic variation. Whereas other organisms which did not grow by moulting could undergo only 'pure' progenesis or neoteny, trilobites could provide, on account of the decoupling of size and shape changes, a wider range of paedomorphs by both terminal and sequential progenesis or by neoteny. The greater familial diversity of the Trilobita compared with other groups of organisms during the Cambrian (Sepkoski 1979) may be explained, in part, by the ability of the Trilobita to evolve such a wide range of morphotypes by these paedomorphic processes.

Acknowledgement. I wish to thank Professor Whittington for his help, advice, and encouragement over the years.

REFERENCES

AIKEN, D. E. 1980. Molting and Growth. *In* COBB, J. S. and PHILLIPS, B. F. (eds.). *The biology and management of Lobsters*, vol. 1, 91–163. Academic Press, New York.

ALBERCH, P., GOULD, S. J., OSTER, G. F. and WAKE, D. B. 1979. Size and shape in ontogeny and phylogeny. *Paleobiology*, **5**, 296–317.

ASHBURNER, M. 1980. Chromosomal action of ecdysone. *Nature*, **285**, 435–436.

BOUNHIOL, J.-J. 1936. Métamorphose après ablation des corpora allata chez le ver à soie (*Bombyx mori*, L.) *C.R. Acad. Sci. Fr.* **203**, 388–389.

—— 1937. Métamorphose prématurée par ablation des corpora allata chez le jeune ver à soie. Ibid. **205**, 175–177.

BRITTEN, R. J. and DAVIDSON, E. H. 1971. Repetitive and non-repetitive DNA sequences and a speculation on the origins of evolutionary novelty. *Q. Rev. Biol.* **46**, 111–133.

DOANE, W. W. 1973. Role of hormones in insect development. *In* COUNCE, S. J. and WADDINGTON, C. H. (eds.). *Developmental systems: insects*, vol. 2, 291–247. Academic Press, London.

ELDREDGE, N. 1977. Trilobites and evolutionary patterns. *In* HALLAM, A. (ed.). *Patterns of evolution as illustrated by the fossil record*. Pp. 305–332. Elsevier, Amsterdam.

EVITT, W. R. 1953. Observations on the trilobite *Ceraurus*. *J. Paleontol.* **27**, 33–48.

FORTEY, R. A. and RUSHTON, A. W. A. 1980. *Acanthopleurella* Groom 1902: origin and life-habits of a miniature trilobite. *Bull. Br. Mus. nat. Hist. (Geol.)*, **33**, 79–89.

GIARD, A. 1887. La castration parasitaire et son influence sur les charactères extérieurs du sexe male chez les crustacés décapodes. *Bull sci. Dept. Nord*, **18**, 1–28.

GILBERT, L. I. 1974. Endocrine action during insect growth. *Rec. Prog. Horm. Res.* **30**, 347–390.

—— and KING, D. S. 1973. Physiology of growth and development: endocrine aspects. *In* ROCKSTEIN, M. (ed.). *The physiology of Insecta*, 2nd edn., 250–370. Academic Press, New York.

GOULD, S. J. 1977. *Ontogeny and phylogeny*. 501 pp. Belknap Press, Cambridge, Massachusetts.

—— and LEWONTIN, R. C. 1979. The spandrels of San Marco and the Panglossian paradigm: a critique of the adaptationist programme. *Proc. R. Soc. Lond.* **B 205**, 581–598.

HUPÉ, P. 1953a. Contribution à l'étûde du Cambrien Inférieur et du Précambrien III de l'Anti-Atlas Morocain. *Notes Mém. Serv. Mines Carte géol. Maroc*, **103**, 1–402.

—— 1953b. Classification des trilobites. *Ann. Paléontol.* **39**, 1–110.

LUDVIGSEN, R. 1979. The Ordovician trilobite *Pseudogygites* Kobayashi in eastern and Arctic North America. *Life Sci. Contrib. R. Ont. Mus.* no. 120, 1–41.

MCNAMARA, K. J. 1978. Paedomorphosis in Scottish olenellid trilobites (early Cambrian). *Palaeontology*, **21**, 635–655.

—— 1981a. Paedomorphosis in Middle Cambrian xystridurine trilobites from northern Australia. *Alcheringa*, **5**, 209–224.

—— 1981b. The role of paedomorphosis in the evolution of Cambrian trilobites. *Open-File Rep. U.S. geol. Surv.* no. 81–743, 126–129.

—— 1982. Heterochrony and phylogenetic trends. *Paleobiology*, **8**, 130–142.

MOORE, R. C. (ed.) 1959. *Treatise on invertebrate paleontology*, pt. **O**, *Arthropoda* 1. Pp. xix + 560. Lawrence, Kansas.

ÖPIK, A. A. 1975. Templetonian and Ordian xystridurid trilobites of Australia. *Bull. Bur. Miner. Resour. Geol. Geophys. Aust.* no. 121, 1–84.

PFLUGFELDER, O. 1937. Bau, Entwicklung und Funktion der Corpora allata und cardiaca von *Dixippus morosus*. *Br. Z. wiss. Zool.* **149**, 477–512.

PHILLIPS, B. F., COBB, J. S. and GEORGE, R. W. 1980. General biology. *In* COBB, J. S. and PHILLIPS, B. F. (eds.). *The biology and management of Lobsters*, vol. 1, 1–82. Academic Press, New York.

ROBISON, R. A. and CAMPBELL, D. P. 1974. A Cambrian corynexochoid trilobite with only two thoracic segments. *Lethaia*, **7**, 273–282.

SEPKOSKI, J. J. 1979. A kinetic model of Phanerozoic taxonomic diversity II. Early Phanerozoic families and multiple equilibria. *Paleobiology*, **5**, 222–251.

SINGH-PRUTHI, H. 1924. Studies on insect metamorphosis: 1, prothetely in mealworms (*Tenebrio mollitor*) and other insects: effects of different temperatures. *Biol. Rev.* **1**, 139–147.

STAAL, G. B. 1968. Experimental evidence on the role of hormone in the larval development of insects. *Proc. Int. Congr. Entomol., 13th, Moscow*, p. 442.

STUBBLEFIELD, C. J. 1936. Cephalic sutures and their bearing on current classifications of trilobites. *Biol. Rev.* **11**, 407–440.

VALENTINE, J. W. 1977. General patterns of metazoan evolution. *In* HALLAM, A. (ed.). *Patterns of evolution as illustrated by the fossil record.* Pp. 27–57. Elsevier, Amsterdam.

—— and CAMPBELL, C. A. 1975. Genetic regulation and the fossil record. *Am. Sci.* **63**, 673–680.

WHITTINGTON, H. B. 1957. The ontogeny of trilobites. *Biol. Rev.* **32**, 421–469.

—— 1981. Paedomorphosis and cryptogenesis in trilobites. *Geol. Mag.* **118**, 591–602.

WIGGLESWORTH, V. B. 1933. The physiology of the cuticle and of ecdysis in *Rhodnius prolixus*; with special reference to the function of the oenocytes and of the dermal glands. *Q. Jl micr. Sci.* **76**, 269–318.

—— 1936. The function of the corpus allatum in the growth and reproduction of *Rhodnius prolixus*. Ibid. **79**, 91–121.

—— 1961. Some observations on the juvenile hormone effects of farnesol in *Rhodnius prolixus*, Stal. *J. Insect Physiol.* **7**, 73–78.

—— 1965. *The principles of insect physiology*, 6th edn. 741 pp. Methuen, London.

WILLIAMS, C. M. and KAFATOS, F. C. 1971. Theoretical aspects of the action of juvenile hormones. *Bull. Soc. Entomol. Suisse*, **44**, 151–162.

WILLIS, J. H. 1974. Morphogenetic action of insect hormones. *Ann. Rev. Entomol.* **19**, 97–116.

WYATT, G. R. 1972. Insect hormones. *In* LITWACK, G. (ed.). *Biochemical actions of hormones*, vol. 2, 385–490. Academic Press, New York.

K. J. MCNAMARA

Department of Palaeontology
Western Australian Museum
Perth
Western Australia 6000

CALYMENE LAWSONI AND ALLIED SPECIES FROM THE SILURIAN OF BRITAIN AND THEIR STRATIGRAPHIC SIGNIFICANCE

by DEREK J. SIVETER

ABSTRACT. Records of *Calymene lawsoni* Shirley, 1962, *C. neointermedia* R. and E. Richter, 1954, and *C. puellaris* Reed, 1920 from the British Silurian have been investigated and the material revised; these form an allied group of species though the occurrence of *C. neointermedia* in Britain is in doubt. *C. lawsoni* and *C. puellaris* are abundant and the latter widespread throughout much of the British Ludlow outcrop; both are represented in the lower Ludfordian, with *puellaris* ranging into the earliest upper Ludfordian and possibly the late upper Gorstian. *C. neointermedia*, the supposed acme of which has previously been widely taken to indicate Upper Leintwardine Formation (lower Ludfordian) and coeval strata, may occur only very rarely in Britain, in the upper Homerian Stage, Wenlock Series, of the Malvern area, the provenance of the material on which this record is based being in doubt. *C. lawsoni* is confined to Britain, while *C. puellaris* and *C. neointermedia* are found on Gotland. The stage in ontogeny at which *C. neointermedia* and *C. puellaris* attain the papillate-buttress structure at 2p lobe precludes them being related to '*C.*' *pompeckji* Kummerow, 1928 from Silurian erratics of north Germany.

THIS paper revises several hitherto poorly understood calymenid trilobites which occur, or were thought to occur, in the Ludlow of Britain. *Calymene lawsoni* has remained essentially undescribed

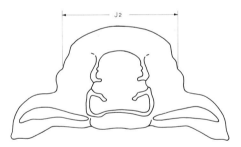

TEXT-FIG. 1. Variate J2 equals width (tr.) of cranidium at anterior extremity of palpebral lobe.

since it was established in 1962, *C. neointermedia* has assumed biostratigraphic importance in many regional accounts yet has commonly been misidentified and *C. puellaris* has been forgotten as a valid taxon. The main purpose of this account is to discuss British examples, though my conclusions regarding some of these species are also necessarily based on an examination of material from Gotland (which I will deal with more fully elsewhere), together with '*C.*' *pompeckji* from north German glacial erratics. The terminology, photographic and preparation techniques are those of Siveter (1977, 1979, 1980), as are the measurements (Siveter 1977, fig. 1), but with the addition of variate J2 (text-fig. 1). Trilobites used or cited are housed in the following institutions: British Museum (Natural History), London (BM); Geological Museum, Institute of Geological Sciences, London (GSM); National Museum of Wales, Cardiff (NMW); Sedgwick Museum, Cambridge (SM); Ludlow Museum, Salop (LM); Leicestershire Museum, Leicester (LEICS); Naturhistoriska

[Special Papers in Palaeontology No. 30, pp. 69–88, pls. 7–10]

Riksmuseet, Stockholm (RM); Paläontologisches Museum, Humboldt-Universität, Berlin (MB); Natur-Museum Senckenberg, Frankfurt am Main (SMF).

SYSTEMATIC PALAEONTOLOGY

Family CALYMENIDAE Milne Edwards, 1840
Subfamily CALYMENINAE Milne Edwards, 1840
Genus CALYMENE Brongniart, 1822

Type species. Calymene blumenbachii Brongniart, 1822; Wenlock Series, Much Wenlock Limestone Formation, Dudley, West Midlands, England.

Calymene lawsoni Shirley, 1962

Plate 7, figs. 1–18; Plate 8, figs. 1–6, 10, 14, 18, 19

?1851   *Calymene subdiademata* (M'Coy); M'Coy, pp. 166, 167, pl. 1F, figs. 9, 9a (specimen lost; see McNamara 1979, p. 71).
1954   *Calymene (Calymene) beyeri*?; R. and E. Richter, p. 19, pl. 2, figs. 25, 26.
?1960   *Calymene* cf. *beyeri* (Richter); Squirrell and Tucker, p. 177 (faunal list; material not seen).
1962   *Calymene* sp. nov.; Whitaker, pp. 345, 348, 350.
1962   *Calymene lawsoni* sp. nov. (*nomen nudum*); Shirley, p. 237.
?1967   *Calymene* cf. *beyeri* Richter; Phipps and Reeve, pp. 345, 367 (faunal list; material not seen).
1968   *Calymene lawsoni*; Shergold and Shirley, pp. 134, 135, pl. 15.
1968   *Calymene lawsoni* Shirley (*nom. nud.*); Haas, pp. 98, 99, text-fig. 16e, f.

Discussion. C. lawsoni has been cited in previous works as a *nomen nudum*, yet the requirements to make this an available name have been met. In 1954 R. and E. Richter figured (pl. 2, figs. 25, 26) as *C. (C.) beyeri*? two incomplete cranidia from the lower Ludfordian 'Mocktree Shales' of Park Farm at the south-western end of Wenlock Edge, and they differentiated them from their new species *beyeri*. Shirley (1962, p. 237) later referred to the Richters' publication and to *C. (C.) beyeri*?—for which he introduced the new name of *lawsoni*. The date and authorship of *C. lawsoni* therefore stems from Shirley's 1962 paper (see I.C.Z.N. Articles 13(a)(i) and (ii)).

Lectotype. Selected herein, internal mould of cranidium, BM It16000, collected Dr. J. H. McD. Whitaker (Univ. Leicester); Plate 7, fig. 14. Shirley (1962) did not formally designate a holotype and the syntypes must be regarded as the two cranidia the Richters collected (SMF 1652a, 1652b), together with those specimens (as then unregistered, now BM It16001–16029) from the Leintwardine area on loan to Shirley (1962, p. 237) from Whitaker. The lectotype 'is, I think, the specimen I pictured at the Bonn Meeting, 1960' (pers. comm. Dr. J. Shirley, Jan. 1982).

Type stratum and type area. Higher part of the Lower Leintwardine Formation, Ludfordian Stage, Ludlow Series, south side of Tatteridge Hill, Leintwardine area, Hereford and Worcester (SO 4209 7259). Field slip locality Le58 of Whitaker.

Paralectotypes. Two internal moulds of incomplete cranidia, SMF 1652a, 1652b, figured R. and E. Richter 1954, pl. 2, figs. 25, 26 respectively. Also several hundred disarticulated exoskeletal parts contained on BM It16001–16029.

Additional material. At least fifty cranidia, fifty pygidia, fifteen hypostomata, and one rostral plate; numerous free cheeks and thoracic segments. All of the material is in mould form and comes mainly from the Leintwardine collections of Whitaker. Repositories include the British Museum (Natural History); National Museum of Wales, Cardiff; Ludlow Museum; Leicestershire Museum, Leicester; Geology Department, Leicester University.

Diagnosis. A *Calymene* species with straight (tr.) to weakly convex (dorsal) outline to anterior cranidial margin and frontal glabellar lobe. Preglabellar area moderately short, about 0·12 to 0·2 times as long as glabella; anterior border (sag. & exsag.) is a fairly short, low, sharply raised convex to subangular rim. Gently convex (sag. & tr.), weakly inflated glabella. Lacks distinct intermediate lobe at inner end of furrow 1p. Posterior border furrow wide (exsag.). Linear (exsag.) and paired (tr.) arrangement of perforate, larger granules on frontomedian glabellar lobe, inner part of fixed cheek, and pygidial axis. Six to eight axial rings, five pleural furrows.

Description. Occipital ring gently to moderately convex (sag. and exsag.), falls very steeply anteriorly, narrows quite sharply behind 1p lateral lobe towards axial furrow. Occipital furrow V-shaped and distinctly incised across central glabellar area, becomes deeper and narrower abaxially. Glabella is bell-shaped, 1·0–1·2 times as long as wide, projects just in front of or is level with (tr.) fixed cheek; in lateral profile (Pl. 7, figs. 2, 17) it is horizontal from occipital furrow to furrow 2p, then falls very gently to anterior margin of frontal lobe, finally vertically for a short distance to bottom of preglabellar furrow. Width of glabella at 1p lobes is 1·4–1·6 times that across frontal lobe.

Subquadrate lobe 1p is just less than one-third as wide as glabella, separated from median lobe by shallow furrow, has a somewhat pointed anterolateral corner. Lateral furrow 1p rather deep and very narrow between 1p and 2p lobes, bifurcates at median lobe, posterior branch turning more backwards and finally running transversely, short anterior branch directed forwards and inwards around inner neck of lobe 2p. Glabella sometimes gently swollen within fork of furrow 1p but no discrete intermediate lobe present. Lobe 2p subcircular or elongate, long axis trending gently forwards and outwards. Furrow 2p runs inwards and slightly backwards. Node-like lobe 3p is slightly elongate (tr.); short (tr.) furrow 3p distinct. Many specimens have a very small 4p lobe and extremely weak 4p furrow (Pl. 7, fig. 18). Subrectangular frontal lobe is three and a half to four times as wide as long, anterior margin straight (tr.) or very weakly convex forwards.

Axial furrow very shallow at occipital ring, narrowest around posterior part of lobe 1p, deepest at furrows 1p and 2p. Some specimens show very weak trace of eye ridge on abaxial side of axial furrow opposite furrow 3p. Anterior pit sited in axial furrow just anterior to 4p lobe. Preglabellar furrow rather short (sag. & exsag.), shallow to moderately deep, its anterior slope rises fairly steeply (less so in small specimens) to slightly longer, convex, or somewhat angular (sag.) anterior border. Anterior margin is straight (tr.) or very weakly arched forwards.

Posterior border widens (exsag.) gradually from axial furrow to fulcrum, abaxially from fulcrum expands more quickly before narrowing towards facial suture. Posterior border furrow rapidly expands in width abaxially, is very wide midway between axial furrow and genal angle, anterior slope is longer and much less steeply inclined than posterior slope. Postocular part of fixed cheek very gently inclined behind palpebral lobe, preocular part slightly more steep and convex. Length of palpebral lobe is in between that of 1p and 2p lobes, mid-length is opposite anterior margin of lobe 2p, it projects outwards and moderately upwards abaxially. Posterior branch of facial suture runs outwards from palpebral lobe before turning moderately then sharply posteriorly to genal angle. Anterior branch curves gently forwards and inwards to anterior margin. Eye socle narrow.

Lateral border furrow rather broad, especially anteriorly, rises quite sharply to lateral border which rolls over and under to meet reflexed doublure (Pl. 7, figs. 8, 11).

Isolated thoracic segments and fragmentary rostral plate are of normal calymenid type. Hypostoma about as wide (tr.) as long (exs.). Narrow, ventrally flexed anterior border; border furrow very weak. Anterior wing has deeply incised pit. Lateral border tightly convex (tr.); border furrow distinct. Posterior border expanded into two short spines; border furrow well marked, transversely directed. Median furrow most distinct where it merges with lateral border furrow, fades very quickly posteriorly towards ovate weakly convex macula, hardly perceptible medially. Anterior lobe has very prominent median protuberance, is three and a half to four times as long as crescent-shaped posterior lobe (Pl. 7, figs. 12, 13, 16).

Axis of pygidium is about 0·3 times pygidial width, tapers gradually and becomes more steeply sided posteriorly, has six to eight axial rings. At least first five ring furrows are complete, sixth sometimes complete, seventh only present abaxially, eighth medially, all except this last being most distinct at axial furrow which is weakest behind well-rounded terminal axial piece. Five well-marked pleural furrows are best impressed at about mid-length, become gradually weaker and narrower adaxially and abaxially, fall well short of lateral margin. Interpleural furrows are very weak on inner half of pleural region, become more distinct abaxially, weaker than pleurals, almost reach lateral margin, fifth furrow forms side of postaxial sector.

Except in preglabellar and axial furrows cranidium has small- to medium-sized granules, in posterior border furrow granules more scattered. On frontomedian glabellar lobe (Pl. 8, fig. 2) there are two parallel rows (exsag.) of larger, perforate granules having the following paired (tr.) arrangement: on occipital ring one-third of its width (tr.) in from axial furrow; opposite inner neck of lobe 1p; opposite (tr.) posterior and anterior margins of lobe 2p; opposite adaxial part of furrow 3p; about opposite lobe 4p. One specimen (Pl. 7, fig. 6) has a further perforation on occipital ring either side of the more central pair, and a possible linear (tr.) arrangement of perforations on posterior border as far as fulcrum. There is at least one row (exsag.) of four large, perforate granules on inner part of fixed cheek running from anterior margin of posterior border furrow to anterior part of palpebral lobe (Pl. 8, fig. 19). Smaller specimens seem to have relatively coarser granulation.

Scattered granules present all over free cheek, become more flattened and scale-like on lateral border. Inner part of pygidial pleural region almost devoid of granules, outer part with them closely spaced. Two parallel rows (exsag.) of perforations on axial rings (Pl. 8, figs. 1, 4, 14). Possible linear (tr.) arrangement of perforations on posterior pleural bands on inner, anterior part of pleural region (Pl. 8, fig. 4).

Ontogeny. Meraspid period. Two small cranidia may be meraspides. Their glabellae are 1·2 mm (Pl. 8, fig. 3) and 1·6 mm (Pl. 8, figs. 5, 6) long and in each it is subparallel sided, having at least three well-developed lateral lobes. Schrank (1970) demonstrated that in small cranidia of certain *Calymene* species the 2p glabellar lobe is not in contact with a genal buttress to form a bridge across the axial furrow, as it is in more adult specimens; both *lawsoni* specimens lack this bridge, or even the beginnings of a genal buttress. One of the cranidia has an external mould, the cast of which (Pl. 8, fig. 6) appears to show the larger paired granules already developed. Both cranidia have a long fixigenal spine.

Contact of 2p lobe and genal buttress. One small internal mould cranidium (Pl. 8, fig. 10), with a glabellar length of 1·9 mm, has a slightly abaxially extended 2p lobe and what may be an extremely weakly incipient genal buttress. The 2p lobe is not in contact with the fixed cheek but it is possible they touched in the testiferous form. Another slightly larger internal mould (Pl. 8, fig. 18), with an estimated glabellar length of 2·3 mm, clearly has a weak genal buttress which almost certainly would have touched the 2p lobe in the testiferous form. Two cranidia (Pl. 8, fig. 19; BM It16036b, unfigured) with glabellae 2·9 mm long (estimated in one), both casts of external moulds, have the 2p lobe and genal buttress joined. The axial furrow was apparently bridged when the glabella was somewhere between 1·6 mm and 2·3 mm long, and certainly by the time it reached 2·9 mm.

Genal spine. One of the cranidia with a glabellar length of 2·9 mm (BM It16036b) has its fixed cheek intact and shows no genal spine as is present on the possible meraspides. This spine seems therefore to be lost earlier in ontogeny than that of *C. puellaris* (see below).

Discussion. This is the first full account of *C. lawsoni*. Variation exists in the nature of the paired, larger granules. Some casts of external moulds appear to lack one or both of the two most anterior pairs on the frontomedian glabellar lobe (Pl. 7, fig. 10) and on others the whole of the paired arrangement is but very weakly discernible (Pl. 7, fig. 18). It is questionable to what extent this is real or a preservational effect. On internal moulds the larger perforate granules are represented by short spikes of sediment which have infilled the cuticle canals (Pl. 8, fig. 2). On some casts of external moulds the larger perforate granules are only represented essentially by the perforations (Pl. 7, fig. 6); this may be a real difference, perhaps a function of size (Pl. 7, figs. 6, 18; Pl. 8, fig. 19), or again

---

EXPLANATION OF PLATE 7

Figs. 1–18. *Calymene lawsoni* Shirley, 1962. All specimens are from the Ludlow Series, Ludfordian Stage. Leintwardine area, Welsh Borderland. 1, 4, 5, 11, 15, 17, 18 are from the Upper Leintwardine Formation, Marlow Lane, about 3 km north of Leintwardine. 2, 6, 7, 9, 12, 13, 16 are from high in the Lower Leintwardine Formation, Marlow Lane. 3, 10 are from high in the Lower Leintwardine Formation, about 400 m south of Haregrove Wood. 8, 14 are from high in the Lower Leintwardine Formation, south side of Tatteridge Hill. 1, 4, 15, cranidium, internal mould, BM It16042 (SO 4086 7673); lateral, dorsal, oblique views, × 3·5. 2, 6, 7, 9, cranidium, cast, and internal mould, BM It16060a, b (SO 4035 7679); lateral, dorsal stereo-pair, frontal views (cast), dorsal view (internal mould), × 3·5. 3, three juxtaposed paralectotype cranidia, internal moulds, BM It16001 (SO 4179 7429); dorsal, oblique views, × 3·5. 5, rostral plate, cast, BM It16032 (SO 4086 7673); ventral view, × 6. 8, topotype free cheek, cast, BM It16035 (SO 4209 7259); ventral view, × 3·5. 10, cranidium, cast, BM It16041; dorsal view, × 7. 11, free cheek, cast, BM It16033 (SO 4086 7673); dorsal stereo-pair, × 3·5. 12, 16, hypostoma, cast, LEICS G222.1982; lateral view, ventral stereo-pair, × 6. 13, hypostoma, internal mould, LEICS G226.1982; ventral view, × 8. 14, lectotype cranidium, internal mould, BM It16000 (SO 4209 7259); dorsal view, × 2. 17, 18, cranidium, cast, BM It16038 (SO 4086 7674); lateral, dorsal views, × 3·5.

All casts are of silicone rubber of external moulds.

PLATE 7

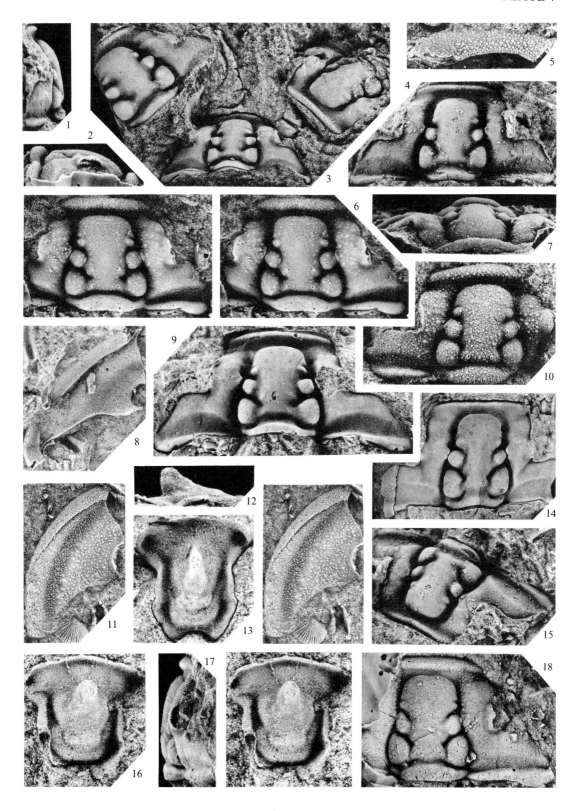

preservational, the rims of the granules being abraded away. Certainly overall smaller cranidia seem to have relatively coarser, more closely spaced, and better developed granulation (Pl. 7, figs. 6, 10, 18; Pl. 8, fig. 19).

Haas (1968, p. 97, pl. 29, figs. 4–9; text-fig. 16a, b) suggested that *C. lawsoni* may be a subspecies of *C. arotia*, his new species from the Ludlow of Turkey. Apart from the fact that *lawsoni* has priority over *arotia* the two taxa are regarded here as specifically distinct. In *C. arotia* the anterior margin is substantially more arcuate, the preglabellar area is relatively longer, and anterior border not so sharply raised up from the preglabellar furrow, there is an obvious intermediate lobe at the inner end of furrow 1p, the posterior border furrow appears less expansive, and it lacks a linear arrangement of the granular sculpture.

The paired larger granules seen in *C. lawsoni* are not unique within the genus. A very small cranidium questionably assigned to *C. orthomarginata* Schrank (1970, p. 128, pl. 6, fig. 5), from north German erratics of upper Wenlock to lower Ludlow age, has a double row of larger granules on the fixed cheek, but there is not the same type of sculpture on its glabella, or anywhere on adult pygidia or cranidia of *orthomarginata*. The latter differs further through its more anteriorly depressed and slightly shorter glabella relative to the fixed cheeks, slightly deeper preglabellar furrow, higher anterior border, and narrower pygidial axis. A pygidium assigned to *C. mimaspera* Schrank (1970, p. 125, pl. 3, fig. 7), also from north German lower Ludlow erratics, has paired larger granules on its axis, but this seems an inconsistent feature of the species (Schrank 1970, pl. 3, figs. 5, 6) which is otherwise easily distinguished from *lawsoni*.

C. puellaris and the similar *C. neointermedia* show, though less obviously, paired larger granules on the posterior part of the median glabellar lobe (Pl. 9, figs. 3, 4), and *puellaris* like *lawsoni* occurs in the lower Ludfordian of the Anglo-Welsh area. It would appear that *C. neointermedia* and *C. puellaris* are closely related to *lawsoni* despite them being clearly distinguished from it on other characters (which are given in the discussions of those species).

Occurrence. C. lawsoni is known from the Ludlow of the central Welsh Borderland. I have seen it from the Upper Leintwardine Formation and most abundantly in the higher part of the Lower Leintwardine Formation of the Leintwardine area, where it forms part of the fauna inhabiting the small submarine canyon-heads which trend westwards from the shelf margin towards the basin (Whitaker 1962). Shergold and Shirley (1968, pl. 15) recorded it from the Upper Leintwardine Formation of Wenlock Edge, the higher part of the Lower Leintwardine Formation of the Ludlow area, the Lower Bodenham, and basal part of the Lower Perton beds of South Woolhope, and they (fig. 2) show its overall range extending into the Upper Bringewood Formation. I can confirm its presence in the Upper Leintwardine Formation of Wenlock Edge, but have not seen material from the other horizons, nor the Ludlow *C.* cf. *beyeri* (? = *C. lawsoni*) of Squirrell and Tucker (1960) and Phipps and Reeve (1967) from the Woolhope, Malverns, and Abberley areas.

Calymene aff. *C. lawsoni* Shirley, 1962

Plate 8, figs. 8, 9, 11, 13, 15

Material. Two cranidia, LM2859, BM It16045; one pygidium, BM It16046.

Occurrence. Middle Elton Formation, Ludlow Series, Upper Millichope, Wenlock Edge, Salop.

Discussion. This form shows the following differences from *C. lawsoni*: the posterior border furrow is less expansive and considerably less deeply impressed; the anterior part of the fixed cheek is exsagitally more convex, descending more steeply forwards; there is no paired arrangement of the granular sculpture; the pygidial pleural furrows are slightly longer. In terms of gross cranidial morphology *C.* aff. *lawsoni* is a more likely ancestor of *C. lawsoni* than any known species.

Calymene neointermedia R. and E. Richter, 1954

Plate 8, figs. 7, 12, 16, 17; Plate 9, figs. 1–3, 8, 10, 13

1885   *Calymene intermedia* n.; Lindström, p. 71, pl. 15, figs. 8–12, ?figs. 5–7.
? 1894   *Calymmene intermedia* Lindström; Schmidt, p. 16, pl. 2, figs. 1–3 (specimens need revision).

1901   *Calymmene intermedia* LDM.; Lindström, p. 48, pl. 3, figs. 6, 7.

non 1909   *Calymmene intermedia*; Moberg and Grönwall, pl. 3, figs. 15, 16.

non 1916   *Calymene intermedia*, Lindström; Reed *in* Gardiner, p. 168.

non 1920   *Calymene intermedia*; Gardiner, pp. 207, 218.

1921   *Calymmene intermedia* Lindström; Hede, pp. 61, 99 (*pars*), ?pp. 58, 59, 60, *non* pp. 68, 69.

1933   *Calymene intermedia* Lindström, 1885; Shirley, p. 64, pl. 1, fig. 16.

non 1933   *Calymene intermedia*; Shirley, p. 64 (SM A3318, SM A3319).

non 1936   *Calymene intermedia* (Lindström); Shirley, p. 400.

non 1937   *Calymene intermedia* Lindström; Straw, p. 452.

1954   *Calymene* (*Calymene*) *neointermedia* n.n.; R. and E. Richter, p. 19.

non 1955   *Calymene intermedia* Lindström; Lawson, p. 112.

non 1959   *Calymene neointermedia* (R. and E. Richter); Walmsley, p. 514.

non 1960   *Calymene neointermedia* (R. and E. Richter); Squirrell and Tucker, p. 177.

? 1960   *Calymene neointermedia* R. and E. Richter; Regnéll *in* Regnéll and Hede, pp. 30, 31 (Faunal list).

1960   *Calymene neointermedia* R. and E. Richter; Hede *in* Regnéll and Hede, p. 82, *non* pp. 80, 81, 83

non 1962   *Calymene neointermedia* R. and E. Richter; Whitaker, p. 345.

non 1963   *Calymene neointermedia* (R. and E. Richter); Holland, Lawson and Walmsley, pp. 117, 118, 126, 145, 147, pl. 6, figs. 4, 7.

? 1967   *Calymene neointermedia* R. and E. Richter; Phipps and Reeve, p. 367.

1968   *Calymene neointermedia* R. and E. Richter; Haas, p. 98, text-fig. 16c, d.

non 1968   *Calymene neointermedia*; Shergold and Shirley, pp. 128, 135.

non 1969   *Calymene neointermedia* R. and E. Richter; Squirrell and Downing, pp. 15, 310, pl. 2, fig. 20.

? 1970   *C. neointermedia*; Männil *in* Kaljo, p. 154 (faunal list).

1970   *C. pompeckji* Kummerow, 1928; Schrank, p. 120, pl. 1, fig. 4; pl. 2, figs. 1–3; ?pl. 1, fig. 6, 6a; pl. 2, fig. 7; *non* pl. 1, fig. 5.

non 1971   *Calymene neointermedia*; Cave and White, pp. 248, 252, 253.

non 1971   *Calymene neointermedia* R. and E. Richter; Shaw, pp. 361, 362, 365.

non 1973   *Calymene neointermedia*; Lawson, pp. 262, 266.

non 1978   *Calymene neointermedia*; Squirrell and White, pp. 16, 42.

1980   *Calymene neointermedia* R. and E. Richter, 1954; Siveter, p. 785.

Holotype. A damaged, enrolled specimen, RM Ar6225; figured Lindström 1885, pl. 15, figs. 11, 12; Shirley 1933, pl. 1, fig. 16; Schrank 1970, pl. 1, fig. 4; herein Plate 8, figs. 12, 16, 17.

Type stratum and type locality. Hemse Beds, Petesvik, Hablingbo Parish, Gotland, Sweden. Martinsson (1962, p. 54) listed *Craspedobolbina percurrens*, *C. robusta*, and *Amphitoxotis curvata* from Petesvik; this ostracode fauna is characteristic of the older, north-western part of the Hemse Beds (see also Martinsson 1967, p. 367). Hede (*in* Regnéll and Hede 1960, p. 82, loc. 44) recorded a concomitant graptolite fauna of *Saetograptus chimaera* and *Neodiversograptus nilssoni*. The Hemse Beds in the Petesvik area are therefore of early lower Gorstian age.

Additional material. One complete specimen (Pl. 9, figs. 1, 2, 8, 10, 13) and one partial cranidium (Pl. 8, fig. 7), both on BM It9145, may come from Britain (see occurrence). I have noted over one hundred specimens from Gotland.

Diagnosis. A *Calymene* species with a long preglabellar area, about 0·25 to 0·3 times as long as glabella, curving forward and moderately upward. Posterior border furrow moderately developed, in section (exsag.) abaxially is like a very open V, has a shallow continuously (tr.) narrow (exsag.) base, tapers gradually towards genal angle. Posterior margins of palpebral lobes are 1·7 to 2·0 times as wide apart as glabellar width at lobe 2p. Main field of free cheek between eye socle and lateral border furrow evenly convex in profile. Thorax has thirteen segments.

Discussion. Together with the material questionably from Britain, many Gotland specimens of *C. neointermedia*, some eighty of which are complete, and about half of which come from the type locality, have been examined. Though *neointermedia* appears closely related to *C. lawsoni* (see discussion of that species), it is easily distinguished, most obviously by the form of the preglabellar area (Pl. 7, fig. 6; Pl. 8, fig. 12) and details of the pygidium (Pl. 8, figs. 1, 4, 14; Pl. 9, fig. 8).

Schrank (1970, p. 120) regarded *neointermedia* as a junior synonym of *C. pompeckji* Kummerow,

1928, a species established on one small specimen (Pl. 9, figs. 5, 11, 12, 14) from a north German glacial erratic believed (Kummerow 1928, p. 9) to be derived from the post-Ludlow Beyrichienkalk, or (Schrank 1970, pp. 120, 123) from a Wenlock–Ludlow horizon within the Graptolithengestein. *C. pompeckji* is similar to *C. neointermedia* in that both have a long, scoop-like preglabellar area, but several differences lead me to regard them as at least separate species:

1. The holotype of *pompeckji* has a glabella which is 4·4 mm long yet has no papillate-buttress structure at lobe 2p; Schrank accounted for this by claiming that it was a juvenile (see ontogeny of *C. lawsoni*). However, all the *neointermedia* specimens I have seen have this structure fully developed, and this includes thirteen Gotland specimens with a shorter glabella (text-fig. 2), the smallest being 3·1 mm long (Pl. 9, fig. 3). In *C. lawsoni*, *C. puellaris*, and *C. tuberculosa* Dalman, 1827 the genal buttress and 2p lobe meet when the glabellar length is between about 1·6 mm and 2·9 mm, 2·4 mm and 2·8 mm, and 2·8 mm and 3·0 mm, respectively. Unless the development of this structure is in a state of arrest in the *pompeckji* holotype, it cannot be considered synonymous with *neointermedia*, nor, moreover, would it belong in *Calymene* or the Calymeninae (Siveter 1977, p. 353; 1979, p. 373; 1980, p. 783).

2. The anterior tip of the preglabellar area is missing in the *pompeckji* holotype, but this area is nevertheless noticeably longer than those of *neointermedia* specimens of similar glabellar length (text-fig. 3).

3. The forwards projecting anterior cranidial margin in *pompeckji* seems to be converging medially in a U or even V-shape rather than in the more rounded outline of *neointermedia* (Pl. 9, figs. 2, 14). This is reflected ventrally in the border sector of the rostral plate (Pl. 9, figs. 12, 13).

4. The glabellar outline in *pompeckji* is more strongly bell-shaped, being relatively narrower in front of lobe 1p (Pl. 9, figs. 2, 3, 14; text-figs. 4, 5).

5. The palpebral lobes in *pompeckji* are not preserved but their distance apart can be estimated by the position of the anterior section of the facial suture, and relative to the glabellar width at lobe 2p they appear more widely separated (Pl. 9, figs. 3, 11, 14; text-fig. 6). Variate J1 cannot be used in *pompeckji*, as it has been in other calymenids—see text-fig. 7 and Siveter 1980, text-fig. 1—to indicate this distance, because adaxially the posterior section of the facial suture is missing.

EXPLANATION OF PLATE 8

Figs. 1–6, 10, 14, 18, 19. *Calymene lawsoni* Shirley, 1962. All specimens are from the Ludlow Series, Ludfordian Stage, Leintwardine area, Welsh Borderland. 1–6, 14 are all from high in the Lower Leintwardine Formation, Marlow Lane, about 3 km north of Leintwardine. 18, 19 are from the Upper Leintwardine Formation, Marlow Lane. 1, pygidium, cast, BM It16044 (SO 4065 7678); posterior view, × 4·5. 2, cranidium, internal mould, LEICS G224.1982, dorsal view, × 5. 3, cranidium, ?meraspis, internal mould, BM It16043 (SO 4065 7678); dorsal view, × 7. 4, pygidium, cast, LEICS G221b.1982; posterior stereo-pair, × 4·5. 5, 6, cranidium, ?meraspis, internal mould, and cast, LEICS G223a,b.1982; dorsal views, × 7. 10, small cranidium, internal mould, BM It16059, high in Lower Leintwardine Formation, about 400 m south of Haregrove Wood (SO 4179 7429); dorsal view, × 7. 14, pygidium, cast, LEICS G221b.1982; posterior view, × 4·5. 18, cranidium, internal mould, BM It16030; dorsal view, × 7. 19, cranidium, cast, BM It16031 (SO 4086 7673); dorsal view, × 7.

Figs. 7, 12, 16, 17. *Calymene neointermedia* R. and E. Richter, 1954. 7, partial cranidium, cast, BM It9145, horizon and locality in question, catalogue data gives Much Wenlock Limestone Formation, Wenlock Series, Malvern, but may be a Gotland specimen (see text); dorsal view, × 3. 12, 16, 17, holotype, complete enrolled specimen, RM Ar6225, Hemse Beds, *Neodiversograptus nilssoni* Biozone, Gorstian Stage, Ludlow Series, Petesvik, Hablingbo Parish, Gotland; dorsal, lateral, frontal views, × 3. Figured Lindström 1885, pl. 15, figs. 11, 12; Shirley 1933, pl. 1, fig. 16; Schrank 1970, pl. 1, fig. 4.

Figs. 8, 9, 11, 13, 15. *Calymene* aff. *lawsoni* Shirley, 1962. All specimens are from the Middle Elton Formation, Gorstian Stage, Ludlow Series, Upper Millichope, Wenlock Edge, Salop. 8, 9, 13, cranidium, LM 2859; dorsal, frontal, lateral views, × 3. 11, cranidium, BM It16045; dorsal view, × 3. 15, pygidium, BM It16046; posterior view, × 2.

All casts are of silicone rubber of external moulds.

PLATE 8

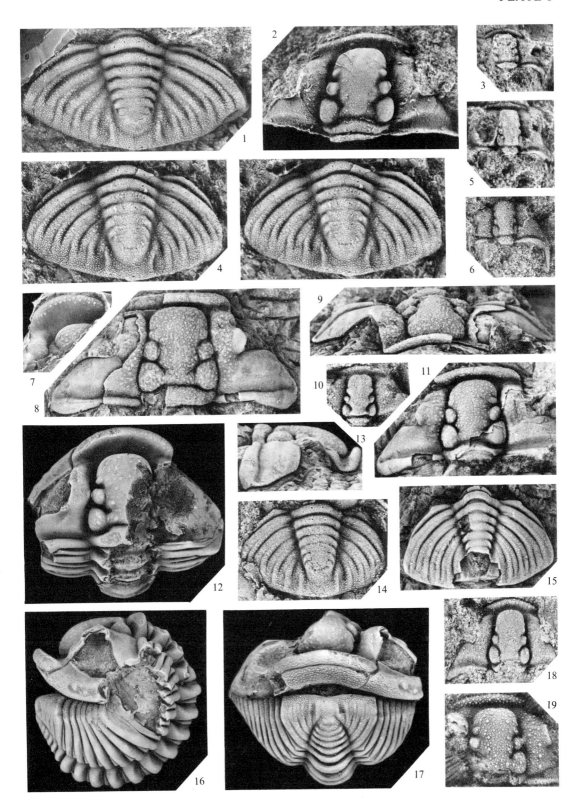

SIVETER, *Calymene*

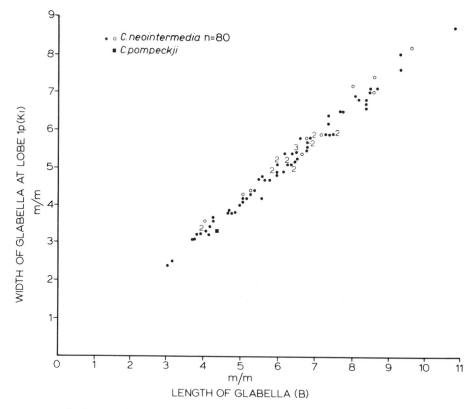

TEXT-FIG. 2. Scatter diagram of glabellar length (B) against glabellar width (K1). All
Calymene neointermedia specimens, 1 questionably from Britain, 79 from Gotland, have the
papillate-buttress structure at 2p lobe developed. Open circle indicates that variate B or K1
has been estimated.

6. The fixed cheeks in *pompeckji*, adaxially opposite 1p lobe, are much more sharply inflated (tr.
and exsag.) and raised above the posterior border and axial furrows (Pl. 9, figs. 1, 2, 5, 14).
   7. The poorly preserved pygidium of *pompeckji* seems to have a more parallel sided, narrower,
relatively longer axis which leaves room for only a very short postaxial sector (Pl. 9, figs. 8, 12).

'*C.*' *pompeckji* appears allied to the three non-buttressed species from the Polish Ludlow which
were assigned (Tomczykowa 1970) to *Spathacalymene* Tillman, 1960, but which represent an
unnamed non-calymenine genus. *C. neointermedia* is most closely related to *C. puellaris* and
differentiated from that species below.

Occurrence. C. neointermedia has for many years been regarded as an index fossil for strata of lower Ludfordian
(specifically Upper Leintwardine Formation) age in Britain, yet it has remained unreported that its type stratum
on Gotland is of early lower Gorstian age, and all the so-called *neointermedia* specimens from the Upper
Leintwardine Formation in Britain I would assign to *C. puellaris*.
   Museum data on the only material of *neointermedia* which is possibly from Britain (Pl. 8, fig. 7; Pl. 9, figs. 1, 2,
8, 10, 13) reads 'Wenlock Limestone, Malvern', and 'old coll. history unknown'; these two specimens are
juxtaposed on BM It9145. Their excellent preservation and the completeness of one of them, together with the
absence of any other *neointermedia* material from Britain and the abundant occurrence of well-preserved
specimens of this species on Gotland, suggested the data on provenance may be erroneous. Dr. David Siveter
comments that 'a reticulate amphitoxotidine beyrichiacean ostracode on BM It9145 has general similarity to
Zorotoxotis Siveter, 1980 from the Wenlock of the Welsh Borderland, but its poor preservation and the lack of a
female precludes firm identification. The subfamily occurs throughout the Wenlock and Ludlow in Britain but

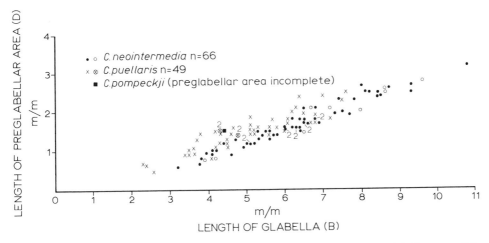

TEXT-FIG. 3. Scatter diagram of glabellar length (B) against length of preglabellar area (D). *Calymene neointermedia* includes 1 specimen questionably from Britain, 65 from Gotland; *C. puellaris* includes 33 from Britain, 16 from Gotland. Open circle and open circle with cross inside indicates that variate B or D has been estimated.

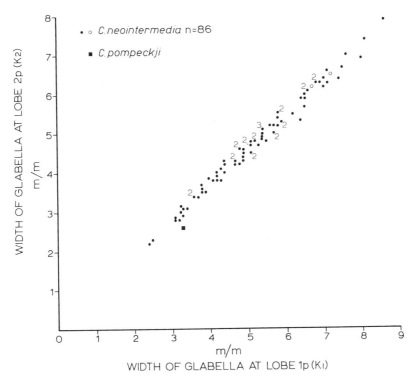

TEXT-FIG. 4. Scatter diagram of glabellar width at lobe 1p (K1) against glabellar width at lobe 2p (K2). *Calymene neointermedia* includes 1 specimen questionably from Britain, 85 from Gotland. Open circle indicates that variate K1 or K2 has been estimated.

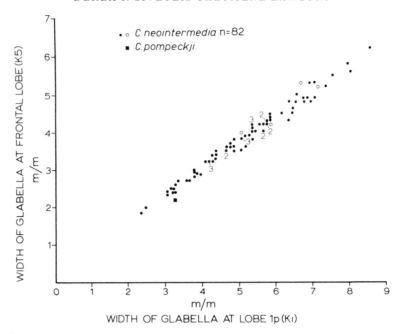

TEXT-FIG. 5. Scatter diagram of glabellar width at lobe 1p (K1) against glabellar width at frontal lobe (K5). *Calymene neointermedia* includes 1 specimen questionably from Britain, 81 from Gotland. Open circle indicates that variate K1 or K5 has been estimated.

on Gotland its genera are, but for very rare exceptions, typical of post-Wenlock strata.' Mr. K. Dorning analysed a small rock sample from BM It9145 and notes that the 'preservation and thermal maturation of the organic material and the type, diversity and abundance of the acritarchs are more like those of the Baltic area than the Welsh Borderland, and the chitinozoa are similar to those recorded from the low Ludlow in Gotland. The sample is unlikely to have come from Malvern.' The suspicion is that BM It9145 is a mislabelled Gotland example. However, if the museum data is correct, the Malvern *C. neointermedia* are only very slightly older than Gotland material. Moreover, Dr. M. G. Bassett (pers. comm., Nov. 1982) informs me that a lot of material from the Malvern area came from Malvern Tunnel, much of it being recorded simply as 'Wenlock Limestone', when in fact the tunnel and adjacent cuttings exposed 'Wenlock Shale', 'Wenlock Limestone', and 'Lower Ludlow Shale'. The Malverns material could even be coeval with that from the type locality.

On Gotland I have seen *neointermedia* from Petesvik and Hablingbo Parish, that is from early lower Gorstian horizons within the Hemse Beds of the west coast. Records of *C. neointermedia* from the Ludlow age Colonus Shale and lower part of the Öved-Ramsåsa Group in Scania (Regnéll *in* Regnéll and Hede 1960, pp. 30, 31), and east Baltic Ludlow–Downton age strata (Schmidt 1894, p. 16, pl. 2, figs. 1–3; Männil *in* Kaljo 1970, p. 154) await investigation, though the species is not listed in the recent distribution table (Männil *in* Kaljo 1977, p. 151) of trilobites from the latter region.

Calymene puellaris Reed, 1920

Plate 9, figs. 4, 6, 7, 9, 15, 16; Plate 10, figs. 1–20

1848   *Calymene tuberculosa*, Salter; Salter (*pars*) *in* Phillips and Salter, p. 342, *non* pl. 12, figs. 1–5.
1849   *Calymene tuberculosa*; Salter (*pars*), p. 8, pl. 8, figs. 8, 8* (GSM 19690), *non* figs. 1–7.
1851   *Calymene tuberculosa* (Salt.); M'Coy (*pars*) *in* Sedgwick and M'Coy, p. 167.
1873   *Calymene tuberculosa*, Salter; Salter (*pars*), p. 166.
1916   *Calymene intermedia* Lindström; Reed *in* Gardiner, p. 168.
1920   *Calymene intermedia*; Gardiner, pp. 207, 218.
1920   *Calymene papillata*, var. nov. *puellaris*; Reed *in* Gardiner, pp. 207, 218, 221, unnumbered text-fig. on bottom left side of p. 221 (holotype).

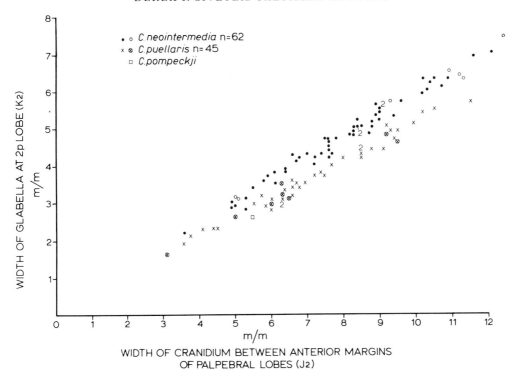

TEXT-FIG. 6. Scatter diagram of cranidial width between anterior margins of palpebral lobes (J2) against glabellar width at lobe 2p (K2). *Calymene neointermedia* includes 1 specimen questionably from Britain, 61 from Gotland; *C. puellaris* includes 31 from Britain, 14 from Gotland. Open circle, open circle with cross inside and open square indicates that variate J2 or K2 has been estimated.

1933   *Calymene* cf. *intermedia* Lindström, 1885; Shirley, p. 63, pl. 1, fig. 15 (holotype of *C. puellaris*).
1933   *Calymene intermedia*; Shirley, p. 64 (SM A3318, SM A3319).
1937   *Calymene intermedia* Lindström; Straw, p. 452.
1955   *Calymene intermedia* Lindström; Lawson, p. 112.
1959   *Calymene neointermedia* (R. and E. Richter); Walmsley, p. 514.
1960   *Calymene neointermedia* (R. and E. Richter); Squirrell and Tucker, p. 177.
? 1960   *Calymene neointermedia* (R. and E. Richter); Regnéll *in* Regnéll and Hede, pp. 30, 31, (faunal list).
1960   *Calymene neointermedia* R. and E. Richter; Hede *in* Regnéll and Hede, pp. 80, 81, 83, *non* p. 82
1962   *Calymene neointermedia* R. and E. Richter; Whitaker, pp. 335, 345.
1963   *Calymene neointermedia* (R. and E. Richter); Holland, Lawson and Walmsley, pp. 117, 118, 126, 145, 147, pl. 6, fig. 7, ?fig. 4.
? 1967   *Calymene neointermedia* R. and E. Richter; Phipps and Reeve, p. 367.
1968   *Calymene neointermedia*; Shergold and Shirley, pp. 128, 135.
1969   *Calymene neointermedia* R. and E. Richter; Squirrell and Downing, pp. 15, 310, pl. 2, fig. 20.
? 1970   *Calymene pompeckji* Kummerow, 1928; Schrank, pl. 1, fig. 6, 6a, pl. 2, fig. 7; *non* pl. 1, figs. 4, 5; pl. 2, figs. 1–3.
1971   *Calymene neointermedia*; Cave and White, pp. 248, 252, 253.
1971   *Calymene neointermedia* R. and E. Richter; Shaw, pp. 361, 362, 365.
1973   *Calymene neointermedia*; Lawson, pp. 262, 266.
1976   *Calymene puellaris* Reed; Siveter *in* Parkin, pp. 597, 598.
1978   *Calymene neointermedia*; Squirrell and White, pp. 14, 42.
1978   *Calymene puellaris* [*neointermedia*] Reed; White and Lawson, p. 8.
1980   *Calymene puellaris* Reed, 1920; Siveter, p. 785.

Holotype. A damaged internal mould of a complete specimen, SM A3320; figured Reed *in* Gardiner 1920, unnumbered text-figure on bottom left-hand side of p. 221; Shirley 1933, pl. 1, fig. 15, herein Plate 10, figs. 13, 15.

Type stratum and type locality. Gardiner (1920, pp. 207, 218) gave the type locality of *C. puellaris* simply as Ludlow Series, locality M24, small quarry east of Longhope church, May Hill inlier. Dr. J. Lawson says (pers. comm., Sept. 1982) that this quarry undoubtedly corresponds to his (1953, Manchester University) Ph.D. thesis locality MY 177 which is in a valley 410 m east of Longhope church (SO 6884 1975), and comprises discontinuous exposure of Upper Blaisdon Beds, Lower Longhope Beds, and Upper Longhope Beds. Throughout the inlier Lawson recorded (1955 and pers. comm.) *C. intermedia* (= *C. puellaris*, see synonymy) from the top Upper Blaisdon Beds and basal Upper Longhope Beds as well as the Lower Longhope Beds, but in the first two of these it occurred only rarely, and in Gardiner's locality M24 he found it only in the latter horizon. It is therefore most likely, but not certain, that the holotype comes from the Lower Longhope Beds, of lower Ludfordian age.

Additional material. From Britain, more than fifty cranidia and pygidia, numerous free cheeks, three hypostomata. Repositories include: Sedgwick Museum, Cambridge; Geological Survey Museum, Institute of Geological Sciences London; British Museum (Nat. Hist.), London; National Museum of Wales, Cardiff.

Diagnosis. A *Calymene* species with a long preglabellar area, about 0·25 to 0·4 times as long as glabella, curving forward and quite strongly upward. Posterior border furrow well developed with fairly steep sides and a sunken base which abaxially is wide (exsag.), tapers sharply towards genal angle. Posterior margins of palpebral lobes are 2·0 to 2·3 times as wide apart as glabellar width at lobe 2p. Main field of free cheek between eye socle and lateral border furrow subangular in profile. Thorax has twelve segments.

Ontogeny. Holaspid period. Some twenty British specimens are complete (I have a further eleven from Gotland). The range of the glabellar length in these is from about 3·7 mm (estimated; Pl. 10, fig. 14) to 6·0 mm (Pl. 10, fig. 9), and all have only twelve thoracic segments, which must be the normal holaspid complement.

Contact of 2p lobe and genal buttress. All testiferous cranidia show the 2p glabellar lobe in contact with a genal buttress; the smallest observed British specimen with this structure (Pl. 9, fig. 7) has a glabella 3·4 mm long,

---

Figs. 1–3, 8, 10, 13. *Calymene neointermedia* R. and E. Richter, 1954. 1, 2, 8, 10, 13, complete specimen, BM It9145, horizon and locality in question, catalogue data gives Much Wenlock Limestone Formation, Wenlock Series, Malvern, but may be a Gotland specimen (see text); lateral, dorsal stereo-pair, and oblique views, × 3, posterior view (pygidium), × 6, ventral view (rostral plate), × 5. 3, complete topotype specimen, RM Ar47789, Hemse Beds, *Neodiversograptus nilssoni* Biozone, Gorstian Stage, Ludlow Series, Petesvik, Hablingbo Parish, Gotland; dorsal view (cephalon), × 5.

Figs. 4, 6, 7, 9, 15, 16. *Calymene puellaris* Reed, 1920. 4, cranidium, RM Ar27115, Hemse Beds, younger southeastern part of Ludlow Series, Alva Kanal, Gotland; dorsal view × 5. 6, cranidium, internal mould, BM It16055a, Lower Llangibby Beds, lower Ludfordian Stage, Ludlow Series, field exposure near spring, 1 km north-north-east of Llangibby Castle, Usk inlier, Gwent (ST 368 982); dorsal view, × 5. 7, cranidium, cast, BM It16047, Upper Leintwardine Formation, Ludfordian Stage, Ludlow Series, Bengry forestry track, Beechenbank Wood, about 0·7 km west of Yatton Court, Aymestrey, Hereford and Worcester; dorsal view, × 5. 9, cranidium, cast, BM It576b, Upper Leintwardine Formation, Ludfordian Stage, Ludlow Series, lane section, The Goggin, Ludlow anticline (SO 465 695); dorsal view, × 5. 15, cranidium, internal mould, BM It16048, Upper Leintwardine Formation, Ludfordian Stage, Ludlow Series, Corfton Bache, hillside exposure 55 m south-west of Halehead Farm, Wenlock Edge, Salop (SO 4890 8612); dorsal view, × 5. 16, complete specimen, cast, NMW 78.52G.20, Lower Llangibby Beds, lower Ludfordian Stage, Ludlow Series, temporary trench west of Cwm-bwrwch Wood, south-west of Llandegfedd Reservoir Dam, Usk inlier, Gwent (ST 322 982–322 985); dorsal view, × 3. See also Pl. 10, fig. 10.

Figs. 5, 11, 12, 14. '*Calymene*' *pompeckji* Kummerow, 1928. Holotype, complete specimen, MB 1969.31, Silurian glacial erratic, Graptolithengestein, found at Butzow/Bezirk Neubrandenburg, Germany; lateral, frontal, ventral views, dorsal stereo-pair, × 5. Figured Kummerow 1928, pl. 1, fig. 5; Schrank 1970, pl. 1, fig. 5.

All casts are of silicone rubber of external moulds, except for figs. 5, 11, 12, 14 which is a glass resin cast.

PLATE 9

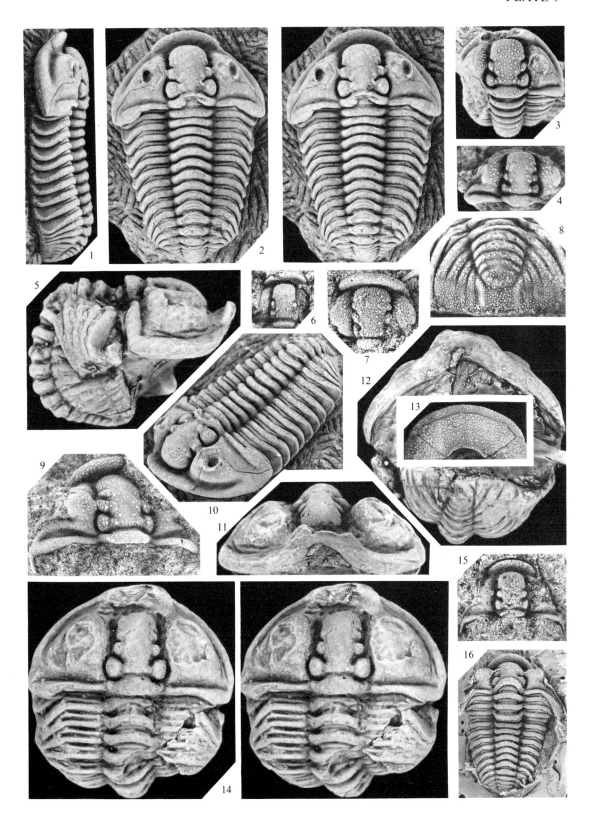

SIVETER, *Calymene*

though in one Gotland example (Pl. 9, fig. 4) it is only 2·8 mm. Internal moulds do not have the axial furrow fully bridged though all of them show a papillate 2p lobe and the remains of a genal buttress. This includes three smaller cranidia with glabellae 2·3 mm (Pl. 9, fig. 6), 2·4 mm (Pl. 9, fig. 15), and 2·6 mm (BM It16052, unfigured) long. It is questionable whether the papillate-buttress structure in these was fully developed, though in one of them (Pl. 9, fig. 15) the buttress may be only incipient. The 2p lobe seems to make contact with the genal buttress when the glabella is about 2·4 to 2·8 mm long.

Genal spine. Several cranidia show a distinct genal spine (Pl. 9, fig. 15), a feature characteristic of small calymenid specimens (see *C. lawsoni*), but two larger undoubtedly holaspid cranidia with glabellae 3·7 mm (Pl. 9, fig. 9) and 4·2 mm long (Pl. 10, fig. 10) still retain a fixigenal spine. How constant this spine is within this size-range is unknown, but it is typical of adult *puellaris* that the posterior cranidial margin does not turn forwards near the genal angle as in many *Calymene* species; instead it is often drawn somewhat backwards (Pl. 10, fig. 6). Apparently *puellaris* loses its genal spine comparatively late in ontogeny.

Discussion. Reed established *C. puellaris* on the basis of one specimen and believed it was closely related to *Papillicalymene papillata* (Lindström, 1885) from the Ludlow of Gotland. Shirley (1933, p. 63) thought *puellaris* 'should be placed with or near to *C. intermedia*' (= *C. neointermedia* R. and E. Richter), but said there was 'no justification for raising a new species or variety'. The holotype of *puellaris* is damaged but is complete and topotype material (Pl. 10, figs. 5, 16) is available. What many authors have recorded as *C. neointermedia* is in fact *C. puellaris*. The two species are very similar, with *neointermedia* doubtless giving rise to *puellaris*, but *puellaris* differs because:

1. As Reed (1920, p. 221) claimed, the holotype of *puellaris* has twelve thoracic segments (Shirley 1933, p. 63 said thirteen), this number for the species being constant (Pl. 8, fig. 16; Pl. 9, figs. 2, 16; Pl. 10, figs. 8, 9, 14, 15).

2. The posterior border furrow is deeper and better developed, particularly abaxially, its sides being steeper and base exsagittally relatively wider (Pl. 9, figs. 2, 9; Pl. 10, figs. 1, 19).

3. The main field of the free cheek is more subangular in section, this profile often continuing anteriorly just on to the fixed cheek (Pl. 9, fig. 10; Pl. 10, figs. 1, 11, 19).

EXPLANATION OF PLATE 10

Figs. 1–16, 18–20. *Calymene puellaris* Reed, 1920. 1, 6, 8–11, 18–20 are all from the Lower Llangibby Beds, lower Ludfordian Stage, Ludlow Series, temporary trench west of Cwm-bwrwch Wood, south-west of Llandegfedd Reservoir Dam, Usk inlier, Gwent (ST 322 982–ST 322 985). 5, 13, 15, 16 are all from the Ludlow Series, almost certainly Lower Longhope Beds, lower Ludfordian Stage, small quarry behind hut in valley, 410 m east of Longhope church, May Hill inlier (SO 6884 1975). 1, 18, almost complete specimen, cast, NMW 78.52G.97b; dorsal stereo-pair, × 3, lateral view, × 5. 2, 12, hypostoma, cast, BM It16049, Upper Leintwardine Formation, Ludfordian Stage, Ludlow Series, lane section, The Goggin, Ludlow anticline (SO 472 700); lateral, ventral views, × 8. 3, 4, 7, cranidium, cast, BM It16056b, Lower Llangibby Beds, lower Ludfordian Stage, Ludlow Series, field exposure near spring, 1 km north-north-east of Llangibby Castle, Usk inlier, Gwent (ST 368 982); dorsal, lateral, frontal views, × 5. 5, topotype cranidium, internal mould, SM A36815; dorsal view, × 3. Listed Gardiner 1920, pp. 207, 218 as *C. intermedia* Lindström. 6, 9, 20, complete specimen, cast, NMW 78.47G.20; oblique, dorsal views, × 2·5, posterior view (pygidium), × 6. 8, complete specimen, cast, NMW 78.52G.32; dorsal view, × 2·5. 10, cephalon of complete specimen, cast, NMW 78.52G.20; dorsal view, × 5. See also Plate 9, fig. 16. 11, free cheek, cast, NMW 78.52G.46; 'dorsal' view, × 5. 13, 15, holotype, complete specimen, internal mould, SM A3320; oblique, dorsal views, × 3. Figured Shirley 1933, pl. 1, fig. 15 as *C.* cf. *intermedia* Lindström. 14, complete specimen, internal mould, NMW 78.35G.485, Lower Leintwardine Formation, Ludfordian Stage, Ludlow Series, Haye Park, 1·6 km north-north-west of Richards Castle, Salop (SO 488 711); dorsal view, × 3. 16, topotype cranidium, internal mould, SM A36814; dorsal view, × 3. Listed Gardiner 1920, pp. 207, 218 as *C. intermedia* Lindström. 19, cephalon and partial thorax, cast, NMW 78.52G.26; dorsal view, × 3.

Fig. 17. *Calymene* cf. *puellaris* Reed, 1920. Complete, distorted specimen, internal mould, GSM 104357, Bannisdale Slates, Ludlow Series, Crook of Lune Flag Quarry, 5 km north-north-west of Sedbergh, Lake District; dorsal view, × 3.

All casts are of silicone rubber of external moulds.

PLATE 10

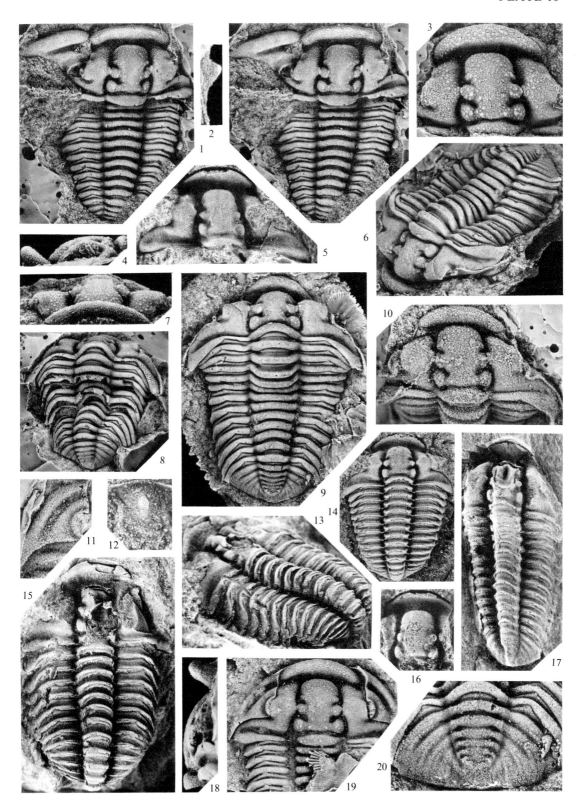

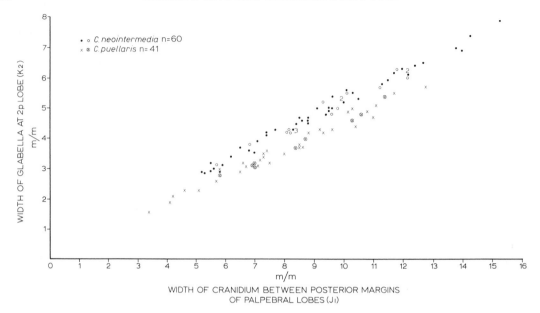

TEXT-FIG. 7. Scatter diagram of cranidial width between posterior margins of palpebral lobes (J1) against glabellar width at lobe 2p (K2). *Calymene neointermedia* includes 1 specimen questionably from Britain, 59 from Gotland; *C. puellaris* includes 28 from Britain, 13 from Gotland. Open circle and open circle with cross inside indicates that variate J1 or K2 has been estimated.

4. The palpebral lobes are almost always wider apart (text-fig. 7).

5. The preglabellar area is more strongly upturned, reaching well above the dorsal surface of the frontal glabellar lobe (Pl. 8, fig. 16; Pl. 9, fig. 1; Pl. 10, figs. 4, 18).

6. The posterior margin and corner of the fixed cheek swing more backward at the facial suture because the genal spine is lost late in ontogeny (Pl. 9, fig. 10; Pl. 10, fig. 6).

Many *puellaris* cranidia have a relatively longer preglabellar area but this distinction does not seem constant (text-fig. 3). Also preglabellar area length plotted against glabellar length shows more variability in *puellaris*, but this may reflect the better preservational and/or geographical homogeneity of the *neointermedia* sample. *C. puellaris*, like *neointermedia*, differs from *C. lawsoni* particularly in the nature of its preglabellar area (Pl. 7, fig. 6; Pl. 10, fig. 3) and pygidium (Pl. 8, figs. 1, 4, 14; Pl. 10, fig. 20). In its relatively long preglabellar area and widely separated palpebral lobes (text-figs. 3, 6), '*C.*' *pompeckji* is more like *C. puellaris* than *C. neointermedia*, but it has thirteen thoracic segments and no papillate-buttress structure (Pl. 9, figs. 5, 14, 16; Pl. 10, fig. 10).

Occurrence. This species is widely distributed. It is particularly characteristic of the late lower Ludfordian (Upper Leintwardine Formation and coeval strata), is less common in the early lower Ludfordian (Lower Leintwardine Formation and correlatives), ranges into the earliest upper Ludfordian (Lake District, at least; basal Kirkby Moor Flags) and possibly the late upper Gorstian (Upper Bringewood Formation). I have identified material from: the Lower and Upper Leintwardine formations, Usk inlier; Lower Longhope Beds, May Hill; Upper Bodenham Beds, Woolhope; *Chonetoidea grayi* Beds, Builth area; base of Cennen Beds, Dyfed; Underbarrow Flags (probably lower division), basal Kirkby Moor Flags and possibly the Bannisdale Slates (Pl. 10, fig. 17), Lake District; Upper Leintwardine Formation of the Newnham (Gloucestershire), Aymestrey, Ludlow, Leintwardine, and Wenlock Edge districts and Lower Leintwardine Formation of the Ludlow anticline. If that Whitcliffe Wood, Ludlow, specimen (SM A3319, Shirley 1933, p. 64) collected by Elles and Slater at the turn of the century is really from the 'Aymestrey Limestone' (= Upper Bringewood Formation), it would be the earliest *Calymene puellaris* yet known. However Dr. M. G. Bassett notes that the combination of ?*Shaleria ornatella*, *Howellella* cf. *elegans*, and ?*Salopina lunata* on SM A3319, and the apparent abundance of the latter (in a shell band), suggests on balance a Leintwardine Formation origin, but the evidence is equivocal. Outside Britain it occurs in the Derrymore Glen Formation, Derrymore Glen inlier, Dingle Peninsula, south-west Ireland, the high Hemse and low Eke Beds of Gotland (work in progress), and possibly Scania (see Regnéll *in* Regnéll and Hede 1960).

Acknowledgements. For the loan of specimens I thank Dr. M. G. Bassett, Dr. P. Crowther, Dr. R. A. Fortey, Dr. V. Jaanusson, Dr. H. Jaeger, Mr. M. Jones, Dr. R. Marsh, Mr. S. F. Morris, Mr. J. Norton, Dr. R. M. Owens, Dr. R. B. Rickards, Dr. A. W. A. Rushton, Dr. J. Shirley, Dr. W. Struve, Mr. and Mrs. C. Taylor, Mr. S. Tunnicliffe, Dr. D. E. White, and Dr. J. H. McD. Whitaker. Drs. M. G. Bassett and J. Lawson kindly provided stratigraphic details on Malvern and May Hill localities, respectively, Drs. Bassett and L. Cherns brachiopod information, and Mr. K. Dorning and Dr. David Siveter comments on microfossils.

REFERENCES

CAVE, R. and WHITE, D. E. 1971. The exposures of Ludlow rocks and associated beds at Tites Point and near Newnham, Gloucestershire. *Geol. J.* **7**, 239–254.

GARDINER, C. I. 1916. The Silurian Inlier of Usk, with a palaeontological appendix by F. R. Cowper Reed. *Proc. Cotteswold Nat. Fld Club,* **19**, 160–172.

—— 1920. The Silurian rocks of May Hill, with a palaeontological appendix by F. R. Cowper Reed. Ibid. **20**, 185–222.

HASS, W. 1968. Trilobiten aus dem Silur und Devon von Bithynien (N.W.-Türkei). *Palaeontographica,* **A130**, 60–207.

HEDE, J. E. 1921. Gottlands Silurstratigrafi. *Sver. geol. Unders.* Ser. C, no. 305, 100 pp.

—— 1960. The Silurian of Gotland. *In* REGNÉLL, G. and HEDE, J. E. The Lower Palaeozoic of Scania. The Silurian of Gotland. *Int. Geol. Congr. XXI Sess. Norden,* 1960, Guidebook d, Sweden. Pp. 89. Stockholm.

HOLLAND, C. H., LAWSON, J. D. and WALMSLEY, V. G. 1963. The Silurian rocks of the Ludlow district, Shropshire. *Bull Br. Mus. nat. Hist.* (Geol.), **8**, 93–171.

KUMMEROW, E. 1928. Beiträge zur Kenntnis der Fauna und der Herkunft der Diluvialgeschiebe. *Jb. preuss. geol. Landesanst. BergAkad.* **48** (for 1927), 1–59.

LAWSON, J. D. 1955. The geology of the May Hill Inlier. *Q. Jl geol. Soc. Lond.* **111**, 85–116.

—— 1973. Facies and faunal changes in the Ludlovian rocks of Aymestrey, Herefordshire. *Geol. J.* **8**, 247–278.

LINDSTRÖM, G. 1885. Förteckning på Gotlands Siluriska Crustacéer. *Öfvers K. Vetensk-Akad. Förh. Stockh.* **6**, 37–100.

—— 1901. Researches on the visual organs of the trilobites. *K. Svensk. Vetensk. Akad. Handl.* **34**, 1–86.

MÄNNIL, R. M. 1970. Trilobita. *In* KALJO, D. (ed.). The Silurian of Estonia. *Eesti NSV Tead. Akad. Inst. Geol.* Tallinn. Pp. 153–157. [In Russian with Estonian and English summaries.]

—— 1977. Distribution of trilobites in the Silurian of the East Baltic area. *In* KALJO, D. (ed.). Facies and Fauna of the Baltic Silurian. *Akad. Nauk. Eston. SSR. Inst. Geol.* Tallinn. Pp. 149–158. [In Russian with Estonian and English summaries.]

MARTINSSON, A. 1962. Ostracodes of the family Beyrichiidae from the Silurian of Gotland. *Bull. Geol. Inst. Univ. Uppsala,* **41**, 1–369.

—— 1967. The succession and correlation of ostracode faunas in the Silurian of Gotland. *Geol. För. Stockh. Förh.* **89**, 350–386.

M'COY, F. 1851. *In* SEDGWICK, A. and M'COY, F. *A synopsis of the classification of the British Palaeozoic rocks, with a systematic description of the British Palaeozoic fossils in the geological museum of the University of Cambridge.* Fasc. I, 1–184. London and Cambridge.

MOBERG, J. C. and GRÖNWALL, K. A. 1909. Om Fyledalens Gotlandium. *Lunds Univ. Årsskr.* N.F. afd. 2, **5**, 1–86.

PARKIN, J. 1976. Silurian rocks of the Bull's Head, Annascaul and Derrymore Glen Inliers, Co. Kerry. *Proc. R. Ir. Acad.* **76**, (B), 577–606.

PHIPPS, C. B. and REEVE, F. A. E. 1967. Stratigraphy and geological history of the Malvern, Abberley and Ledbury Hills. *Geol. J.* **5**, 339–368.

REED, F. R. C. 1916. Palaeontological appendix. *In* GARDINER, C. I. The Silurian Inlier of Usk. *Proc. Cotteswold Nat. Fld Club,* **19**, 160–172.

—— 1920. Description of two trilobites. *In* GARDINER, C. I. The Silurian rocks of May Hill. Ibid. **20**, 219–222.

REGNÉLL, G. 1960. The Lower Palaeozoic of Scania. *In* REGNÉLL, G. and HEDE, J. E. The Lower Palaeozoic of Scania. The Silurian of Gotland. *Int. Geol. Congr. XXI Sess. Norden,* 1960, Guidebook d, Sweden. Pp. 89. Stockholm.

RICHTER, R. and RICHTER, E. 1954. Die Trilobiten des Ebbe-Sattels. *Abh. Senckenb. naturforsch. Ges.* no. **488**, 1–76.

SALTER, J. W. 1848. *In* PHILLIPS, J. and SALTER, J. W. Palaeontological appendix to Professor John Phillips' Memoir on the Malvern Hills, compared with the Palaeozoic districts of Abberley etc. *Mem. geol. Surv. U.K.* **2**, 331–386.

—— 1849. Figures and descriptions illustrative of British organic remains. Ibid. 2 Dec.

—— 1873. *A catalogue of the collection of Cambrian and Silurian fossils contained in the Geological Museum of the University of Cambridge.* Pp. 204. Cambridge.

SCHMIDT, F. 1894. Revision der Ostbaltischen Silurischen Trilobiten, Abt. 4. Calymmeniden, Proetiden, Bronteiden, Harpediden, Trinucleiden, Remopleuriden und Agnostiden. *Mém. Acad. imp. Sci. St-Pétersb.* (7), **42**, (5), 93 pp.

SCHRANK, E. 1970. Calymeniden (Trilobita) aus Silurischen Geschieben. *Ber. Deutsch. Ges. geol. Wiss., A, Geol.-Paläont.* **15**, 109–146.

SHAW, R. W. L. 1971. The faunal stratigraphy of the Kirkby Moor Flags of the type area near Kendal, Westmorland. *Geol. J.* **7**, 359–380.

SHERGOLD, J. H. and SHIRLEY, J. 1968. The faunal stratigraphy of the Ludlovian rocks between Craven Arms and Bourton, near Much Wenlock, Shropshire. Ibid. **6**, 119–138.

SHIRLEY, J. 1933. A redescription of the known British Silurian species of *Calymene* (s.l.). *Mem. Proc. Manchester Lit. Phil. Soc.* **77**, 51–67.

—— 1936. Some British trilobites of the family Calymenidae. *Q. Jl geol. Soc. Lond.* **92**, 384–422.

—— 1962. Review of the correlation of the supposed Silurian strata of Artois, Westphalia, the Taunas and Polish Podolia. *Symposiums—Band der 2 internationalen Arbeitstagung über die Silur/Devon-Grenze und die Stratigraphie von Silur und Devon*, 234–242.

SIVETER, DEREK J. 1977. The middle Ordovician of the Oslo region, Norway, 27. Trilobites of the family Calymenidae. *Norsk geol. tiddskr.* **56** (for 1976), 335–396.

—— 1979. *Metacalymene* Kegel, 1927, a calymenid trilobite from the Kopanina Formation (Silurian) of Bohemia. *J. Paleont.* **53**, 367–379.

—— 1980. Evolution of the Silurian trilobite *Tapinocalymene* from the Wenlock of the Welsh Borderlands. *Palaeontology*, **23**, 97–101.

SQUIRRELL, H. C. and DOWNING, R. A. 1969. Geology of the South Wales Coalfield. Part 1. The country around Newport (Mon.), 3rd edn. *Mem. geol. Surv. U.K.* 333 pp.

—— and TUCKER, E. V. 1960. The geology of the Woolhope Inlier (Hertfordshire). *Q. Jl geol. Soc. Lond.* **116**, 139–185.

—— and WHITE, D. E. 1978. Stratigraphy of the Silurian and Old Red Sandstone of the Cennen Valley and adjacent areas, south-east Dyfed, Wales. *Rep. Inst. Geol. Sci.* no. 78/6, 1–45.

STRAW, S. H. 1937. The higher Ludlovian rocks of the Builth district. *Q. Jl geol. Soc. Lond.* **96**, 406–456.

TOMCZYKOWA, E. 1970. Silurian *Spathacalymene* Tillman, 1960 (Trilobita) of Poland. *Acta palaeont. pol.* **15**, 63–94.

WALMSLEY, V. G. 1959. The geology of the Usk Inlier (Monmouthshire). *Q. Jl geol. Soc. Lond.* **114**, 483–521.

WHITAKER, J. H. MCD. 1962. The geology of the area around Leintwardine, Herefordshire. Ibid. **118**, 319–351.

WHITE, D. E. and LAWSON, J. D. 1978. The stratigraphy of new sections in the Ludlow Series of the type area, Ludlow, Salop, England. *Rep. Inst. Geol. Sci.* no. 78/30, 1–10.

DEREK J. SIVETER
Department of Geology
University of Hull
Hull HU6 7RX

ADDITIONAL FAUNAL DATA FOR THE BEDINAN FORMATION (ORDOVICIAN) OF SOUTH-EASTERN TURKEY

by W. T. DEAN

ABSTRACT. The Bedinan Formation of the Derik–Mardin area is of Caradoc and possibly Ashgill age, and is divisible into three members. The lowest beds of the lower shale member, exposed only at Şip Dere, between Derik and Bedinan, indicate quiet conditions of deposition during the earliest part of an extensive marine transgression. Increasing depth of water is suggested by successive faunules containing: *a*, primitiid ostracods, small brachiopods (*Aegiromena*), and rare fragments of graptolites but only a single trilobite hypostoma (*Dalmanitina*); *b*, dalmanitids (*Dalmanitina*), calymenids (*Colpocoryphe*), and rare asaphids (*Nobiliasaphus*); and *c*, abundant trinucleids (*Deanaspis*) with only rare *Dalmanitina* and *Colpocoryphe*. *Kloucekia* appears still higher and, together with *Dalmanitina* and *Deanaspis*, persists into the highest unit of the lower shale member. No recognizable macrofossils were found in the sandstone member. The upper shale member commences with sandstones (probably late Caradoc or early Ashgill) which contain the trilobites *Calymenella* and *Dreyfussina*?, indicating close affinities with southern France and Iberia, and ends with grey-green shales in which no macrofossils were found.

THE present paper marks the latest stage in work on Cambrian and Ordovician stratigraphy and trilobites of the Mediterranean region *sensu lato* and the Near East that has continued intermittently since the 1960s. In 1965, while working in the Montagne Noire, southern France, I was visited in the field by Harry and Dorothy Whittington at a time when Harry was accumulating data for a paper which appeared later (Whittington, 1966) as a presidential address of the Paleontological Society. Although this contained much useful information on Ordovician trilobites, their regional relationships could not then be fully assessed because at that time the concept of 'Continental Drift' was out of favour, in spite of widespread appreciation that the distribution of similar sediments and fossils, both Cambrian and Ordovician, on opposite sides of the present-day Atlantic could be more easily explained by invoking theories of horizontal plate movements such as now form part of undergraduate teaching. The position was partly rectified by Whittington and Hughes (1972) in a paper where the name Selenopeltis Province was used to encompass a region occupied by the *Selenopeltis* Fauna of Whittington (1966) and what have been variously and informally termed Bohemian, Mediterranean, or Tethyan faunas. Finding a single representative trilobite genus for such a province is difficult and the eponymous odontopleurid has not yet been found throughout the Ordovician succession in any one area. Whichever of the above terms one prefers, it certainly covers south-eastern Turkey, including the Derik–Mardin area (part of the Border Folds of Rigo de Righi and Cortesini 1964) which, together with the Amanos Mountains (text-fig. 1) and Taurus Mountains to the west and the Arab Platform to the east, constituted an integral part of Gondwanaland in Cambrian and Ordovician times. The location of the South Pole in various parts of West Africa throughout the Ordovician lends support to suggestions that most of the Gondwanaland Ordovician rocks and faunas represent cool-water deposits and assemblages.

The close resemblance of Arenig strata in south and south-eastern Turkey to corresponding rocks in western and south-western Europe has been noted elsewhere, as has the widespread absence of Llandeilo rocks (Dean 1980, p. 1). The new material from the Bedinan Formation emphasizes the remarkably close similarity not only of Caradoc (or early Ashgill?) trilobite faunas but also of the corresponding stratigraphic record in areas as widely separate as southern France and south-eastern Turkey. The base of the Bedinan Formation also affords a good example of how, on the margins

[Special Papers in Palaeontology No. 30, pp. 89–105, pls. 11–13]

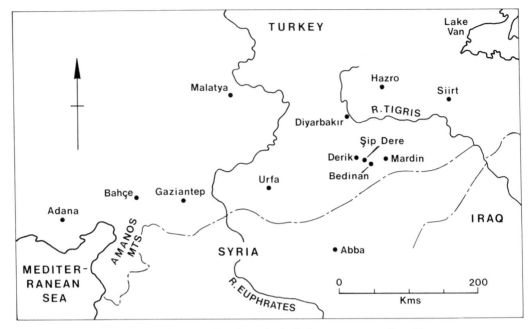

TEXT-FIG. 1. Outline map showing principal place-names mentioned in text.

of a continental mass such as Gondwanaland, an unconformity of very large proportions may be represented in the field by nothing more than a minor change in lithology with no basal conglomerate.

STRATIGRAPHIC SUCCESSION AND FAUNAL RELATIONSHIPS

Review of previous work. Although of geographically limited extent, the inliers of Cambrian and Ordovician rocks in the region between Derik and Mardin (text-figs. 1, 2) are of key importance in deciphering the Lower Palaeozoic history of south and south-eastern Turkey. The only Ordovician rocks there are those of the Bedinan Formation, which crops out in three inliers. The largest is in the vicinity of Bedinan village (text-fig. 2); the others lie immediately east of Sosink village and in the isolated valley Şip Dere (text-fig. 2). The formational name was introduced by Kellogg (1960) who divided the rocks into two parts: a shale member at least 502 m thick, followed by a sandstone member 121 m thick. The latter was succeeded, possibly unconformably, by 28 m of dark-brown sandstone and 59 m + of olive-grey, silty shales containing well-rounded quartz grains; both these units were together termed Dadaş Formation and their age was stated to be Lower? Silurian. Kellogg claimed that the Bedinan Formation rests unconformably on the Sosink Formation (Cambrian) at Şip Dere but no break was indicated by Schmidt (1965) in a correlation chart which showed the Bedinan Formation extending through both Ordovician and Silurian and containing in its upper half successive, discrete horizons termed Handof 'B' Sand and Handof 'A' Sand. Handof Formation was introduced by Rigo de Righi and Cortesini (1964, p. 1913) for 900 m of shales, some black, in the Handof-1 well in the Hazro area, north of Mardin (text-fig. 1) and a recent review (Dean 1980, p. 8) suggested that the strata are probably Silurian with no evidence of Ordovician. Fontaine (1981) has shown that strata in the Hazro area previously incorrectly termed Bedinan Formation belong to the Dadaş Formation, which is of upper Silurian and lower Devonian age; use of Handof Formation or Handof Sand in the Derik–Mardin region is considered here to be inappropriate.

More recently the stratigraphic nomenclature for the Bedinan Formation has been slightly modified (Dean *et al.* 1981, p. 271) and three members are now recognized. The lower shale member follows Kellogg's (1960) usage with the addition of the overlying 36 m unit of fine-grained sandstones and sandy shales that formed the lowest part of his sandstone member, originally 121 m thick. Sandstone member as now used comprises the remaining 85 m of thick-bedded sandstones of Kellogg's sandstone member. Upper shale member, 87 m +, is equivalent to the Dadaş Formation as used in the area east of Bedinan by Kellogg.

Lower shale member: Şip Dere. The oldest rocks at Şip Dere are of Cambrian age and belong to the Koruk Formation, followed by the Sosink Formation (text-fig. 3). The latter, in its type area 6 km to the west, has been divided into a shale member approx. 225 m and a sandstone member more than 850 m thick (Kellogg 1960; Dean *et al.* 1981). At Şip Dere similar subdivision proved impracticable but locs. Yo. 236–238 in the north part of the valley (text-fig. 3) yielded the late Middle Cambrian trilobites *Paradoxides (Eccaparadoxides) remus* Dean, 1982, *Peronopsis* sp., and *Conocoryphe (Conocoryphe) caecigena* Dean, 1982 which indicate correlation with the shale

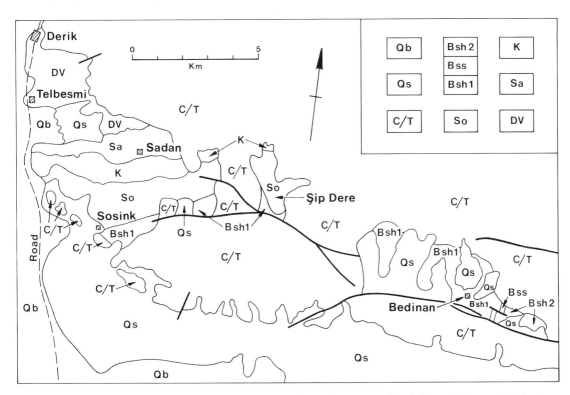

TEXT-FIG. 2. Geological map of the Derik–Bedinan area, slightly modified from Kellogg (1960). Key: DV = Derik Volcanics Formation; Sa = Sadan Formation; K = Koruk Formation; So = Sosink Formation; Bsh1, Bss, and Bsh2 = lower shale, sandstone, and upper shale members of Bedinan Formation; C/T = undifferentiated Cretaceous and Tertiary limestones; Qs = Quaternary sediments; Qb = Quaternary basalts.

member of the type area. Higher strata exposed on the west side of Şip Dere to the south of Kaniah (text-fig. 3) were found to contain *Peronopsis* sp. and *Holasaphus mesopotamicus* Dean, 1972 and are correlated with approximately the middle part of the sandstone member; the close proximity of these beds to the outcrop of the overlying Bedinan Formation suggests that the highest part of the sandstone member at Sosink has been cut out by the sub-Ordovician unconformity.

The lowest strata of the lower shale member are exposed only at Şip Dere (text-fig. 3); there is only a small discrepancy of dip between the two formations and the precise line of junction is difficult to distinguish. As indicated by Kellogg (1960) and confirmed by Dean *et al.* (1981, p. 272) no basal conglomerate is developed, though the break in sedimentation involves a time gap of at least fifty million years.

The lowest fossiliferous horizon in the Bedinan Formation was found 0·5 m above its base at loc. Yo. 242, where the sample included common (7–10) primitiid ostracods, bryozoans, and small brachiopods (*Aegiromena* sp.), rare (1–3) fragments of diplograptid graptolites, and a single hypostoma of *Dalmanitina* sp. About 46 m higher in the succession, at loc. Yo. 243A, the fauna was more varied, and suggests that deepening marine conditions permitted the establishment of a more diverse benthos which included abundant (>10) undetermined small bivalves and bellerophontid gastropods, several (4–6) small brachiopods (*Aegiromena* sp.), abundant

Dalmanitina proaeva (Emmrich, 1839) and *Colpocoryphe grandis* (Šnajdr, 1956), and two specimens of *Nobiliasaphus* cf. *nobilis* (Barrande, 1846). The absence of trinucleid trilobites at Yo. 243A was noteworthy, particularly by comparison with loc. Yo. 243B, 34 m higher in the lower shale member, where *Deanaspis orthogonius* (Dean, 1967) and primitiid ostracods were abundant, whilst *D. proaeva, C. grandis*, and *N.* cf. *nobilis* proved to be rare.

Lower shale member: Sosink area. Beds of the lower shale member which crop out just east of Sosink village (text-fig. 2) were considered by Kellogg (1960) to be faulted against sandstones of the Sosink Formation, an

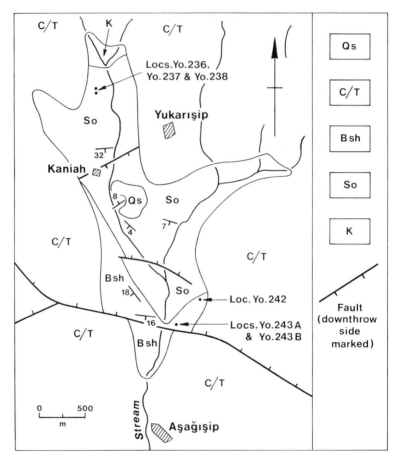

TEXT-FIG. 3. Geological map (after Kellogg 1960) of the inlier of Cambrian and Ordovician rocks at Şip Dere, 12 km south-east of Derik. For key, see text-fig. 2.

interpretation followed by Dean (1967, p. 89, fig. 4). As the result of a visit to the section in 1980 the boundary was considered by Dean *et al.* (1981, p. 272) to be insufficiently well exposed for certain interpretation but probably an unconformity analogous to that at Şip Dere. Support for such a view may derive from the occurrence at loc. Yo. 243B of *Deanaspis orthogonius*, a species originally described from the higher part of the incomplete Sosink section in association with, *inter alia, Colpocoryphe* and *Dalmanitina proaeva*, and stratigraphically higher than shales containing gastropods, *Aegiromena, Lasiograptus, Dalmanitina*, and rare *Selenopeltis* but no trinucleids (Dean 1967, p. 89). *Nobiliasaphus* has been reported with certainty only at Şip Dere but may also be represented by an undetermined asaphid hypostoma at the section east of Sosink (Dean 1967, p. 121). In both cases the horizon is in the lowest part of the lower shale member.

Lower shale member: Bedinan area. Both east of Sosink and at Şip Dere the outcrop of the lower shale member is truncated by the profound sub-Cretaceous unconformity. Conversely, higher beds of the member are exposed only in the immediate vicinity of Bedinan village (text-figs. 2, 5), where the lower boundary of the Bedinan Formation is not seen owing to faulting. *Colpocoryphe* (possibly *C. grandis*) was found in only the lowest beds south-west of Bedinan (Dean 1967, p. 89) and the section there may overlap slightly with those at Şip Dere and east of Sosink. *Kloucekia phillipsii euroa* Dean, 1967 appears slightly higher in the section and extends upwards to the top of the lower shale member; *Dalmanitina proaeva* ranges through the whole of the member. The trilobites in the middle third (approx.) of the lower shale member are predominantly trinucleids, particularly *Deanaspis bedinanensis* (Dean, 1967) for which this is the type area, which occur in abundance at and immediately west of Bedinan. South-east of the village successively higher strata become progressively siltier and there is a corresponding diminution in the number of all groups of fossils.

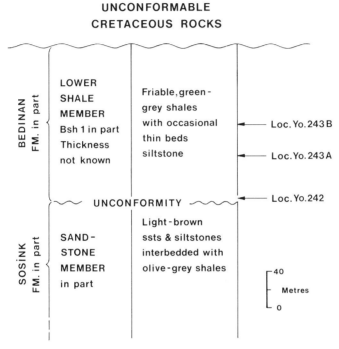

UNCONFORMABLE
CRETACEOUS ROCKS

BEDINAN FM. in part — LOWER SHALE MEMBER Bsh 1 in part Thickness not known — Friable, green-grey shales with occasional thin beds siltstone — Loc. Yo. 243 B, Loc. Yo. 243 A

UNCONFORMITY — Loc. Yo. 242

SOSİNK FM. in part — SAND-STONE MEMBER in part — Light-brown ssts & siltstones interbedded with olive-grey shales — 40 Metres 0

TEXT-FIG. 4. Stratigraphic levels of fossiliferous samples from the lowest part of the Bedinan Formation at Şip Dere (see also text-fig. 3).

In an earlier account of the Bedinan faunas (Dean 1967) those from the highest part of the formation could not be fully investigated owing to time limitations, a deficiency partly remedied during the summers of 1979 and 1980 (Dean *et al.* 1981). It was then confirmed that a 36 m unit of buff-weathering, fine-grained sandstones and sandy shales at the top of the lower shale member (text-fig. 6), though largely unfossiliferous, contained several *D. bedinanensis* and *K. phillipsii euroa*, rare *Dalmanitina proaeva*, abundant small brachiopods (especially *Aegiromena* sp.), and echinoderm debris at loc. Yo. 217B (text-figs. 5, 6). Fossils proved rare in the underlying shales, only *K. phillipsii euroa* and primitiid ostracods being recorded at loc. Yo. 217A.

Lower shale member: age and correlation. A previous assessment (Dean 1967, pp. 91, 92) broadly equated the lower shale member with the Černin Beds and Chlustina Beds of Bohemia, strata since renamed, respectively, Vinice Formation and Zahořany Formation. The appropriate stratigraphy was reviewed by Havlíček and Vaněk (1966) who assigned the Vinice Formation and Zahořany Formation to the upper half of the Caradoc, excluding the topmost part of that series, and gave detailed faunal lists which are used in the present appraisal.

In Bohemia both *Nobiliasaphus nobilis* and *D. proaeva* are recorded from the Vinice Formation, Zahořany Formation and the Bohdalec Formation, while *K. phillipsii* is reported from the whole of the Caradoc Series, from the Liben Formation to the Bohdalec Formation (material from the Králův Dvůr Formation, Ashgill Series, assigned to the species by Havlíček and Vaněk 1966, pl. 14, fig. 4 appears to be specifically distinct). *C. grandis* occurs in the Llandeilo (Dobrotivá Formation) of Bohemia and the succeeding Liben Formation (horizon of the type material) and Letná Formation, that is to say its range in Bohemia precedes that of *N. nobilis* and *D. proaeva*. It seems likely that the species ranges slightly higher in south-eastern Turkey, into strata of the lower shale member approximately equivalent to the Vinice Formation. Some support for this view may derive from the occurrence in the lower shale member east of Sosink of *Dionide formosa* (Barrande, 1846) *anatolica* Dean (1967, pp. 91, 109), subspecies of a species recorded only from the Vinice Formation (Havlíček and Vaněk 1966, p. 55).

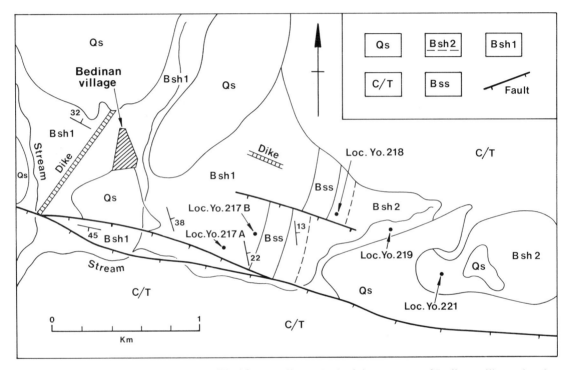

TEXT-FIG. 5. Geological map (slightly modified from Kellogg 1960) of the area east of Bedinan village, showing location of fossiliferous samples from the Bedinan Formation noted in text. For key, see text-fig. 2.

Correlation of the highest beds of the lower shale member with the Bohemian succession is as yet tenuous but the presence of both *K. phillipsii euroa* and *Dalmanitina proaeva* at loc. Yo. 217B suggests a horizon no higher than the upper Caradoc as represented by the Bohdalec Formation or even the underlying Zahořany Formation.

The wide distribution of *C. grandis* and *D. proaeva* in the western Mediterranean region is shown by records from the Caradoc of the Anti-Atlas, Morocco (Destombes 1966, p. 35; 1972, p. 40), and, in the case of the former species, of central Spain (Gil Cid 1970, p. 287). *N. nobilis* also is described from the Caradoc of Spain (Gil Cid 1972, p. 92). Wider distribution of the lower shale member's fauna in the Middle East is suggested by Stubblefield's record (in Sudbury 1957) of *C. arago* (Rouault) and *Pseudobasilicus* cf. *nobilis* in the lowest part of the Abba Group, described from a borehole in northern Syria, 150 km south-south-west of Bedinan (see Dean 1975, p. 369 for review). The Abba borehole did not reach the base of the unit, but the strata described, though said to be of Llandeilo age, are likely to belong with the Caradoc.

Work now in preparation by the author on the sparse Ordovician fauna of the Bahçe Formation (as used by Yalçın 1980) in the northern Amanos Mountains (text-fig. 1) suggests that it represents that of the lower shale member of the Bedinan Formation.

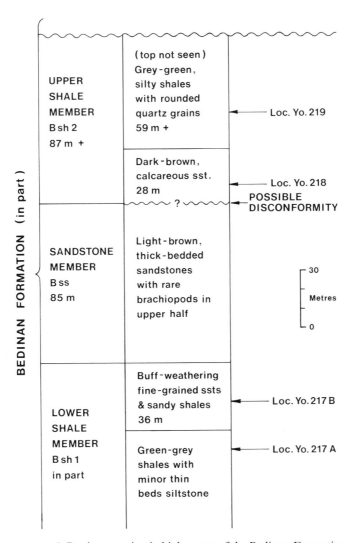

TEXT-FIG. 6. Rock succession in higher part of the Bedinan Formation, east of Bedinan village, showing stratigraphic levels of fossiliferous samples noted in text.

Sandstone member. In its new, restricted, interpretation the sandstone member comprises 85 m of thick- or very thick-bedded, fine-grained sandstones which weather light brown and form a conspicuous feature on the hill slope east of Bedinan (text-fig. 5) where the outcrop is truncated by unconformable Cretaceous rocks. The junction with the underlying lower shale member is normal and gradational. Fossils are confined to occasional unidentifiable brachiopod fragments, and no satisfactory evidence of age is available.

Upper shale member. The colour change from light yellow-brown, fine-grained sandstones of the sandstone member to dark-brown, often coarse-grained, calcareous, and ferruginous sandstones of the 28 m unit at the base of the upper shale member is clearly seen in the topography east of Bedinan village. Kellogg (1960) described the poorly exposed lower boundary as appearing to rest on an irregular surface but considered the 'unconformity?' to be 'probably minor'.

Macrofossil evidence of age is sparse but a faunule from loc. Yo. 218, 9 m above the base of the dark-brown sandstone, includes several fragments of the trilobite *Calymenella boisseli* Bergeron, 1890 and rare examples of *Dreyfussina?* sp. together with poorly preserved brachiopods and echinoderm debris. Up to this point in the Bedinan Formation the closest faunal relationships are with Bohemia and, to a lesser degree, parts of the western Mediterranean and North Africa. In this instance there is a strong link with the Montagne Noire, southern France, where *C. boisseli* was first described (Bergeron 1890) from the Grès du Grand Glauzy of Hérault, and with southern Spain, where *C.* cf. *boisseli* was described from the upper Caradoc of the eastern Sierra Morena by Hammann (1976, p. 54).

There is some disagreement as to the precise age of *C. boisseli* in the Grès du Grand Glauzy, a unit from which the type species of *Dreyfussina, D. exophtalma* (Dreyfuss 1948) and *Dalmanitina proaeva* were also described (Dreyfuss 1948). According to Hammann (1974, p. 96) *Dreyfussina exophtalma exophtalma* is of Caradoc age, as is its Spanish subspecies *D. exophtalma castiliana* (Hammann 1971). On the other hand, Henry (1980*b*, p. 74) puts the age of the Grès du Grand Glauzy at 'Caradoc?, Ashgill?'. The evidence from Bedinan suggests a horizon at least high in the Caradoc, possibly corresponding to part of the Bohdalec Formation of Bohemia, but the magnitude of the inferred break below the basal sandstone of the upper shale member is not yet known and an early Ashgill age cannot be excluded.

No macrofossils have yet been found in the grey-green shales, at least 59 m thick (top not seen), which constitute the topmost unit of the upper shale member. Age evidence provided by acritarchs (Rauscher *in* Dean *et al.* 1981, p. 273) is equivocal, but an unspecified horizon in the Ashgill Series is probably indicated.

SYSTEMATIC PALAEONTOLOGY

Family TRINUCLEIDAE Hawle and Corda, 1847
Subfamily MARROLITHINAE Hughes, 1971
Genus DEANASPIS Hughes, Ingham and Addison, 1975

Type species. Cryptolithus? bedinanensis Dean, 1967.

Deanaspis orthogonius (Dean, 1967)

Plate 12, fig. 8

1967   *Marrolithus orthogonius* sp. nov., Dean, p. 96, pl. 1, figs. 1–9.
1975   *Deanaspis orthogonius* (Dean, 1967); Hughes, Ingham and Addison, pp. 575, 595.

Figured material. It. 16084.

Horizon and locality. Lower shale member of the Bedinan Formation, loc. Yo. 243B, Şip Dere, where the species

---

EXPLANATION OF PLATE 11

Figs. 1, 7, 9, 10. *Nobiliasaphus* cf. *nobilis* (Barrande, 1846). All specimens from lower shale member of Bedinan Formation at Şip Dere. 7, large pygidium, loc. Yo. 243A, It. 15775, latex cast, × 1·3. 1, posterior part of axis of same specimen, × 3·2. 9, incomplete pygidium, loc. Yo. 243B, It. 15777, partly internal, partly external mould, × 3. 10, almost complete, small, dorsal exoskeleton, loc. Yo. 243A, It. 15776, internal mould, × 3.
Figs. 2, 4, 6, 8, 11, 12. *Dreyfussina?* sp. Basal sandstone of upper shale member, Bedinan Formation, loc. Yo. 218, east of Bedinan village. 2, 4, incomplete cranidium, It. 16079, internal mould, left lateral and dorsal views, × 3. 6, fragment of cranidium showing rounded genal angle, It. 16081, internal mould, oblique right lateral view, × 3. 8, 11, 12, incomplete cephalon, It. 16080, internal mould, dorsal, anterior, and right lateral views, × 3·5.
Figs. 3, 5. *Deanaspis bedinanensis* (Dean, 1967). Lower shale member of Bedinan Formation, loc. Yo. 217B, east of Bedinan village. 3, group of three small cranidia, It. 16085, internal mould, × 5. 5, incomplete large cranidium, It. 16086, internal mould, × 4.

PLATE 11

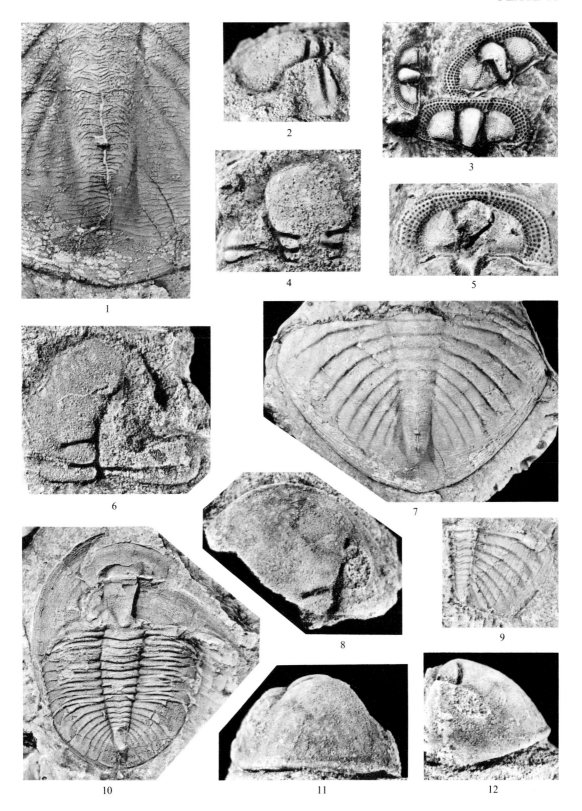

was abundant (>10 specimens) and included both articulated and disarticulated examples. Elsewhere the species occurs, at probably the same horizon, in the section east of Sosink, near Derik (text-fig. 2), from which the type material was described.

Deanaspis bedinanensis (Dean, 1967)

Plate 11, figs. 3, 5

1967   *Cryptolithus?* bedinanensis sp. nov., Dean, p. 104, pl. 3, figs. 1–7, 9; pl. 4, figs. 2–9.
1975   *Deanaspis bedinanensis* (Dean, 1967); Hughes, Ingham and Addison, pp. 573, 591, pl. 9, figs. 100, 101.

Figured Material. It. 16085 (Pl. 11, fig. 3), It. 16086 (Pl. 11, fig. 5).

Horizon and localities. The type material came from the section in the lower shale member west of Bedinan village, where specimens are abundant. The species ranges through approximately the upper two-thirds of the member and the present figured specimens are from loc. Yo. 217B, east of Bedinan, in the highest unit of the lower shale member (text-figs. 5, 6).

Family DALMANITIDAE Vogdes, 1890
Subfamily DALMANITININAE Destombes, 1972
Genus DALMANITINA Reed, 1905

Type species. Phacops socialis Barrande, 1846.

Dalmanitina proaeva (Emmrich, 1839)

Plate 12, figs. 2, 4, 9; Plate 13, fig. 3

1967   *Dalmanitina proaeva proaeva* (Emmrich); Dean, p. 112, pl. 6, figs. 1–9, 11–13; pl. 7, figs. 4, 5. Includes partial synonymy.
1972   *Dalmanitina (Dalmanitina) proaeva* (Emmrich, 1839); Destombes, p. 40, pl. 7, figs. 1–6.

Figured material. It. 15771 (Pl. 12, fig. 2), It. 15772 (Pl. 13, fig. 3), It. 15773 (Pl. 12, fig. 9), It. 15774 (Pl. 12, fig. 4).

Horizon and localities. All the known material from the Derik–Mardin region is from the lower shale member of the Bedinan Formation. For localities in the lower and middle parts of the member near Sosink and west of Bedinan village, see Dean 1967, pp. 89, 91. In the present account It. 15771–15774 are from loc. Yo. 243A, Şip Dere, where other material includes nine fragmentary cranidia and nine pygidia. Only two specimens were found at loc. Yo. 243B, where the fauna consists mostly of the trinucleid *Deanaspis orthogonius*.

Subfamily ACASTINAE Delo, 1935
Genus KLOUCEKIA Delo, 1935

Type species. Phacops Phillipsii Barrande, 1846.

Kloucekia phillipsii (Barrande, 1846) *euroa* Dean, 1967

1967   *Kloucekia phillipsii* (Barrande) *euroa*, Dean 1967, p. 113, pl. 6, fig. 10; pl. 7, figs. 1–3, 6, 7, 9, 12.

Description and discussion. Kloucekia has not been recorded from the lowest part (*c*. 440 m) of the lower shale member in the section east of Sosink (Dean 1967, pp. 89–91), nor from the lowest beds exposed west of Bedinan village (Dean 1967, pp. 86–88) but occurs abundantly in the approximately middle part of the member there. *K. phillipsii* was redescribed by Whittington (1962, p. 7) using material from the Zahořany Formation of Bohemia, the horizon from which the lectotype of the species, designated by Šnajdr (1956, p. 34), was obtained. The Turkish subspecies differs only in minor details of the pygidium and is of similar age.

Subfamily Uncertain
Genus DREYFUSSINA Hupé *in* Choubert *et al.*, 1956

Type species. Dalmania exophtalma Dreyfuss, 1948.

Dreyfussina? sp.

Plate 11, figs. 2, 4, 6, 8, 11, 12

Figured material. It. 16079 (Pl. 11, figs. 2, 4), It. 16080 (Pl. 11, figs. 8, 11, 12), It. 16081 (Pl. 11, fig. 6). All are from the basal sandstone unit of the upper shale member of the Bedinan Formation at loc. Yo. 218, 1·5 km east of Bedinan village. No other specimens found.

Dimensions (in mm). IM = internal mould.

	It. 16079 (IM)	It. 16081 (IM)
Length of cranidium	8·8	12·7 estd
Maximum breadth of cranidium	17·6 estd	24·0 estd
Length of glabella	7·2	10·6 estd
Maximum breadth of glabella	8·2 estd	11·2 estd
Basal breadth of glabella	5·8	8·2 estd

Description and discussion. The fragmentary Turkish specimens, comprising one cephalon and two cranidia, are insufficient for certain generic identification but agree in most respects with topotypes of *Dalmania exophtalma* from southern France illustrated and discussed by Henry (1980, p. 169, pl. 44, figs. 5, 10). One of the French cephala (Henry 1980, pl. 44, fig. 10a) has a small, pointed, median projection on the front of the glabella but in another (Henry 1980, pl. 44, fig. 10b) the corresponding structure is broadly rounded in plan, as shown in Plate 11, fig. 4 and Plate 11, fig. 8. Henry drew particular attention to the presence of a narrow marginal rim on the cephalon of *D. exophtalma*, and a similar one is visible in Plate 11, figs. 8, 11, 12. None of the material from Bedinan has the eyes preserved but their location apparently opposite the 2p and part of, if not all, the 3p glabellar lobes at least approximates to that in *D. exophtalma*. Two cranidia (Pl. 11, figs. 4, 6) show rounded genal angles, in contrast to the genal points noted by Henry (1980, p. 169) for *D. exophtalma*, though not clearly visible in his illustrations, and to the slim fixigenal spines shown by Hammann (1974, text-fig. 35) for *D. exophtalma castiliana* (Hammann 1971) from the Caradoc of Spain. Though the evidence is inconclusive there appears to be no vincular furrow on the only partially visible ventral surface of It. 16080 (Pl. 11, figs. 11, 12).

Family CALYMENIDAE Burmeister, 1843
Subfamily COLPOCORYPHINAE Hupé, 1955
Genus COLPOCORYPHE Novák in Perner, 1918

Type species. Calymene arago Rouault, 1849.

Colpocoryphe grandis (Šnajdr, 1956)

Plate 12, figs. 1, 3, 5–7, 10–12

1956   *Calymene (Colpocoryphe) grandis* nov. sp., Šnajdr, pp. 501, 529, pl. 3, figs. 1–9.
1980   *Colpocoryphe grandis* (Šnajdr, 1956); Henry, p. 64, pl. 8, figs. 1–3, 5, 7, 8; pl. 45, fig. 3; text-fig. 21. Includes previous synonymy.
1981   *Colpocoryphe* sp.; Dean, Monod and Perinçek, p. 272.

Figured material. It. 15778 (Pl. 12, figs. 1, 6); It. 15779 (Pl. 12, fig. 11); It. 16071 (Pl. 12, figs. 10, 12); It. 16072 (Pl. 12, fig. 5); It. 16073 (Pl. 12, figs. 3, 7). These specimens, together with eight fragments of cranidium and three of pygidium, came from the lower shale member of the Bedinan Formation at loc. Yo. 243A, Şip Dere. A single fragment of cranidium was found at loc. Yo. 243B.

Dimensions (in mm). All internal moulds.

	It. 15779	It. 16071
Median length of cranidium	—	19·0
Maximum breadth of cranidium	—	47·0
Length of glabella and occipital ring	10·0	16·5
Basal breadth of glabella	8·5	14·7
Distance across palpebral lobes	14·1	—

	It. 15778	It. 16072	It. 16073
Length of pygidium (excluding half-ring and measured parallel to surface of axis)	4·4	5·8	10·4
Maximum breadth of pygidium	5·7 estd	8·0 estd	14·0
Frontal breadth of axis	2·6	3·6	6·3 estd

Description and discussion. Most of Šnajdr's illustrations of the type material, from the Drabov Beds and Letná Beds of Bohemia, show close agreement with the Turkish material, though much of the latter is slightly compressed and its proportions may vary. His illustrations of the pygidium (Šnajdr 1956, pl. 3, figs. 1, 7) are less clear, but the original description stated that five or six rings occupy the anterior two-thirds of the axis. The larger Turkish pygidia show six well-developed axial rings and traces of a seventh. Henry's (1980, p. 64) revision of the species using well-preserved material from Brittany shows minor differences from the Turkish specimens. The pygidium has up to eight axial rings, with evidence of paired apodemes in most of the ring furrows (Henry 1980, pl. 8, figs. 2a, 5b) and the terminal piece appears more swollen in dorsal view due to the slightly greater development of flange-like protuberances on either side of it which form part of the vincular structure. One pygidium figured by Henry (1980, pl. 8, fig. 2c) shows what appears to be a narrow ridge running from the posterolateral margins to form a narrow, almost pointed, arch with apex at the tip of the axis. It. 16072 (Pl. 12, fig. 5) carries a similar structure, formed by a line of small, pointed tubercles, which marks a break in ornamentation and separates the densely granulate ventral surface from the apparently more sparsely granulate dorsal surface.

Only three specimens of the genus are known from the lower shale member of the Bedinan Formation in the Derik and Bedinan areas, where they were recorded as *Colpocoryphe* sp. (Dean 1967, p. 120, pl. 9, figs. 5–10). The most complete cranidium there is very small and its apparent differences from *C. grandis* may simply reflect changes during ontogeny.

Subfamily REEDOCALYMENINAE Hupé, 1955

Genus CALYMENELLA Bergeron, 1890

Type species. Calymenella boisseli Bergeron, 1890.

Calymenella boisseli (Bergeron, 1890)

Plate 13, figs. 1, 2, 4–12

    1890   *Calymenella boisseli* n.g., n.sp., Bergeron, p. 365, pl. 5, figs. 1–7.
    1980   *Calymenella boisseli* Bergeron; Henry, p. 74, pl. 12, figs. 6–10; text-figs. 28, 30.
    1981   *Calymenella* aff. *boisseli* Bergeron; Dean, Monod and Perinçek, pp. 273, 278.

Figured material. It. 16074, cranidium (Pl. 13, figs. 2, 6, 7, 12); It. 16075, right librigena (Pl. 13, fig. 8); It. 16076, pygidium (Pl. 13, fig. 11); It. 16077, pygidium (Pl. 13, fig. 5); It. 16078, pygidium (Pl. 13, figs. 1, 4, 9); It. 16089 (Pl. 13, fig. 10). Other material comprises three fragments of pygidium.

EXPLANATION OF PLATE 12

Figs. 1, 3, 5–7, 10–12. *Colpocoryphe grandis* (Šnajdr, 1956). All specimens from lower shale member of Bedinan Formation at loc. Yo. 243A, Şip Dere. 1, slightly disarticulated, small exoskeleton, It. 15778, internal mould, × 4. 6, pygidium of same specimen, × 5. 3, 7, large pygidium, It. 16073, internal mould, posterior and dorsal views, × 4. 5, incomplete small pygidium, It. 16072, internal mould, × 5. 10, 12, incomplete, large cranidium, It. 16071, internal mould, dorsal and anterior views, × 2. 11, incomplete cranidium showing palpebral lobes, It. 15779, internal mould, × 4.

Figs. 2, 4, 9. *Dalmanitina proaeva* (Emmrich, 1839). All specimens from lower shale member of Bedinan Formation at loc. Yo. 243A, Şip Dere. 2, fragmentary, slightly distorted, small cranidium, It. 15771, × 4. 4, pygidium with broken posterior spine, It. 15774, internal mould, × 2. 9, hypostoma, It. 15773, internal mould, × 4.

Fig. 8. *Deanaspis orthogonius* (Dean, 1967). Lower shale member of Bedinan Formation, loc. Yo. 243B, Şip Dere. Incomplete cranidium showing part of enrolled thorax beneath, It. 16084, internal mould, × 5.

PLATE 12

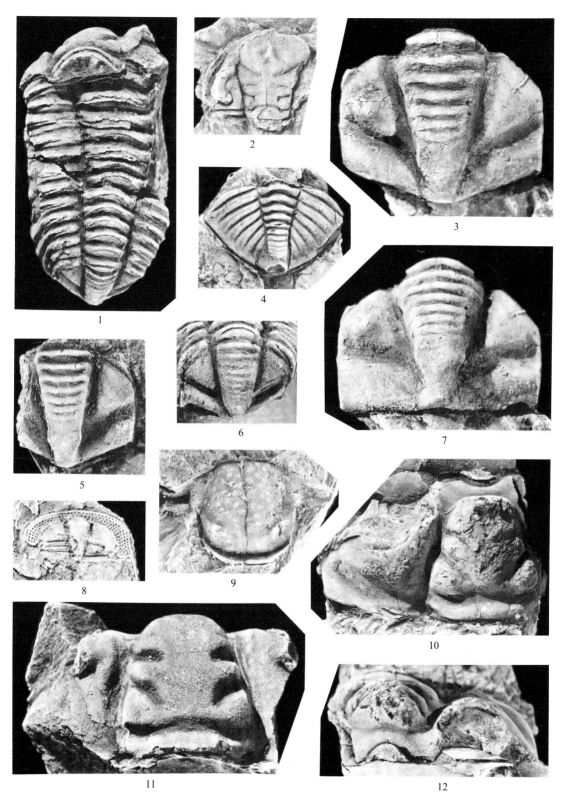

1

2

3

4

5

6

7

8

9

10

11

12

DEAN, Ordovician trilobites

Dimensions (in mm). IM = internal mould, EM = external mould.

	It. 16077 (EM)	It. 16078 (IM)
Projected length of pygidium (excluding half-ring)	10·0 estd	8·4 estd
Maximum breadth of pygidium	15·6 estd	15·0
Frontal breadth of axis	9·0 estd	7·6
Length of axis (measured parallel to dorsal surface)	11·0 estd	10·7

Horizon and locality. All the specimens are from the dark-brown sandstone at the base of the upper shale member of the Bedinan Formation, loc. Yo. 218, east of Bedinan village.

Description and discussion. The large Turkish cranidium now figured (Pl. 13, figs. 2, 6, 7, 12), though incomplete, agrees closely with topotype material from southern France figured by Dreyfuss (1948), Hammann and Henry (1978), and Henry (1980). Particularly conspicuous is the long, slightly convex preglabellar field (compare Pl. 13, fig. 2 with Henry 1980, text-fig. 30B); the small anterior border is slightly more developed on the external surface (Pl. 13, fig. 6) than on the internal mould (Pl. 13, fig. 7). Henry (1980, p. 75) drew attention to the pair of large tubercles sited at the ends of each axial ring on the internal mould of the pygidium, and the same feature is seen here in Plate 13, figs. 4, 9.

Pygidia from the Bedinan Formation have seven, possibly eight, axial rings and six and a half pairs of pleural ribs, numbers in agreement with those given by Henry (1980). Interpleural furrows, though not preserved on the internal moulds, are visible on a latex cast (Pl. 13, fig. 5) of the external surface. The post-axial area is not well preserved and the 'parallélépipédique' terminal piece noted by Henry (1980, p. 75) is not evident.

Family ASAPHIDAE Burmeister, 1843
Subfamily ASAPHINAE Burmeister, 1843
Genus NOBILIASAPHUS Přibyl and Vaněk, 1965

Type species. Asaphus nobilis Barrande, 1846.

Junior objective synonym. Pamirotchechites Balashova, 1966.

Nobiliasaphus cf. *nobilis* (Barrande, 1846)

Plate 11, figs. 1, 7, 9, 10

Figured material. It. 15775, external mould of large pygidium, figured here as latex cast (Pl. 11, figs. 1, 7); It. 15776, almost complete, small exoskeleton (Pl. 11, fig. 10); It. 15777, incomplete pygidium (Pl. 11, fig. 9). No other specimens found.

Dimensions (in mm). IM = internal mould, EM = external mould.

	It. 15775 (EM)	It. 15776 (IM)
Length of pygidium (excluding half-ring)	36·7	6·2
Maximum breadth of pygidium	60·0 estd	12·0
Length of axis (measured parallel to dorsal surface)	28·0 estd	4·5 estd
Frontal breadth of axis	12·2	2·5

Horizon and localities. Lower shale member of the Bedinan Formation at Şip Dere. It. 15775 and It. 15776 are from loc. Yo. 243A; It. 15777 is from Yo. 243B.

EXPLANATION OF PLATE 13

Figs. 1, 2, 4–12. *Calymenella boisseli* Bergeron, 1980. All specimens from basal sandstone of upper shale member, Bedinan Formation, loc. Yo. 218, east of Bedinan village. 1, 4, 9, pygidium, It. 16078, internal mould, posterior, dorsal, and left lateral views, ×3. 2, 7, 12, incomplete cranidium, It. 16074, internal mould, ×2·5. 6, latex cast of same specimen, ×2·5. 5, pygidium, It. 16077, latex cast, ×3. 8, right librigena, It. 16075, latex cast, ×3. 10, fragment of cranidium showing median ridge on glabella, It. 16089, internal mould, ×5. 11, incomplete pygidium, It. 16076, latex cast, ×4.

Fig. 3. *Dalmanitina proaeva* (Emmrich, 1839). Lower shale member of Bedinan Formation at loc. Yo. 243A, Şip Dere. Incomplete, uncompressed cephalon, It. 15772, internal mould, ×3.

PLATE 13

DEAN, Ordovician trilobites

Description and discussion. Though not yet revised in detail, *Nobiliasaphus* has been the subject of comments by Přibyl and Vaněk (1965, p. 277; 1968, p. 192), Kříž and Pek (1974), and Hughes (1979, p. 117). The most conspicuous and easily recognized feature is the ornamentation of the pygidium, with transverse ridges (termed 'middle ridges' by Kříž and Pek 1974, p. 22) on each axial ring curving backwards both at the axial furrows and at the sagittal line where they form a posterior projection, sometimes pointed. Similar ornamentation is less strongly developed on the thoracic axis.

Nobiliasaphus is uncommon and fragmentary in the lower shale member of the Bedinan Formation. The largest pygidium (Pl. 11, fig. 7) has a median length three-fifths of the maximum breadth and the front of the axis is slightly more than one-fifth of the overall breadth. The axial furrows appear slightly concave abaxially in dorsal view; seven axial rings occupy 0·55 of the length of the axis and are delimited by transversely straight ring furrows which become successively weaker. No further ring furrows can be clearly distinguished on this large specimen, but at least four or five more may be present judging from the ornamentation, and ten rings are visible on the far from complete axis of the small pygidium in Plate 11, fig. 9, which shows also the very wide doublure with terrace lines. The tip of the axis declines posteriorly to merge gradually with the border. Excluding furrows the dorsal surface is ornamented with anastomosing terrace-lines which are approximately transverse in direction on the pleural fields but become subparallel to the margin upon attaining the unfurrowed border.

The smallest Turkish pygidium (Pl. 11, fig. 10) has, in addition to the anterior half-ribs, eight pairs of ribs and the largest (Pl. 11, fig. 7) has nine pairs, numbers in broad agreement with the nine or ten pairs seen in Barrande's (1852, pl. 31, figs. 1, 4, 5; pl. 35, fig. 8) somewhat idealized illustrations of material from the Zahořany Formation. Of the latter, the original of Barrande 1852, pl. 31, fig. 1 was chosen as lectotype by Kielan (1960, p. 76) who noted fourteen to fifteen axial rings, though Barrande's other illustrations show numbers in agreement with the Turkish material. The large pygidium from Şip Dere (Pl. 11, fig. 7) has a length/breadth ratio of 0·61, which falls within the range of 0·6–0·7 given for *N. nobilis* by Kielan (1960, p. 76), and its outline closely resembles that of smaller pygidia from Zahořany figured by Barrande. It is smaller than the largest illustrated Bohemian pygidia, in which the length/breadth ratio is about 0·7, and it is possible that the proportions of the pygidium changed during ontogeny.

Acknowledgements. My field-work in the Derik–Mardin area owes much to the help and co-operation of T.P.A.O. (Turkish Petroleum Corporation) and its geologists, especially Doğan Perinçek, and of Olivier Monod and Philippe Janvier, Université de Paris. Financial assistance from the Natural Environment Research Council is gratefully acknowledged.

REFERENCES

BARRANDE, J. 1846. *Notice préliminaire sur le Systême silurien et les Trilobites de Bohême.* vi + 97 pp. Leipsig.
—— 1852. *Systême silurien du centre de la Bohême. I ère partie. Recherches paléontologiques.* xxx + 935 pp. Prague and Paris.
BERGERON, J. 1890. Sur une forme nouvelle de Trilobite de la famille des Calymenidae [genre *Calymenella*]. *Bull. Soc. géol. Fr.* **18**, 365–371.
DEAN, W. T. 1967. The correlation and trilobite fauna of the Bedinan Formation (Ordovician) in south-eastern Turkey. *Bull. Br. Mus. nat. Hist.* (Geol.), **15**, 81–123.
—— 1975. Cambrian and Ordovician correlation and trilobite distribution in Turkey. *Fossils and Strata*, **5**, 353–373.
—— 1980. The Ordovician System in the Near and Middle East. *Int. Union Geol. Sci.* Publ. 2, 1–22.
—— MONOD, D. and PERINÇEK, D. 1981. Correlation of Cambrian and Ordovician rocks in southeastern Turkey. Petroleum Activities at the 100th Year (100 Yilda Petrol Faaliyeti). *T.C. Petrol Isleri Gen. Müd. Der.* **25**, Ankara, 269–291 (English), 292–300 (Turkish).
DESTOMBES, J. 1966. Quelques Calymenina (Trilobitae) de l'Ordovicien moyen et supérieur de l'Anti-Atlas (Maroc). *Notes Serv. géol. Maroc*, **26**, 33–53.
—— 1972. Les trilobites du sous-ordre des Phacopina de l'Ordovicien de l'Anti-Atlas (Maroc). *Notes Mém. Serv. Mines Carte géol. Maroc*, no. 240, 1–111.
DREYFUSS, M. 1948. Contribution à l'étude géologique et paléontologique de l'Ordovicien supérieur de la Montagne Noire. *Mem. Soc. géol. Fr.* **58**, 1–62.
FONTAINE, J.-M. 1981. La Plate-forme Arabe et sa marge passive au Mésozoïque: l'exemple d'Hazro (S.E. Turquie). Thèse Université de Paris-Sud, Orsay.

GIL CID, M. D. 1970. Contribución al estudio de la fauna del Ordovicico de Montes de Toledo (Trilobites). *Estudios geol. Inst. Invest. geol. Lucas Mallada*, **26**, 285–295.

—— 1972. Sobre algunos Asaphidae (Trilobites) del Ordovicico de los Montes de Toledo (España). Ibid. **28**, 89–101.

HAMMANN, W. 1974. Phacopina und Cheirurina (Trilobita) aus dem Ordovizium von Spanien. *Senckenberg. leth.* **55**, 1–151.

—— 1976. Trilobiten aus dem oberen Caradoc der östlich Sierra Morena (Spanien). Ibid. **57**, 35–85.

—— and HENRY, J.-L. 1978. Quelques espèces de *Calymenella, Eohomalonotus* et *Kerfornella* (Trilobita, Ptychopariida) de l'Ordovicien du Massif Armoricain et de la Péninsule Ibérique. Ibid. **59**, 401–429.

HAVLÍČEK, V. and VANĚK, J. 1966. The biostratigraphy of the Ordovician of Bohemia. *Sb. geol. Ved. Praha*, **P 8**, 7–69.

HENRY, J.-L. 1980. Trilobites ordoviciens du Massif Armoricain. *Mém. Soc. géol. minéral. Bretagne* **22**, 1–250.

HUGHES, C. P. 1979. The Ordovician trilobite faunas of the Builth–Llandrindod Inlier, central Wales. Part III. *Bull. Br. Mus. nat. Hist.* (Geol.), **32**, 109–181.

—— INGHAM, J. K. and ADDISON, R. 1975. The morphology, classification and evolution of the Trinucleidae (Trilobita). *Phil. Tr. Roy. Soc. Lond.* **B 272**, 537–607.

KELLOGG, H. E. 1960. The geology of the Derik–Mardin area, southeastern Turkey. Rep. Explor. Divn Am. Overseas Petr. Ltd. (unpublished). Ankara.

KIELAN, Z. 1960. Upper Ordovician trilobites from Poland and some related forms from Bohemia and Scandinavia. *Palaeont. pol.* **11**, 1–198.

KŘÍŽ, J. and PEK, I. 1974. *Dysplanus, Nobiliasaphus* and *Petrbokia* (Trilobita) in the Llandeilo of Bohemia. *Věst. ústřed. Úst. geol.* **49**, 19–27.

PŘIBYL, A. and VANĚK, J. 1965. Neue Trilobiten des böhmischen Ordoviziums. Ibid. **40**, 277–282.

—— —— 1968. Einige Trilobiten aus dem böhmischen Ordovizium. Ibid. **43**, 191–197.

RIGO DE RIGHI, M. and CORTESINI, A. 1964. Gravity tectonics in the foothills structure belt of southeast Turkey. *Bull. Am. Petrol. Geol.* **48**, 1911–1937.

SCHMIDT, G. C. 1965. Chart I. Proposed rock unit nomenclature, Petroleum District V, S.E. Turkey. Revised edn. *Stratig. Comm. Turkish Assoc. Petrol. Geol.*, Ankara (unpublished).

ŠNAJDR, M. 1956. Trilobiti drabovských a letenských vrstev českého ordoviku. *Sb. Ústřed. Úst. Geol.*, Paleont. **22**, 477–533.

SUDBURY, M. 1957. *Diplograptus spinulosus* sp. nov., from the Ordovician of Syria. *Geol. Mag.* **94**, 503–506.

WHITTINGTON, H. B. 1962. A monograph of the Ordovician trilobites of the Bala area, Merioneth, I. *Palaeontogr. Soc.* [*Monogr.*], 1–32.

—— 1966. Phylogeny and distribution of Ordovician trilobites. *J. Paleont.* **40**, 696–737.

—— and HUGHES, C. P. 1972. Ordovician geography and faunal provinces deduced from trilobite distribution. *Phil. Trans. roy. Soc. Lond.* **B 263**, 235–278.

YALÇIN, N. 1980. Amanosların litolojik karakterleri ve Güneydoğu Anadolu'nun tektonik evrimindeki anlamı. *Türk. Jeol. Kur. Bült.* **23**, 21–30.

W. T. DEAN
Department of Geology
University College
Cardiff CF1 1XL
U.K.

TRILOBITES FROM THE UPPER CAMBRIAN
OLENUS ZONE IN CENTRAL ENGLAND

by A. W. A. RUSHTON

ABSTRACT. The *Olenus* Zone is well developed in the Outwoods Shales Formation of the Nuneaton district, central England. The trilobite fauna of twenty four species is comparatively rich: seventeen are identified or compared with existing taxa, six are under open nomenclature, and one, *Olenus veles*, is new. Four *Olenus* subzones are recognized and correlated with those in Scandinavia. The base of the *Olenus* Zone is correlated approximately with the base of the *Aphelaspis* Zone of North America and the base of the Idamean Stage of Australia. The top of the *Olenus* Zone is correlated approximately with the top of the *Elvinia* Zone of North America, the top of the Sakian Stage of the Maly Karatau, Kazakhstan, and with the lower part of the post-Idamean stage in Australia.

THE *Olenus* Zone represents a substantial part of the Upper Cambrian Merioneth Series as developed in north-west Europe. Although much of the fauna has a restricted distribution in Europe the approximate limits of the zone can be widely recognized by the occurrence of *Glyptagnostus reticulatus* near the base and by species of *Irvingella* near the top. These limits show that the *Olenus* Zone—the oldest but one of the eight zones in the north-west European sequence—approximates to five of the ten or twelve Upper Cambrian assemblage zones in North America, and to five zones of the more finely subdivided Australian sequence.

Despite the occurrence of some widely ranging trilobites, intercontinental correlation of the Upper Cambrian remains difficult, as is shown by the different solutions suggested in various correlation charts; thus the upper boundary of the *Olenus* Zone is correlated at higher levels by Shergold (1980, p. 16; 1982, p. 14) and especially Ergaliev (1980, p. 51) than was suggested by Cowie *et al.* (1972, pl. 5). The correlation of the *Olenus* Zone is reviewed below.

Considering the bewildering multiplicity of Upper Cambrian trilobite taxa in the shallow-water deposits of the cratonic areas of North America, Asia, and Australia, the trilobite faunas of the *Olenus* Zone in north-west Europe are restricted, consisting mainly of *Olenus* species and agnostids. Fifteen or sixteen species have been described from the zone in Scandinavia (Westergård 1947, p. 22) and seven from Wales (Lake 1906, 1908). The Outwoods Shales of the English Midlands, however, has a rather richer trilobite fauna of twenty-four species, but of these only *Homagnostus obesus*, *Sulcatagnostus securiger*, and *Irvingella nuneatonensis* have been figured. The present account surveys all the species and completes the description of the Cambrian trilobites of the Nuneaton district, following Illing (1916), Rushton (1966, 1967, 1978, 1979), and Taylor and Rushton (1972).

Divisions of the Olenus *Zone*. In Scandinavia the *Olenus* Zone, sometimes known as the Zone of *Olenus* with *H. obesus*, has been divided into six subzones (Westergård 1947); the lower part of the Upper Cambrian is thus subdivided as follows:

Parabolina spinulosa	{ *P. spinulosa*
	{ *P. brevispina*
	{ *Olenus scanicus* and *Cyclotron angelini*
	{ *O. dentatus*
Olenus with	{ *O. attenuatus*
Homagnostus obesus	{ *O. wahlenbergi*
	{ *O. truncatus*
	{ *O. gibbosus*
Agnostus pisiformis	

[Special Papers in Palaeontology No. 30, pp. 107–139, pls. 14–19]

The lower subzones are widely recognized but the upper subzones are absent over most of Scandinavia (Westergård 1947, p. 21; Martinsson 1974, p. 206); the succession is complete only in Scania, and there Westergård (1944, p. 29) showed an unfossiliferous thickness of strata above the highest characteristic *Olenus* species and below fossils typical of the overlying *Parabolina spinulosa* Zone, such as *P. brevispina* Westergård, *Protopeltura aciculata* (Angelin), and *Orusia lenticularis* (Wahlenberg).

In the Nuneaton district *Olenus gibbosus*, *O. truncatus*, and *O. Wahlenbergi* occur in succession and are taken to indicate the presence of the same subzones as in Scandinavia. The higher Scandinavian subzones cannot be identified satisfactorily in Britain, so the upper part of the *Olenus* Zone is referred to a local *cataractes* Subzone (Allen *et al.* 1981, p. 308) which overlies the *wahlenbergi* Subzone in the Nuneaton district.

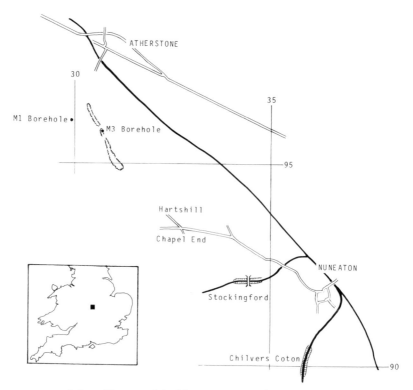

TEXT-FIG. 1. Locality map of the Nuneaton area. The numbered lines refer to the National Grid within square SP.

THE *OLENUS* ZONE IN BRITAIN

The *Olenus* Zone is proved in North and South Wales and near Nuneaton in the English Midlands. It may be present in the Malvern Hills (although no trilobites of the zone are known) but is unknown in Shropshire. Rushton (1974) reviewed the Cambrian of these areas.

In North Wales the zone is present in the Maentwrog Formation, a thick sequence of mudstones and siltstones deposited mainly as turbidites and contourites (Allen *et al.* 1981, pp. 303, 321). The *gibbosus* Subzone is recognized 140 m to 250 m above the base of the formation, and is overlain by the *truncatus* and *cataractes* Subzones. The *Olenus* Zone persists into the base of the overlying Ffestiniog Flags Formation (Allen *et al.* 1981, p. 311), a division of shallow-water sandstone and siltstone. The trilobite fauna consists of *O. gibbosus*, *O. truncatus*, *O. cataractes*, *O. micrurus*, *H. obesus*, and *Glyptagnostus reticulatus*, with records of single specimens of *O. austriacus*? (recorded below) in the *gibbosus* Subzone, *Olentella rara* (Orłowski) in the *cataractes* Subzone (Allen *et al.* 1981, p. 308) and *Olenus* cf. *solitarius* at the top of the Maentwrog Formation (see below).

In South Wales the *Olenus* Zone is present in the '*Lingula* Flags' of the St. David's district, Dyfed. They form a 600 m thickness of silty and sandy alternations, similar to the Ffestiniog Flags of North Wales. The trilobites recorded from the local Treffgarne Bridge Beds are *O. cataractes*, *O. mundus*, and *H. obesus* (Lake 1906, 1908), and the *cataractes* Subzone is the only subdivision proved there.

In the Malvern Hills the *Olenus* Zone is inferred from the presence of *Cyclotron lapworthi* in a thin shale unit (Rushton 1969), but this is not yet supported by other evidence. In Shropshire the *Olenus* Zone is unknown and may be absent, since the overlying *P. spinulosa* Zone lies directly on the Middle Cambrian Upper Comley Sandstone locally (Stubblefield 1930; Rushton 1974, p. 100).

Finally, the *Olenus* Zone is well developed in the Outwoods Shales of the Midlands, as discussed below.

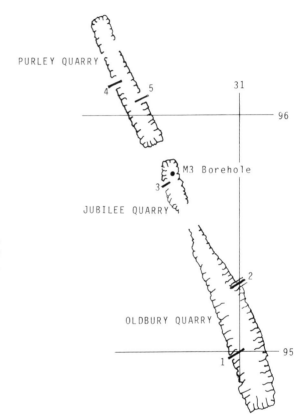

TEXT-FIG. 2. Sketch-map of the Mancetter Quarries to show localities 1–5 and the position of the Merevale No. 3 Borehole (M3), simplified from Taylor and Rushton 1972, p. 14.

The Outwoods Shales Formation. The Outwoods Shales Formation, the thickest division of the Stockingford Shales Group, is about 230 m thick, discounting intrusive sills. The formation is composed of innumerable alternations of pale grey bioturbated mudstone and dark grey laminated pyrite-rich mudstones, with infrequent thin sandy or silty beds. The formation was described in detail by Taylor and Rushton (1972, p. 9), who interpreted it as an outer-shelf deposit.

The main exposures of the Outwoods Shales are in the Mancetter Quarries (localities 1–5 in text-fig. 2). The higher beds are best seen in Stockingford Railway Cutting (text-fig. 1), where trilobites were mainly collected in the north bank at localities 6, 7, and 8, respectively 140 m, 157 m, and 166 m west of the bridge over the cutting. There is a number of spot localities to the north-west and south of Nuneaton, several of which were mentioned by Taylor and Rushton (1972, p. 12), and Mr A. F. Cook has collected a few specimens from a pipe-line trench just north of Stockingford Cutting. Much of the information on the Outwoods Shales was derived from the Merevale Nos. 1 and 3 Boreholes (hereafter referred to as M1 and M3 Boreholes) which respectively penetrated the upper third and lower half of the formation. The Outwoods Shales has also been encountered farther afield in a few exploratory boreholes, for instance at Trickley Lodge, 7 km south-west of Tamworth (SP 1603 9844), and at Kineton, 12 km north-north-west of Banbury (SP 3844 5015).

The Outwoods Shales are more or less fossiliferous throughout. The distribution of trilobites, based on

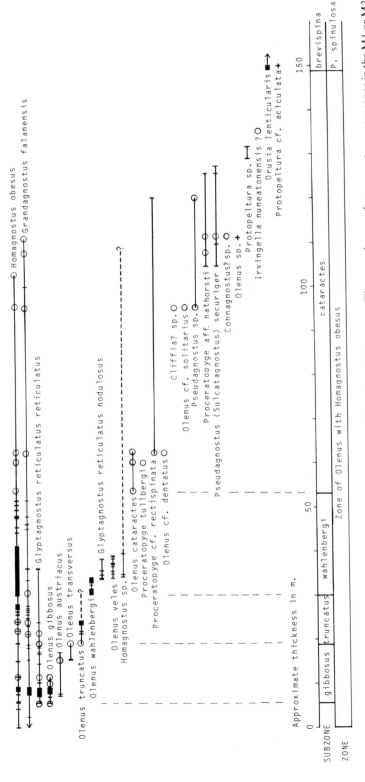

TEXT-FIG. 3. Distribution of trilobites in the *Olenus* Zone. Continuous lines show their ranges. The cross-bars each represent an occurrence in the M1 or M3 Boreholes; the open circles represent records from surface exposures positioned so far as possible in relation to the borehole depths. The numbers refer to estimated stratigraphical thicknesses in metres above the base of the zone. The appearance of the brachiopod *Orusia lenticularis* is included because it contributes to recognition of the base of the overlying *Parabolina spinulosa* Zone.

specimens from all localities that could be related stratigraphically, is shown in text-fig. 3. The lower parts of the succession are richly fossiliferous, especially with species familiar from Scandinavia; the upper parts are much less so, but include more cosmopolitan elements, some of which are not otherwise recorded from Europe. There is a poorly known interval in the middle of the *Olenus* Zone at a higher level than the exposures around Oldbury Quarry (loc. 1) and apparently much affected by igneous intrusions in Stockingford Cutting: this interval would repay future study.

BIOSTRATIGRAPHY

The lowest 9 m of the Outwoods Shales is referred to the Middle Cambrian *Lejopyge laevigata* Zone and the succeeding 58 m to the *Agnostus pisiformis* Zone (Rushton 1978). The base of the *Olenus* Zone is taken at the appearance of *H. obesus* at a depth of 200 ft 8 in. (61·16 m) in the M3 Borehole. The lowest part of the zone is poorly fossiliferous and yields only the long-ranging species *H. obesus* and *Grandagnostus falanensis*, together with a single fragment of *Proceratopyge nathorsti* from loc. 5 (Rushton 1978, p. 247).

The *gibbosus* Subzone is recognized about 6 m higher by the appearance of *O. gibbosus* and *Glyptagnostus reticulatus reticulatus* at a depth of 180 ft 4 in. (54·96 m) in the M3 Borehole, and at loc. 5. A little higher *O. austriacus* appears (at loc. 4 and M3 Borehole) and *O. transversus* occurs in the upper part of the Subzone in the M3 Borehole.

The *truncatus* Subzone is recognized by the appearance of *O. truncatus* at loc. 4 where it is associated with *O. transversus*, and in the M3 Borehole. No other species make their first appearance in this subzone apart from doubtful fragments of *O. wahlenbergi* and *O. veles* sp. nov. in the M3 Borehole. *G. r. reticulatus* is much less common than in the *gibbosus* Subzone.

The *wahlenbergi* Subzone is characterized by the appearance of definite *O. wahlenbergi* at a depth of 90 ft (27·43 m) in the M3 Borehole. The base of the subzone also lies within the section at loc. 2 but was not located precisely. *H. obesus* is very abundant in this subzone and *G. r. reticulatus* is rare; *G. r. nodulosus*, *O. veles*, and *Homagnostus* sp. appear above the base in the M3 Borehole.

The *cataractes* Subzone is recognized in Oldbury Quarry (loc. 1), some 20 m above the highest record of the *wahlenbergi* Subzone; the intervening strata are baked by a sill and fossils have not been found in them. The trilobites are *O. cataractes*, *O.* cf. *dentatus*, *P.* cf. *rectispinata*, *P. tullbergi*, *H. obesus*, and *G. falanensis*, but *G. reticulatus* has not been found. In the higher parts of the *cataractes* Subzone, exposed in Stockingford Railway cutting (locs. 6–8) and cored in the M1 Borehole, the fauna differs: olenids and *H. obesus* are rare but species of *Pseudagnostus* appear and *Proceratopyge* is comparatively common; *O.* cf. *solitarius* occurs at loc. 6 in beds about 80 m below the top of the Outwoods Shales. As the highest beds of the Outwoods Shales reach above the top of the *Olenus* Zone they are to be correlated with the lower parts of the Ffestiniog Flags Formation. Thus the occurrence of *O.* cf. *solitarius* at the top of the Maentwrog Formation, just below the local base of the Ffestiniog, may be coeval with the occurrence in the Outwoods Shales. When more is known of the ranges of the trilobites in the upper part of the *cataractes* Subzone it may be possible to separate an overlying subzone, corresponding partly to the barren interval in the Scanian succession.

The top 12 m of the Outwoods Shales in the M1 Borehole yielded *Orusia lenticularis* and fragments probably of *Protopeltura aciculata*, they are referred to the *Parabolina spinulosa* Zone (Taylor and Rushton 1972, p. 21). *Irvingella nuneatonensis* occurs near the top of the Outwoods Shales in Stockingford Cutting, probably within the top 60 m of the Formation, and specimens of *Irvingella* referred to the species with doubt were collected from beds apparently about 30 m below the top of the Outwoods Shales in Parkhill Outwoods, 1·5 km south-west of Atherstone (Rushton 1967, p. 354). Thus *I. nuneatonensis* is likely to have originated from the upper part of the *cataractes* Subzone, and although its horizon is not fully established it cannot lie higher than the very base of the *P. brevispina* Subzone.

WORLDWIDE CORRELATION OF THE *OLENUS* ZONE

Correlation within the *Olenus* Zone is best achieved by means of *Olenus* species which appear in a consistent sequence in the Baltic and British areas. The boundaries between the subzones are sharp because the *Olenus* species tend to abound in some layers but are absent or very rare in the intervening beds. There is generally no gradation between the species, which thus conform to a pattern of sequential invasions by allopatrically evolved species, as already shown by Kaufmann (1933). *Olenus* species favoured outer-shelf sites and are associated with oxygen-deficient facies; their areal and facial distribution is thus generally restricted and they are therefore of limited correlative value.

There are certain trilobites characteristic of shallow-shelf seas which are found in the olenid biofacies as 'occasional invaders'. Such forms are of some correlative value but, being rare, little is known of their stratigraphical ranges and morphological variation. Examples are:

1. The damesellid *Drepanura eremita* Westergård (1947) is known from three fragments from the *pisiformis* Zone of Sweden, and has been recorded from the Tuorian of Siberia (Ivshin and Pokrovskaya 1968, p. 98). Damesellidae are characteristic of the early Upper Cambrian in northern China and Australia.

2. The pterocephaliid *Pedinocephalus peregrinus* (Henningsmoen 1957b; 1958, p. 180), known from two specimens from the *gibbosus* Subzone of Västergötland, Sweden, is related to, and may be a synonym of, *Acrocephalites? rarus* Westergård (1922, p. 119), which is based on one specimen from the *gibbosus* Subzone of Sweden. *Pedinocephalus* occurs in the early Upper Cambrian (upper Tuorian Stage) of Kazakhstan.

3. One specimen of another pterocepaliid, *Olentella rara*, has been found in the *cataractes* Subzone of North Wales; *O. rara* was described from shallow-water quartzites in Poland (Orłowski 1968, p. 268) but the genus typically occurs in the upper Tuorian of Kazakhstan.

Most valuable for correlation are the trilobites which are found regularly in both on- and off-shelf sites. These are the supposedly pelagic agnostid trilobites *Glyptagnostus* and *Homagnostus* and the polymerids *Irvingella* and perhaps *Proceratopyge*.

The presence of *G. r. reticulatus* in the lower subzones of the *Olenus* Zone allows a general intercontinental correlation: the subspecies occurs in the base of the pterocephaliid Biomere (in the *Aphelaspis* Zone) in Texas and Nevada, U.S.A. (Palmer 1962) and *G. reticulatus* gives its name to a zone in the Tuorian Stage in Siberia (Ivshin and Pokrovskaya 1968, p. 99) and the Sakian Stage in Kazakhstan (Ergaliev 1980), a zone of the Machari Shale of South Korea (Kobayashi 1962), and to a zone or zones in the Idamean Stage in Queensland, Australia (Öpik 1963; Shergold 1982); it also occurs in the *Chuangia-Prochuangia* Zone in Hunan and Guizhou provinces, China (Yang 1978). All these occurrences seem to be approximately contemporaneous. However, in Europe there is no record of the earlier forms *G. r. angelini* Resser or *G. stolidotus* Öpik and it is possible that the earliest European examples of *G. reticulatus* s.s. are not as old as the earliest occurrences in other continents. In Europe the ranges of *Agnostus pisiformis* (Wahlenberg) and *G. reticulatus* s.s. are not known to overlap whereas Ergaliev (1980, p. 39) recorded *A. pisiformis* in his *G. reticulatus* Zone, together with several Idamean species and a few Mindyallan forms (*Pseudagnostus sericatus* Öpik, *Innitagnostus innitens* Öpik). *G. reticulatus* has a short but definite stratigraphical range and there may therefore be some discrepancy in the correlation; but the presence of rare exotic trilobites in the *pisiformis* Zone shows that part at least of that zone lies lower than the *G. reticulatus* Zone, contrary to Ergaliev's correlation chart (1980, pp. 50, 51).

The presence of *Irvingella nuneatonensis* at a horizon near the top of the *Olenus* Zone suggests a correlation with the world-wide dispersal of *Irvingella* species. In the Great Basin of the U.S.A. Palmer (1965, pls. 21–23) showed an ascending succession of *Irvingella* species through the upper part of the *Elvinia* Zone, namely *I. angustilimbata* (Kobayashi), *I. flohri* Resser, *I. transversa* Palmer, and *I. major* Ulrich and Resser. The last is so prevalent that it has been used to designate a thin subzone at the top of the *Elvinia* Zone in parts of the U.S.A. Rushton (1967) distinguished *I. nuneatonensis* from the early form *I. angustilimbata* (which has an anterior border on the mature cranidium) and found it more comparable with, though specifically distinct from, *I. flohri* and *I. major*. It is clear that the horizon of *I. nuneatonensis* underlies all or nearly all the *P. brevispina* Subzone, so Rushton (1967, p. 347) correlated the top of the *Olenus* Zone approximately with the top of the *Elvinia* Zone. *Irvingella* species characterize a zone at the base of the Shidertan Stage in Kazakhstan and the Siberian Platform (Ivshin and Pokrovskaya 1968, pp. 99, 101) and at the base of the 'post-Idamean' in Queensland (Shergold 1982, p. 16). *Irvingella* species occur in the upper two zones of the Sakian Stage of the Maly Karatau (Ergaliev 1980); in Tien-Shan and South Korea *Irvingella* appears to occur with *Proceratopyge rectispinata*, suggesting a correlation of the *Iwayaspis* Zone of the Machari Shale and the *Eochuangia* Zone of Kobayashi (1962, pp. 8, 11) with the upper part of the *Olenus* Zone. Finally, the zonal position of the Swedish *I. suecica* Westergård remains uncertain but probably lies in or below the *brevispina* Subzone (Westergård 1949, p. 606). Westergård noted that the original material of *I. suecica* was associated with a pygidium of *Olenus* type (Westergård 1947, pl. 3, fig. 4).

These considerations show that Shergold's (1980, p. 16) correlation of the upper part of the European *Parabolina spinulosa* Zone with the *I. tropica* Zone and the lower part of the *Elvinia* Zone makes the *spinulosa* Zone too old. His later revision (Shergold 1982, p. 14) is an improvement but still implies that *I. nuneatonensis* is

SWEDEN	CENTRAL ENGLAND	NORTH WALES	GREAT BASIN, U.S.A.	QUEENSLAND	MALY KARATAU, KAZAKHSTAN
P. spinulosa	P. spinulosa	Briscoia? celtica	Conaspis	post-Irvingella	Pseudagnostus pseudangustilobus
P. brevispina	P. brevispina	P. spinulosa / Parabolinoides bucephalus	Parabolinoides spp.		**I. major / I. perfecta**
[unfossiliferous]	**Irvingella nuneatonensis**		**Irvingella major / Irvingella flohri**	Irvingella tropica	Ivshinagnostus ivshini
O. scanicus & C. angelini			Elvinia		**I. tropica**
O. dentatus	O. cataractes	O. cataractes / O. micrurus / Olentella rara	Dunderbergia	Stigmatoa diloma	Pseudagnostus curtare
O. attenuatus			**Erixanium**		
O. wahlenbergi	O. wahlenbergi		Prehousia	Erixanium sentum	Erixanium
O. truncatus	O. truncatus	**O. truncatus**	Dicanthopyge	Proceratopyge cryptica	Homagnostus longiformis
			Aphelaspis		
O. gibbosus	O. gibbosus	**O. gibbosus / Glyptagnostus reticulatus**	**G. reticulatus**	Glyptagnostus reticulatus	Glyptagnostus reticulatus
Pedinocephalus rarus			**G. stolidotus**		**A. pisiformis**
		[no fossils]			
Agnostus pisiformis	Agnostus pisiformis		Crepicephalus	Glyptagnostus stolidotus	Glyptagnostus stolidotus

Zonal / stage schemes:

- SWEDEN: *Parabolina spinulosa* zone; *Olenus* and *Homagnostus obesus* zone.
- CENTRAL ENGLAND: MONKS PARK SHALES / MOOR WOOD; OUTWOODS SHALES FORMATION. *Parabolina spinulosa* zone; *Olenus* and *Homagnostus obesus* zone.
- NORTH WALES: CWM-HESGEN; FFESTINIOG FLAGS FM.; MAENTWROG FORMATION.
- GREAT BASIN, U.S.A.: FRANCONIAN; DRESBACHIAN.
- QUEENSLAND: post-IDAMEAN; IDAMEAN; MINDYALLAN.
- MALY KARATAU, KAZAKHSTAN: SAKIAN.

TABLE 1. Suggested correlation of the *Olenus* Zone in England and Wales with zonal schemes in Scania, Sweden (Westergård 1947), the Great Basin, U.S.A. (Palmer 1965), Queensland, Australia (Shergold 1982), and Maly Karatau, Kazakhstan (Ergaliev 1980). The names of selected key taxa are added in the zones in which they occur.

older than the supposedly early species *I. tropica* Öpik and *I. angustilimbata*. Incidentally, these correlations also make the Welsh species *Parabolinoides bucephalus* (Belt), which occurs with and just below *Parabolina spinulosa* itself (Allen *et al.* 1981, p. 311) older than the typical occurrences of *Parabolinoides* in North America which are confined to the *Conaspis* Zone. This is of course not impossible, and it could be argued that *P. bucephalus* was the extracratonic ancestor of the North American species; but, on the other hand, the correlation proposed in Table 1 brings the known generic ranges more closely into line (cf. Rushton 1968, p. 414). Furthermore, if the Welsh *Briscoia*? *celtica* (Salter) from the upper part of the *spinulosa* Zone (see Allen *et al.* 1981, p. 314) is related to the typical North American Dikelocephalidae, it is not unreasonable to regard the upper *spinulosa* Zone as about the same age as the upper Franconian in which the earlier North American *Briscoia* and *Dikelocephalus* appear. The correlation chart by Ergaliev (1980, pp. 50, 51) shows the *Leptoplastus* Zone correlated with the higher *Irvingella* horizons, which makes the former much too old. He may have been influenced by Westergård's earlier supposition as to the stratigraphical position of *I. suecica*, without considering his later correction (Westergård 1949). The correlation chart by Lu and Lin (1981, p. 119) places *I. tropica* higher yet, equating it with the *Protopeltura praecursor* Zone. Based on the correlation of the Chinese and Australian successions, I here suggest that the *Olenus* Zone is approximately equivalent to the *Chuangia* Zone together with the *Changshania* Zone of northern China; to the upper part of the Huayansi Formation in western Zhejiang, viz. from the *G. reticulatus* Zone to the *clavatus-kiangshanensis* Zone (Lu and Lin 1981); and to the *Chuangia-Prochuangia* Zone of Hunan and Guizhou (where *G. reticulatus* and *O. austriacus* are found), together with the *Lopnorites orthogonialis* Zone (Yang 1978).

SYSTEMATIC PALAEONTOLOGY

Discussion. The trilobites of the *Olenus* Zone consist mainly of known species and for these only salient points are discussed. A few species are considered more fully. The descriptive terms for agnostids follow Öpik (1963, 1967) and Shergold (1977) and for polymerids follow Henningsmoen (1957*a*) where appropriate, but I prefer 'rhachis' to 'axis' (which term can then be reserved for its geometrical sense). Note that the 'glabella' is restricted to the preoccipital part of the 'cephalic rhachis'. Nearly all the cited specimens are in the collections of the Sedgwick Museum, Cambridge (abbreviated SM) or the Institute of Geological Sciences, London (IGS).

Family AGNOSTIDAE M'Coy
Subfamily AGNOSTINAE M'Coy
Genus HOMAGNOSTUS Howell, 1935*b*

Discussion. This genus was discussed briefly by Rushton (1978, p. 259) but fuller consideration of the validity of the genera *Homagnostus*, *Rudagnostus*, and *Micragnostus* requires evaluation of many species, and is not attempted here.

EXPLANATION OF PLATE 14

All figures are of internal moulds unless stated otherwise. All the specimens are from the *Olenus* Zone in the Outwoods Shales Formation of the Nuneaton district, except for Plate 14, fig. 1, and Plate 18, figs. 2 and 3, which are from North Wales.

Figs. 1–10. *Homagnostus obesus* (Belt). All × 8. 1, lectotype, British Museum (Natural History) I.7646; *gibbosus* Subzone, Afon Mawddach, Dolmelynllyn, Gwynedd, North Wales (SH 7295 2373). 2–9, showing variation, especially in the pygidial rhachis; upper part of *wahlenbergi* Subzone, east side of Jubilee Quarry. Respectively SM A59109, IGS Zl 8310, SM A59114 (juvenile) and 59115, SM A59106, IGS Zl 8310, SM A59110, A59108, A59107; 10, IGS BDA508, *wahlenbergi* Subzone, M3 Borehole.

Figs. 11–13, 17? *Homagnostus* sp. 11–13 from *wahlenbergi* Subzone, M3 Borehole, respectively IGS BDA265, BDA380, BDA463, all × 6. 17, doubtfully referred specimen IGS RU1626, × 8; *cataractes* Subzone, M1 Borehole.

Figs. 14–16. *Connagnostus*? sp. Pygidium from sandstone lenticle, *cataractes* Subzone, Stockingford Cutting, loc. 8, × 8. 14 is latex cast of SM A59412; 15, 16, top and left side views of SM A59412a.

PLATE 14

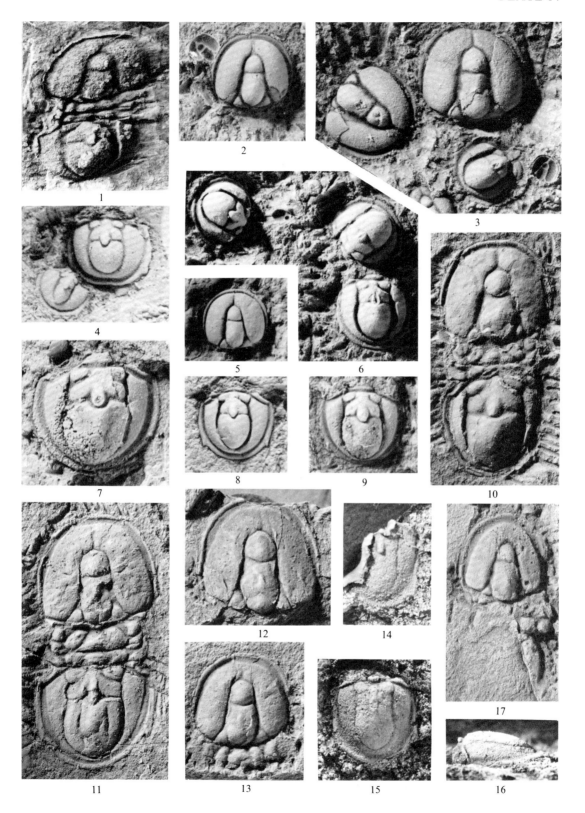

Homagnostus obesus (Belt, 1867)

Plate 14, figs. 1–10; text-fig. 4

Material. The lectotype (selected Rushton *in* Allen *et al.* 1981, pl. 1, fig. 2) shows the diagnostic features but is not well preserved (Pl. 14, fig. 1) and the species is interpreted here by reference to better preserved material from the Outwoods Shales (cf. Rushton 1978, pl. 25, fig. 4), especially numerous adults and some meraspids from a silty bed cropping out just east of Jubilee Quarry (SP 3075 9579).

Remarks. Immature cranidia have a narrower rhachis and show stronger lateral glabellar furrows than when fully grown (text-fig. 4*a*, *b*). Pygidia show a widening of the rhachis which continues to the gerontic stage (Pl. 14, figs. 7, 10); the posterior lobe becomes plump, tends to become blunter behind, and extends backwards with growth, but it never reaches the posterior border furrow. An important feature is the development of the anterior lobes and furrows on the pygidial rhachis: in meraspid pygidia the anterior furrow arches forwards across the rhachis rather as in mature *Ptychagnostus* (text-fig. 4*c*); with growth the median part of this furrow becomes effaced and, concurrently, faint longitudinal furrows develop exsagittally tending to divide the anterior rhachial lobe into three parts (text-fig. 4*d*), much as in species of *Innitagnostus* (Shergold 1982, pl. 5, figs. 3–5); with further growth the lateral furrows curve forward into the longitudinal furrows and join the articular furrow, giving the structure characteristic of *Homagnostus* (text-fig. 4*e*). These observations are based on few specimens only and require further investigation from better preserved material, but they suggest (*a*) that *Homagnostus* is related to and may be descended from the Ptychagnostinae, (*b*) that the furrows on the glabella and pygidial rhachis in *Innitagnostus* are primitive in comparison with *Homagnostus*, and (*c*) that the structure of the anterior part of the pygidial rhachis in adult *Micragnostus* (Fortey 1980, pl. 1, figs. 15, 16, 18) could have been developed in *Micragnostus* by paedomorphosis in the *Homagnostus* stock.

Occurrence. Throughout the *Olenus* Zone at many localities in the Outwoods Shales. Common in the *gibbosus* and especially the *wahlenbergi* Subzone in the M3 Borehole; present in the *cataractes* Subzone at loc. 1 but rare in the upper part of the Subzone (four specimens from Stockingford cutting, none found in the M1 Borehole).

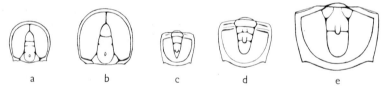

a   b   c   d   e

TEXT-FIG. 4. Sketches of meraspid and early holaspid shields of *Homagnostus obesus* (Belt) to show changes in ontogeny, all × 20. Drawn from specimens (SM) on the prolific slabs from Jubilee Quarry (locality as for Pl. 14, figs. 2–9).

Homagnostus sp.

Plate 14, figs. 11–13, 17?

Material. One complete specimen and two cephala from the *wahlenbergi* Subzone, M3 Borehole; one doubtfully referred specimen (Pl. 14, fig. 17) from the *cataractes* Subzone, M1 Borehole.

Remarks. The present form shows definite differences from gerontic *Homagnostus obesus* (Pl. 14, fig. 10) because the median glabellar node is stronger and elongate, the preglabellar median furrow is weaker, the border furrows are wider, the cephalic acrolobe is weakly constricted (expressed as an abrupt narrowing of the border furrow) and faintly scrobiculate, and the pygidial rhachis shows distinct notular furrows. The structure of the anterior part of the pygidial rhachis, however, is like that of *H. obesus* and is taken to be a shared derived character justifying reference to *Homagnostus*. As discussed below, the distinct notular furrows distinguish the present form from *H. laevis*.

Rushton (in Taylor and Rushton 1972, pl. 2) noted 'cf. *Homagnostus obesus laevis*' in the upper part of the Outwoods Shales in the M1 Borehole. This was based on the fragmentary specimen here tentatively referred to *Homagnostus* sp. (Pl. 14, fig. 17). The cephalon somewhat resembles *H. o. laevis* Westergård (1947) but the counterpart of the fragmentary pygidium attached has the interanotular axis clearly expressed and this suggests that reference to *H. o. laevis* is inappropriate because, though the rhachis of the holotype of *laevis* (Westergård 1922, pl. 1, fig. 5) is imperfect, the figured paratype (ibid., fig. 6) shows no intranotular axis.

Agnostus captiosus Lazarenko (1966, pl. 1, figs. 13–23) shows some similarity to the present species, although only one pygidium (her fig. 19) shows notular furrows; the cephalic border furrows are narrower, however, and the anterior part of the pygidial rhachis is divided into three lobes, as in *Innitagnostus* or *Cyclagnostus quasivespa* Öpik (1967, pl. 59, fig. 5). The cephalon of *A. scrobicularis* Ergaliev (1980, pl. 17, figs. 1, 2) has narrower border furrows and an unconstricted acrolobe and the pygidium has a shorter, narrower rhachis without notular furrows, but has a scrobiculate acrolobe. The structure of the anterior part of the pygidial rhachis distinguishes the present species from *Lotagnostus* species such as *L. trisectus* (Salter) (Westergård 1922, pl. 1, figs. 11, 12) and *L. asiaticus* (Troedsson 1937, pl. 1, figs. 10–16). Known *Lotagnostus* species have narrower cephalic borders and an unconstricted acrolobe.

Subfamily QUADRAGNOSTINAE Howell
Genus GRANDAGNOSTUS Howell, 1935a
Grandagnostus falanensis (Westergård, 1947)

Plate 15, figs. 2–5, 8, 9

1978 *Grandagnostus falanensis* (Westergård); Rushton, p. 256, pl. 24, figs. 6–14, fig. 3a [synonymy].

Material. Besides the material listed by Rushton (1978) from the *laevigata* and *pisiformis* Zones there are specimens from all the subzones of the *Olenus* Zone in the Merevale Nos. 1 and 3 Boreholes and from Purley and Oldbury Quarry; also several specimens from the *cataractes* Subzone in the Stockingford Cutting, locs. 6–8.

Remarks. Although the specimens show considerable apparent variation due to compaction in shale there is no consistent difference observed between the material described in 1978 and that figured here. Specimens from a calcareous sandstone lenticle (Pl. 15, figs. 4, 5, 8, 9) show the natural convexity. None of the specimens show the narrow rhachial structure and subcircular pygidium with subdued 'shoulders' which characterize *Phaldagnostus* (Ergaliev 1980, pl. 12, figs. 5, 6) or *Peratagnostus* (Shergold 1982, pl. 6, figs. 10, 11). Shergold (1982, p. 22) discussed the identity of these two genera. He figured juveniles attributed to *P. nobilis* (1982, pl. 6, figs. 4–7) which have a pygidial structure sharply distinct from the young of *G. falanensis* attributed by Rushton (1978, text-fig. 3a) but very similar to *Phalagnostus nudus* (Beyrich), as discussed by Rushton (1978, p. 255), who considered that the marginal flange was not merely a border but pleural regions with or without the border. In *Phalagnostus* this juvenile structure persists to the adult stage whereas in *Peratagnostus* the adult form has the rhachis, pleural lobes, and border all developed (Palmer 1968, pl. 10, fig. 24; Shergold 1982, pl. 6, fig. 11). If Shergold is correct in attributing the juvenile forms to *Peratagnostus* some of the unnamed pygidia compared by Rushton (1978, p. 256) to *Phalagnostus* may be referable to *Peratagnostus* which genus must, in such a case, show a remarkable ontogenetic metamorphosis in pygidia 1·0 mm to 1·6 mm long.

Subfamily GLYPTAGNOSTINAE Whitehouse
Genus GLYPTAGNOSTUS Whitehouse, 1936
Glyptagnostus reticulatus reticulatus (Angelin, 1851)

Plate 15, figs. 1, 6

Remarks. Shergold (1982, p. 24, pl. 4) has commented on the ontogenetic changes in *G. reticulatus*, among which is an increase in the density of the distinctive reticulation or blistering of the surface.

G. reticulatus s.l. has a wide distribution and a restricted stratigraphic range as discussed fully by Palmer (1962, pp. 1, 11). The Outwoods Shale material is typical of *G. reticulatus reticulatus* and proved abundant in the *gibbosus* Subzone and rare in the *truncatus* and *wahlenbergi* Subzones. In North Wales Rushton (*in* Allen *et al.* 1981, p. 307) observed *G. reticulatus* only in the *gibbosus* Subzone. In Sweden *G. r. reticulatus* occurs in the *gibbosus* and *truncatus* Subzones and it is known from the *wahlenbergi* Subzone in Bornholm (Westergård 1947, pp. 6, 22).

Glyptagnostus reticulatus nodulosus Westergård, 1947

Plate 15, fig. 7

Remarks. This subspecies is distinguished by a finer and denser reticulation than in specimens of subsp. *reticulatus* of comparable size. In the example illustrated the surface blistering is more subdivided than in Westergård's specimens (1947, pl. 1, figs. 7–9) presumably because it is a large specimen. In the Outwoods Shales it is confined to the *wahlenbergi* Subzone at loc. 2 and M3 Borehole; in Sweden it occurs in the *truncatus* and *wahlenbergi* Subzones and in Bornholm it occurs at the base of the *Olenus* Zone (Westergård 1947, p. 7).

Family DIPLAGNOSTIDAE Whitehouse
Subfamily DIPLAGNOSTINAE Whitehouse
Genus CONNAGNOSTUS Öpik, 1967
Connagnostus? sp.

Plate 14, figs. 14–16

Material. One pygidium from the *cataractes* Subzone, loc. 8, in Stockingford Cutting.

Remarks. The fragmentary external mould shows an elongate median rhachial node and one of the marginal spines; the internal mould shows the pygidial rhachis divided into two short anterior segments which are traversed by the median node and a posterior lobe which is longer than wide and much longer than the two anterior segments. The front of the rhachis is embayed, and the articulating device appears to be of diplagnostoid type, as in *Connagnostus venerabilis* Öpik (1967, pl. 54, fig. 11). The comparative shortness of the anterior segments is in contrast with the species of *Connagnostus* described by Öpik (1967, pl. 54, figs. 11–13; pl. 55, figs. 1, 2) and by Shergold (1980, pl. 12, figs. 2–4,

---

EXPLANATION OF PLATE 15

Figs. 1, 6. *Glyptagnostus reticulatus reticulatus* (Angelin). IGS BDA992, BDA1064; both × 8, *gibbosus* Subzone, M3 Borehole.

Figs. 2–5, 8, 9. *Grandagnostus falanensis* (Westergård). 2, flattened, SM A59125, × 6; *cataractes* Subzone, Stockingford Cutting, loc. 6. 3, exuvia, in relief, showing traces of the cephalic and pygidial doublures, SM A59120, × 6; *wahlenbergi* Subzone, Oldbury Quarry, loc. 2. 4, 5, top and right side views of cephalon, SM A59127, associated with cephalon of *Pseudagnostus* (*Sulcatagnostus*) *securiger* (Lake)?, SM A59128, × 6. Sandstone lenticle, *cataractes* Subzone, Stockingford Cutting, loc. 8. 8, 9, top and right side views of pygidium, SM A59126, × 10; occurrence as for figs. 4, 5.

Fig. 7. *Glyptagnostus reticulatus nodulosus* Westergård. Large cephalon, IGS BDA459, × 4; *wahlenbergi* Subzone, M3 Borehole.

Figs. 10, 12–14, 17–18. *Pseudagnostus* sp. 10, IGS CS338 (latex), × 5; *cataractes* Subzone, Stockingford Cutting, loc. 6. 12, 13, latex cast of pygidium and side view of partly enrolled specimen, SM A59099b and a, both × 6; top of *cataractes* Subzone, Parkhill Outwoods (SP 2980 9654). 14, 17, small complete specimen and large enrolled specimen, both latex casts, IGS CS337, CS341, × 5; locality as for fig. 10. 18, latex cast of cephalon, IGS RU1433, × 5; *cataractes* Subzone, M1 Borehole.

Figs. 11, 15, 16. *Pseudagnostus* (*Sulcatagnostus*) *securiger* (Lake). See also figs. 4, 5. 11, pyritized cephalon, IGS RU1530, × 5; *cataractes* Subzone, M1 Borehole. 15, holotype (monotypy), IGS GSM57650, × 5; *cataractes* Subzone?, Chapel End (see text); figured Lake 1906, pl. 2, fig. 11. 16, fragmentary specimen, IGS RU1673, × 4; *cataractes* Subzone, M1 Borehole.

PLATE 15

RUSHTON, Cambrian trilobites

7–10). Shergold's specimens have a pygidial rhachis that is *Homagnostus*-like anteriorly and with agnostoid rather than diplagnostoid articulation; it seems probable that they are offshoots of the *Homagnostus* stock rather than diplagnostids. The rachis of the present specimen resembles that of *Innitagnostis inexpectans* (Kobayashi) (Shergold 1982, p. 20 gives figures and a synonymy), but the border furrow appears wider and the articulation appears to be of diplagnostid rather than agnostid type.

Subfamily PSEUDAGNOSTINAE Whitehouse
Subgenus PSEUDAGNOSTUS (PSEUDAGNOSTUS) Jaekel, 1909

Discussion. The classification used here follows Shergold's (1977) valuable review of the genus, to which the reader is also referred for comments on morphology, distribution, and terminology (Shergold 1977, p. 97).

Pseudagnostus (*Pseudagnostus*) sp.

Plate 15, figs. 10, 12–14, 17, 18

Material. A small complete specimen, a crushed enrolled specimen, two cephala, a thoracic segment, and two pygidia from locality 6, Stockingford Cutting. One poorly preserved but typical specimen (IGS JR 3009, 3010) from Outwoods Shales 'near junction of Griff and Coventry canals, $2\frac{1}{4}$ miles south of Nuneaton', grid ref. SP 364 883 approx. One partly enrolled specimen (SM A59099) from Park Hill Outwoods (grid ref. SP 2983 9654). One cephalon from the M1 Borehole. All are from the upper part of the Outwoods Shales, *cataractes* Subzone, but the exact horizon of JR 3009–3010 is not certain.

Remarks. The present material comes closest to Shergold's *cyclopyge* group of *Pseudagnostus*. Salient features are: parallel-sided glabella with elongate node, the transverse glabellar furrow weak or absent; preglabellar median furrow distinct; borders deliquiate; two anterior segments of pygidial rhachis not differentiated; subplethoid; marginal spines long, thin, inwardly curved at both early and late stages of growth, though their early position is more retral. The cephalon resembles those of *P. primus* Kobayashi (especially Kobayashi 1962, pl. 5, fig. 9) but pygidia of *P. primus* 'invariably lack the [marginal] spines' (ibid., p. 32). Comparable also is a cephalon assigned to *Neoagnostus* [*Pseudorhaptagnostus*] *simplex* (Lermontova 1951, pl. 2, fig. 16) though in that the glabellar node is in a more forward position. The holotype pygidium of *N. simplex* has a shorter anterior part to the rhachis and very small marginal spines. Long marginal spines are present in *N. canadensis* (Billings; Rasetti 1944, pl. 36, figs. 8–13), but in that species the cephalon shows the papillionate condition characteristic of *Neoagnostus* and the pygidium has a more definitely plethoid deuterolobe. Shergold (1980, pl. 8, figs. 10–12) illustrated small pygidia of '*Pseudagnostus* sp. V' with curved pygidial spines. His species has a proportionally longer pygidium at 2 mm stage of growth and the acrolobe is constricted. Shergold suggested that his *Pseudagnostus* sp. V might be juveniles of *P. aulax* Shergold (1980, pl. 5); if so the adult form resembles that of the present species in having the spines in a more advanced position, but differs in many other respects, among them the reduction of the spines with growth of the animal.

Subgenus PSEUDAGNOSTUS (SULCATAGNOSTUS) Kobayashi, 1939
Pseudagnostus (*Sulcatagnostus*) *securiger* (Lake, 1906)

Plate 15, figs. 4?, 5?, 11, 15, 16

Material. Holotype IGS GSM 57650 (Pl. 15, fig. 15), and a small associated cephalon (GSM 57651) referred to the species, from '40 feet below the unconformity' at Chapel End, 4 km north-west of Nuneaton (?grid ref. SP 321 936). Two cephala and one pygidium from Stockingford Cutting (locality 8). Two cephala and fragment of a complete specimen from the M1 Borehole.

Remarks. Since Lake's original description (1906, p. 20, pl. 2, fig. 11) the holotype has been cleaned, revealing a posterior median spine on the pygidium. The horizon at which it was collected has been re-assessed as Outwoods Shales, upper part of the *Olenus* Zone (Taylor and Rushton 1972, p. 20). The glabella has a distinct anterior transverse furrow and an elongate node; the cheeks are strongly scrobiculate on the holotype but less so on smaller specimens. The pygidium is plethoid; the anterior part of the rhachis is more strongly divided into a shorter anterior lobe and longer second lobe than in the pygidium in Plate 15, fig. 16; this second specimen confirms the presence of weak rugae on the pygidial flanks but there is no confirmation of the longitudinal furrow seen on the right flank of the holotype which may therefore be an accident of preservation. Two pygidia confirm the constricted pygidial acrolobe and the presence of a posterior median spine.

Sulcatagnostus ququqensis Lu (*in* Lu *et al.* 1974, p. 84, pl. 1, fig. 10) from the late Middle Cambrian of Xingjiang (Singkiang) has unconstricted acrolobes in both shields, and no median pygidial spine is observed or described. The pygidial rhachis is narrower posteriorly; it appears not to be deuterolobate, and the species may therefore represent a caecate diplagnostine species.

S. rugosus Ergaliev (1980, p. 112, pl. 17, figs. 3, 4) is a caecate, deliquiate, plethoid *Pseudagnostus* with a constricted cephalic (but unconstricted pygidial?) acrolobe. Although very like *P. (S.) securiger*, the pygidium has an entire posterior border. If the subgenus *Sulcatagnostus* is to be maintained on the presence of the posterior median border spine, *S. rugosus* should be transferred to the subgenus *Pseudagnostus*. The only other form possibly referable to the subgenus *Sulcatagnostus* remains the undescribed pygidium cited by Shergold (1977, p. 86).

Family OLENIDAE Burmeister

Discussion. Henningsmoen (1957a, p. 91) suggested that in olenids generally the free cheeks were united anteriorly by a thin ventral doublure, and he demonstrated this with an illustration of united free cheeks in *Parabolina spinulosa* and *Olenus attenuatus*. This suggestion has been supported by evidence from Arenigian olenids (Fortey 1974, p. 14). Palmer, however, has inferred from well-preserved free cheeks and cranidia that the early *O. gibbosus* had a rostral plate 'comparable to that of *Aphelaspis buttsi*' (Palmer 1962, p. 35, pl. 6, fig. 15). A small rostral plate (Pl. 17, figs. 2, 3), found with parts of *O. wahlenbergi* and referred to that species, is much narrower sagittally than the rostrum of *Aphelaspis*, but the ends show that the connective sutures took a very similar course. It appears that the early *Olenus* had a rostral plate and that the fusion of the connective sutures to unite the free cheeks took place in the upper part of the *Olenus* Zone, after *O. wahlenbergi* had evolved but before its successor *O. attenuatus*.

Subfamily OLENINAE Burmeister
Genus OLENUS Dalman, 1827

Remarks. Since Henningsmoen's thorough review (1957a, p. 96) several new species have been proposed:

Olenus altaicus Ivshin, 1962, *Trudy SNIIGGIMS*, vol. 19, p. 229 [description not seen].
Olenus austriacus Yang MS *in* Zhou *et al.*, 1977 (see below).
Olenus delicatus Öpik, 1963, p. 62, pl. 4, fig. 12.
Olenus guizhouensis Lu and Chien *in* Yin and Lee 1978, p. 475, pl. 163, fig. 3.
Olenus ogilviei Öpik, 1963, p. 59, pl. 1; pl. 2, figs. 2–5.
Olenus proximus Lazarenko, 1966, p. 55, pl. 5, figs. 8, 9. This species was described from the *stolidotus* Zone in Siberia, which is a low horizon for an *Olenus*. It is about contemporaneous with *O. alpha* Henningsmoen (*pisiformis* Zone of Norway) but differs notably in the rounded glabellar front and convex ('pelturoid') postocular sutures.
Olenus rarus Orłowski, 1968, p. 268, pl. 4, figs. 6–19. This species was transferred to *Olentella* by Rushton (*in* Allen *et al.* 1981, p. 308).
Olenus sinensis Lu, 1964, re-figured in Lu *et al.* 1974, p. 84, pl. 1, fig. 14; pl. 2, fig. 10. It closely resembles *O. asiaticus* Kobayashi but originates from a much higher horizon.
Olenus veles sp. nov. (below).

Beltella solitaria Westergård, 1922, is here transferred to *Olenus*. Of the doubtful species in Henningsmoen's list *O.? bucephalus* (Belt) was transferred to *Parabolinoides* by Rushton (1968).

O.? wilsoni Henningsmoen was made the type of a new genus *Simulolenus* by Palmer (1965, p. 55), to which he also referred *O. granulatus* Palmer, 1960 and *S. quadrisulcatus* Palmer, 1965. *Simulolenus* was stated to be characterized by a large glabella with three or four pairs of furrows, a short preglabellar field, and narrow fixed cheeks, small and forwardly placed eyes, an obtuse inner spine angle, and a transversly subquadrate pygidium composed of few (one or two) segments which has a poorly defined border and no marginal spines, and a weak indentation behind the rhachis. I consider the generic status of *Simulolenus* to be doubtful. In all cephalic features *Simulolenus* agrees with such species as *O. micrurus*, *O. cataractes*, and *O. solitarius* so that *Simulolenus* can only be recognized by the peculiarities of the pygidium. *O. cataractes* has a spinous pygidium typical of *Olenus*. Those of *O. micrurus* and *O. solitarius*, though small, being composed of two to three segments, are not as transverse as the pygidia of species assigned to *Simulolenus*, nor are they indented behind the rhachis; thus they are morphologically intermediate between *Olenus* and *Simulolenus*. Therefore, although Palmer's suggestion (1965, p. 56) that the species represent a local development of the olenid stock may well be valid, I do not think that *Simulolenus* can be sustained on morphological grounds as a distinct genus. If *Simulolenus* is reduced to a subgenus of *Olenus* (distinguished by a small transverse pygidium with a median indentation), or is suppressed, *O. (S.) quadrisulcatus* (Palmer) is in danger of becoming a junior homonym of Westergård's (1922) *Olenus*? species which Henningsmoen (1957a, p. 128) made the type of *Parabolina*? *quadrisulcata*, depending on the generic reference preferred for that species (see Öpik 1963, p. 64).

Ten species of *Olenus* have been found in the Outwoods Shales. Six of these are close to species originally described from Scandinavia and they occur in the same ascending sequence: *O. gibbosus*, *O. transversus*, *O. truncatus*, *O. wahlenbergi*, *O.* cf. *dentatus*, *O.* cf. *solitarius*. Most of these species are well known and illustrated (Westergård 1922; Henningsmoen 1957a) and only brief notes are appended here.

O. gibbosus (Wahlenberg) (Pl. 16, figs. 1–3). See Henningsmoen 1957a, p. 105. The Outwoods Shale material has a short pygidium, corresponding to Kaufmann's (1933) 'anterior' variant, and includes both narrow and broad forms of the species (Westergård 1922, p. 195). Abundant at the base of the *gibbosus* Subzone, Purley Quarry (locs. 5, 4) and the M3 Borehole. Allen *et al.* (1981, p. 307) commented on its distribution in North Wales.

O. transversus Westergård (Pl. 16, fig. 6). See Henningsmoen 1957a, p. 108. The pygidia from the Outwoods Shales do not show marginal spines, in contrast to some figured by Kaufmann (1933, figs. 12, 17). In Norway *O. transversus* occurs with an *O. gibbosus* 'posterior' (i.e. late) variant but in Scania it occurs above *O. gibbosus*, as in the M3 Borehole where it is rare. In Purley Quarry it occurs rarely with *O. truncatus* at loc. 4.

O. truncatus (Brünnich) (Pl. 16, figs. 4, 5, 9, 10). See Henningsmoen 1957a, p. 109. The figured material includes a cranidium with an unusually long glabella (Pl. 16, fig. 4). The free cheeks from the Outwoods Shales generally show a more acute flip (Henningsmoln 1957a, p. 13) than in Swedish specimens, but the cheek spine is continuous with the lateral margin. The pygidium is typical of the

---

EXPLANATION OF PLATE 16

Figs. 1–3. *Olenus gibbosus* (Wahlenberg), all *gibbosus* Subzone, M3 Borehole. 1, large pygidium, latex cast of IGS BDA962, × 6. 2, cranidium, BDA 1057, × 6. 3, free cheek, BDA1068, × 8.

Figs. 4, 5, 9, 10. *Olenus truncatus* (Brünnich). 4, cranidium with unusually long glabella, latex cast of IGS BDA729, × 4; *truncatus* Subzone, M3 Borehole. 5, axial shield showing 13 thoracic segments, SM A59052, × 4; *truncatus* Subzone, Oldbury Quarry, loc. 2. 9, free cheek with acute flip, latex cast of IGS BDA728, × 4; occurrence as fig. 4. 10, pygidium, latex cast of BDA735, × 6; occurrence as fig. 4.

Fig. 6. *Olenus transversus* Westergård. Axial shield, SM A59045, × 4; *truncatus* Subzone, Purley Quarry, loc. 4.

Figs. 7, 8, 11, 12. *Olenus austriacus* Yang. 7, cranidium with glabella flattened by compaction, IGS BDA859, × 6; *gibbosus* Subzone, M3 Borehole. 8, fragmentary dorsal shield, BDA1085, × 4; occurrence as fig. 7. 11, 12, dorsal shields, SM A59074, A59075, both × 3; *gibbosus* Subzone, Purley Quarry, loc. 4.

PLATE 16

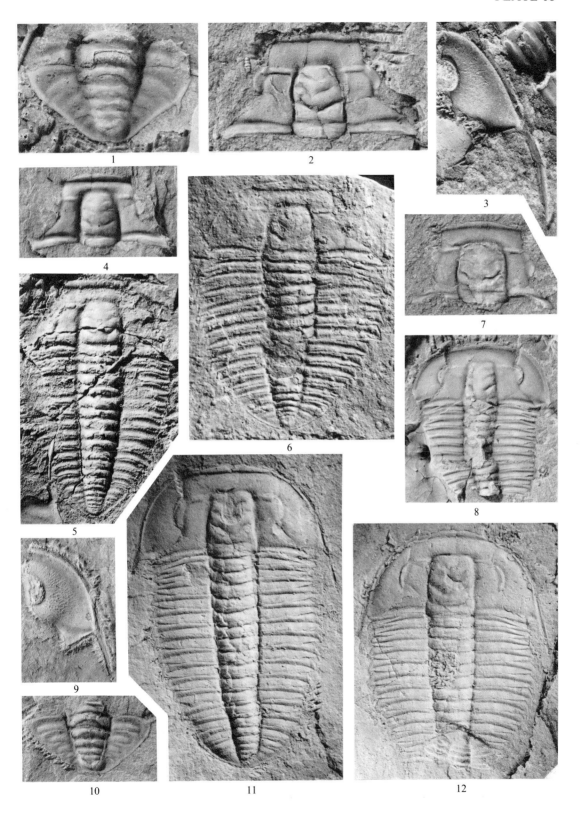

1

2

3

4

6

7

5

8

9

10

11

12

RUSHTON, *Olenus*

species. Not common in the *truncatus* Subzone, Purley Quarry, loc. 4; common in Oldbury Quarry, loc. 2. Not common in the M3 Borehole. Allen *et al.* (1981, p. 308) commented on the distribution in North Wales.

O. wahlenbergi Westergård (Pl. 17, figs. 1–6; text-fig. 5*g*). See Henningsmoen 1957*a*, p. 110. Part of Illing's (1913, p. 452) record of *O. truncatus* is referable here, the free cheek being distinctive (Pl. 17, fig. 5). Several specimens show fifteen thoracic segments, but one doubtfully referred specimen has sixteen. The rostral plate shown in Plate 17, figs. 2, 3 (here seen for the first time in *Olenus*) is 0·35 mm long sagittally and is 2·0 mm wide anteriorly, 1·3 mm wide posteriorly. The connective sutures make an inwardly concave curve. There are four terrace lines on the anterior part of the rostral plate but the posterior band appears smooth. Common in the *wahlenbergi* Subzone, Oldbury Quarry, loc. 2, and the M3 Borehole.

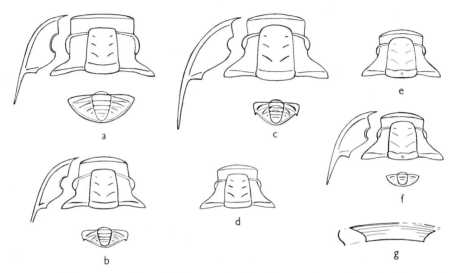

TEXT-FIG. 5. Reconstructions of *Olenus* species based on specimens from the Outwoods Shales; *a*, *O. austraicus*, × 2; *b*, *O. veles*, × 2; *c*, *d*, *O. cataractes*, *c*. × 1½ and × 3; *e*, *f*, *O.* cf. *solitarius*, *c*. × 1½ and × 3; *g*, *O. Wahlenbergi*, rostral plate seen from the ventral side and cross-section, × 25, reconstructed from IGS BDA 416, 417 (see Pl. 17, fig. 3).

O. cf. *dentatus* Westergård (Pl. 18, figs. 4, 8). See Henningsmoen 1957*a*, p. 104. The material, all from loc. 3, is fragmentary, but the cranidium agrees in proportion with shale-preserved specimens figured by Westergård. The pygidium is twice as wide as long, is rounded posteriorly, and has a widely conical rachis, but the strong marginal spines typical of *O. dentatus* are not well seen. Scarce in the *cataractes* Subzone, Jubilee Quarry, loc. 3.

O. cataractes Salter (Pl. 18, figs. 1, 2, 5–7, 9; text-fig. 5*c*, *d*). See Henningsmoen 1957*a*, p. 104. The material from the Outwoods Shales, found at loc. 1 in Oldbury Quarry, corresponds closely with the description by Lake (1908, p. 56) based mainly on material from South Wales. The small specimen figured here (Pl. 18, fig. 1) has faint rhachial nodes and stronger eye-ridges and shorter pleural spines than in the full-grown specimens, and in these resembles *O. mundus* Lake, 1908. As similar differences were observed between smaller and larger specimens of *O.* cf. *solitarius* it seems likely that *O. mundus* represents a younger stage of *O. cataractes*, as Lake had suspected. *O. cataractes* resembles *O.* cf. *solitarius* but has a longer pygidium with marginal spines. The pygidium resembles that of *O. dentatus* but its rhachis is less strongly tapered; in *O. dentatus* the preglabellar field is longer. Loc. 1, Oldbury Quarry, *cataractes* Subzone; not very common. South Wales, North Wales (Allen *et al.*, 1981, p. 308).

Olenus cf. *solitarius* (Westergård, 1922)

Plate 18, figs. 3, 10–14; text-fig. 5*e, f*

cf. 1922 *Beltella solitaria* n. sp.; Westergård, p. 140, pl. 14, fig. 1, *non* fig. 2.
cf. 1957*a* *Protopeltura? solitaria* (Westergård 1922); Henningsmoen, p. 230 [synonymy, remarks].

Material. A large cranidium and thorax, a small axial shield, fragments of a cephalon, cranidia, and thoraces, all from Stockingford Cutting, locality 6. Also referred to the same form are a dorsal shield, cranidium, and other fragments (SM A55470–55473), all on one block from 'east of Craig Ddu' (= Graig Ddu), near Criccieth, Gwynedd, North Wales, i.e. from the top beds of the Maentwrong Formation approximately at grid reference SH 524 375.

Remarks. Westergård's doubtful specimen of *Beltella solitaria* (1922, pl. 14, fig. 2) has a spinose pygidium and is referable to *Protopeltura* (Henningsmoen 1957*a*, p. 231). The lectotype (Westergård 1922, pl. 14, fig. 1) is transferred to *Olenus* by comparison with the more complete material from England and Wales.

The cranidium shown in Plate 18, fig. 10 is nearly the same size as the lectotype and is similarly proportioned except that the postocular cheeks are four-fifths as wide as the occipital ring, compared with two-thirds as wide in the type. The occipital node is clearly seen in fig. 13. The free cheek (not known from Sweden) has a rather broad border for *Olenus*; the inner spine angle is slightly obtuse and the flip is right-angled (fig. 12). The thorax has fourteen segments in the presumed exuvia in figs. 10, 11 and also in the specimen (SM A55470) from east of Graig Ddu (Pl. 18, fig. 3). In these specimens there are weak median nodes on most of the segments, though not in the large specimen shown in fig. 14. The pleurae are slightly wider than the rhachis whereas in the lectotype they are slightly narrower. The pygidium, known from the exuvia in fig. 11 and SM A55470, is nearly three times wider than long, has a blunt rhachis composed of two rings and a terminal part and pleural fields, slightly narrower than the rhachis, with one or two pleural grooves; the border is indistinct, has no marginal spines and is convex posteriorly.

The pygidium differs from those referred to *Simulolenus* by Palmer (1965, pl. 8, figs. 4, 10, 12) which are at least three times wider than long and are concave in outline behind the rhachis.

Despite the long glabella and rather small and forwardly placed eyes, this species is referred to *Olenus* rather than *Parabolina* which has twelve thoracic segments and usually a spinose pygidium; it differs from *Protopeltura* which has ten to twelve segments and an outwardly deflected cheek-spine. *P. solitaria* resembles *O. cataractes* in the cranidium but at a comparable stage of growth the eyes are closer to the glabella and the preglabellar field is shorter (text-fig. 5*c*–*f*). In *O. cataractes* the pygidium is longer, with four rhachial rings, and bears a pair of marginal spines.

Compared with *O. (S.) wilsoni* of the same size (Palmer 1965, pl. 8, fig. 11) *O.* cf. *solitarius* has a shorter preglabellar field (one-eighth rather than one-sixth of glabellar length) and slightly narrower interocular cheeks. *O. (S.) quadrisulcatus* (Palmer 1965, pl. 8, fig. 2) has a short preglabellar field but differs from *O.* cf. *solitarius* in the square cephalic rhachis and more posteriorly placed eyes.

Occurrence. The lectotype originated from the *brevispina* Subzone of the *Parabolina spinulosa* Zone at Andrarum, Scania, Sweden. The British material compared with it is from the upper part of the *cataractes* Subzone at Graig Ddu, North Wales, and from loc. 6 in the Outwoods Shales, Stockingford Cutting, where it is scarce.

Olenus austriacus Yang *in* Zhou *et al.*, 1977

Plate 16, figs. 7, 8, 11, 12; text-fig. 5*a*

?1922 *Olenus transversus?*; Westergård, p. 126, pl. 5, figs. 16, 17 only.
1972 *Olenus* sp. 1; Taylor and Rushton, p. 20, pl. 4 [borehole record].
1977 *Olenus austriacus* Yang MS *in* Zhou *et al.* p. 152, pl. 47, fig. 5.
1978 *Olenus austriacus* sp. nov.; Yang, p. 38, pl. 6, figs. 1, 2.

Material. Two complete and two incomplete dorsal shields, a cephalon, a cranidium, and a pygidium.

Diagnosis. An *Olenus* species with a long preglabellar field, long eyes, a small obtuse flip, thirteen thoracic segments with short pleural spines, and a relatively large pygidium which is rounded behind and lacks marginal spines.

Description. The present material agrees closely with the holotype (Yang 1978, pl. 6, fig. 1). Glabella rectangular, with three pairs of glabellar furrows which are convex forwards and oblique backwards and inwards. Preglabellar field between a quarter and a third as long as cephalic rhachis. Palpebral lobes long for *Olenus*, nearly one-third cranidial length; eye-ridges transverse. Preocular sutures very slightly divergent forwards. Interocular cheeks over half as wide as glabella. Postocular cheeks slender (exsag.), about as wide as occipital ring. Flip small, obtuse, inner spine angle obtuse. Thorax of thirteen segments with short pleural spines. Pygidium rather large, width more than twice length, rounded behind, without spines. Rhachis about a quarter of total width, composed of four or five rings, and a terminal piece; flanks with four pairs of pleural grooves.

Remarks. *O. austraicus* differs from *O. ogilviei* Öpik (1963, pl. 1, pl. 2, figs. 2–5) because the glabella is more truncate anteriorly, the preglabellar field is longer, the eyes shorter, the genal spines longer, there are thirteen rather than fourteen thoracic segments, and the pygidium is larger. The short pleural spines recall *O. transversus* Westergård (1922) which, however, is a broader species with a shorter pygidium; the longer pygidia which Westergård doubtfully referred to *O. transversus* may be referable to *O. austriacus* but have an extra rhachial ring. The cranidium shows some resemblance to *Hancrania brevilimbata* Kobayashi (1962, pl. 9, figs. 2–6), of which only cranidia and a hypostome were figured. In *H. brevilimbata* the glabella shows only two pairs of furrows, the preocular sutures diverge forwards slightly more and the greater length of the eye results in a slenderer postocular cheek.

Occurrence. Purley Quarry, loc. 4; M3 Borehole; temporary section north of Stockingford Railway Cutting (A. F. Cook coll.); all from the *gibbosus* Subzone, rare. A doubtful cranidium from the *gibbosus* Subzone at Cefndeuddwr, North Wales (Allen *et al.* 1981, p. 308). The types are from Tingziguan, Fenghuang County, Hunan Province, China, associated (as in Britain) with *Glyptagnostus reticulatus*.

Olenus veles sp. nov.

Plate 17, figs. 7–13; text-fig. 5*b*

Name. 'Light-armed soldier', the pleural spines being comparatively short.

> 1913 *Olenus truncatus*; Illing (*pars*), p. 452.
> 1923 *Parabolinella triarthra*; Pringle *in* Eastwood *et al.*, p. 35.
> 1972 *Olenus* sp. 2; Taylor and Rushton, p. 20, pl. 4 [borehole record].

Holotype. IGS BDA155, 156, counterparts of an axial shield (Pl. 17, fig. 11) from the *wahlenbergi* Subzone, M3 Borehole, depth 58 ft 4 in. (17.78 m).

EXPLANATION OF PLATE 17

Figs. 1–6. *Olenus wahlenbergi* Westergård. 1, free cheek showing acute flip and deflected genal spine, IGS BDA430, × 8; *wahlenbergi* Subzone, M3 Borehole. 2, 3, rostral plate associated with inverted hypostome, BDA410, × 10 and × 20 (see text-fig. 5*g*); occurrence as fig. 1. 4, moulted exoskeleton with the eighth thoracic segment missing, latex cast of SM A59056b, × 6; *wahlenbergi* Subzone, Oldbury Quarry, loc. 2. 5, free cheek (specimen now damaged), SM A10778, × 4; Oldbury Quarry (exact locality unknown). 6, pygidium, latex cast of IGS BDA574, × 6; occurrence as fig. 1.
Figs. 7–13. *Olenus veles* sp. nov. 7, cranidium, latex cast of SM A10777, × 4; Oldbury Quarry (exact locality unknown). 8, 10, pygidium of axial shield, SM A10780, × 6, and latex cast of counterpart, IGS GSM57917, × 4; occurrence as fig. 7. 9, pygidium of IGS TE11, × 6; trench near Old Wharfe Inn, Chilvers Coton (grid ref. SP 3623 9047), exact horizon unknown. 11, holotype, axial shield, latex cast of IGS BDA156, × 4; *wahlenbergi* Subzone, M3 Borehole. 12, 13, dorsal shields, IGS TE42, Zi4824, both × 3; occurrence as fig. 9.

PLATE 17

Other material. IGS GSM 57917, SM A10,780 (counterparts), A10,777 (all Illing collection, from Oldbury Quarry, exact locality and horizon uncertain); IGS TE11, 13, 15, 42, 43, Zi4824, from trench at Old Wharfe Inn, Chilvers Coton, Nuneaton (Grid ref. SP 3623 9047), associated with *Homagnostus obesus*; about ten specimens from the *wahlenbergi* Subzone in the M3 Borehole, including IGS BDA140, 173–174, 249–250, 251–252, 268, 269, 271–272, 341, 344, 369–370.

Diagnosis. An *Olenus* species with medium-sized palpebral lobes, oblique eye ridges, a small obtuse flip, thirteen thoracic segments with short pleural spines, and a wide triangular pygidium with a pair of small marginal spines.

Description. Glabella (excluding occipital ring) nearly square, with three pairs of simple glabellar furrows but S3 are short and faint. SO straight, simple, occipital ring without node. Preglabellar field about one-fifth to one-sixth as long as cephalic rhachis; preocular sutures parallel in front of the eyes (converge forewards in juveniles). Interocular cheeks just over a third as wide as glabella at eye-line. Palpebral lobes placed opposite S2, about a quarter of the cranidial length (proportionately more in juveniles), connected to glabella by oblique ocular ridges. Postocular cheeks nearly as wide as occipital ring, postocular sutures sinuous. Free cheeks with obtuse flip and obtuse inner spine angle; spine continuous with lateral margin, extending back to the fifth thoracic segment. Hypostome unknown.

Thorax of thirteen segments; rhachis without nodes; pleurae about as wide as rhachis, their tips with short spines only.

Pygidium subtriangular, length about two-fifths of width. Rhachis about one-third of pygidial width, divided into four or five rings and a terminal piece. Pleural regions with three or four pairs of grooves and a tiny pair of marginal spines. The anterior segment of the pygidium looks well developed and not well fused with the rest of the pygidium.

Measurements (in mm).

	Sagittal length	
	Cranidium	Pygidium
IGS Zi4824	7·9	—
IGS BDA156 (holotype)	5·2	2·1
BDA140	3·0	—
BDA250	1·7	—
SM A10,780	—	2·3

Remarks. The short pleural spines of *O. veles* distinguish it from several *Olenus* species, such as *O. gibbosus*, *O. truncatus*, *O. wahlenbergi* (Westergård 1922), *O. cataractes*, and *O. micrurus* (Lake 1908), in which the posterior thoracic segments have longer spines. The obliquity of the eye-ridges in *O. veles* distinguishes it from all the Scandinavian species (Westergård 1922) except *O. alpha* Henningsmoen (1957a) which has a shorter preglabellar field, wider interocular cheeks, less transverse postocular sutures, a composite occipital furrow and no pygidial spines. In *O. veles* the preglabellar field is shorter than in *O. asiaticus* Kobayashi and *O. sinensis* Lu, and the preglabellar

---

EXPLANATION OF PLATE 18

Figs. 1, 2, 5–7, 9. *Olenus cataractes* Salter. All from *cataractes* Subzone. 1, external mould of small axial shield, SM A59041b, × 6; Oldbury Quarry, loc. 1. 2, one of Salter's specimens, × 1½; Maentwrog Formation, Treflys, *c.* 2 km east of Criccieth, Gwynedd, North Wales; IGS GSM8946, figured Lake 1908, pl. 5, fig. 13. 5, 6, internal and external moulds of fragmentary cranidia SM A59039a and A59044b, both × 3; Oldbury Quarry, loc. 1. 7, posterior thoracic segments and pygidium, SM A59040a, × 4; Oldbury Quarry, loc. 1. 9, external mould of cephalon, SM A59043a, × 4; Oldbury Quarry, loc. 1.

Figs. 3, 10–14. *Olenus* cf. *solitarius* (Westergård). 3, dorsal shield and cranidium, SM A55472, A55471, × 2; *cataractes* Subzone, upper beds of Maentwrog Formation, east of Graig Ddu, *c.* 2 km East of Criccieth, Gwynedd, N. Wales. 10–14, all from *cataractes* Subzone, Stockingford Cutting, loc. 6. 10, 11, parts of one exuvia, SM A59092a, b, × 6. 12, fragmentary cephalon showing free cheek, SM A59090a, × 6. 13, fragmentary cranidium, SM A59089a, × 3. 14, large cranidium and partial thorax, SM A59091, × 2.

Figs. 4, 8. *Olenus* cf. *dentatus* Westergård. Pygidium and fragmentary cranidium, SM A59063 (latex cast), A59062, both × 4; *cataractes* Subzone, Jubilee Quarry, loc. 3.

PLATE 18

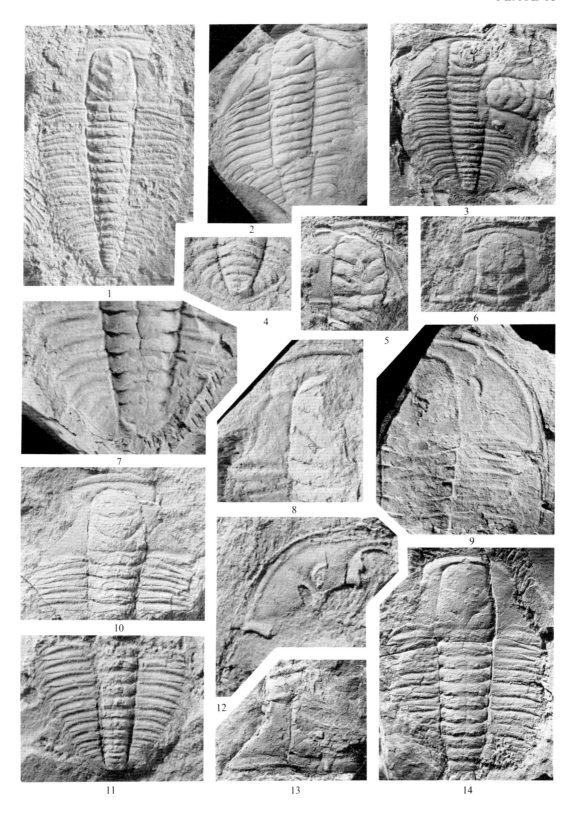

field is longer and the interocular cheeks are wider than in *O. cataractes* and *O. solitarius* (text-fig. 5). *O. micrurus* differs in having fourteen thoracic segments and a smaller, almost semi-circular pygidium (Lake 1908, pl. 5, fig. 12; Allen *et al.* 1981, pl. 1, figs. 3, 4). *O. austriacus* differs from *O. veles* in having a longer preglabellar field, longer eyes, transverse eye ridges, and a larger, rounded, and spineless pygidium. *O. transversus* differs in having a longer preglabellar field, wider interocular cheeks, and wider pleural regions. *O. ogilviei* has longer eyes, shorter genal spines, fourteen thoracic segments, and a smaller pygidium.

Occurrence. In the M3 Borehole *O. veles* occurs at the top of the local range of *O. wahlenbergi* and ranges up. Rushton (*in* Taylor and Rushton 1972, pl. 4) extended the range (as '*Olenus* sp. 2') down somewhat on the basis of a juvenile specimen here regarded as of doubtful identity. The exact horizons of the material from Oldbury Quarry and Chilvers Coton is unknown.

Olenus sp.

Plate 19, fig. 1.

Remarks. One cranidium from Merevale No. 3 Borehole, depth 623 ft 11 in. (190·17 m), resembles *O. cataractes* but has wider interocular cheeks. The sagittal length of the frontal border is great for *Olenus* but is less than in *Parabolinoides bucephalus* (Belt) (Allen *et al.* 1981, pl. 1, fig. 10).

Subfamily PELTURINAE Hawle and Corda
Genus PROTOPELTURA Brögger, 1882
Protopeltura sp.

Plate 19, figs. 2, 3

Material. Two specimens from the top of the *cataractes* Subzone in the M1 Borehole, depths 544 ft (165·81 m) and 556 ft (169·47 m).

Remarks. The glabella is subquadrate and has two pairs of weak furrows. SO composite; occipital ring without distinct node. The frontal area shows only a trace of the anterior border furrow. Eyes small, opposite L3. Postocular sutures nearly straight. Free cheeks with acute flip and short spine. Thorax of at least ten segments, pleural tips pointed but not spined. Pygidial rhachis with three rings and terminal part, pleural regions without a spine.

The figured dorsal shield (Pl. 19, fig. 3) together with its counterpart, seems to show ten thoracic segments but the pygidium is not very distinct. The well-preserved fragment in Plate 196, fig. 2, has ten thoracic segments but the facet on the anterior pleura is less extensive than the anterior pleura in Plate 19, fig. 3; this suggests that there may be one or two segments missing from the fragmentary shield.

These are among the earliest pelturines known. Although the glabella and free cheek are typical of the subfamily the postocular suture is nearly straight instead of convex as in typical Pelturinae. Most *Protopeltura* species (Henningsmoen 1957*a*, p. 220; Ivshin 1962, p. 28) differ from the present form in having a somewhat tapered glabella, distinct glabellar furrows, and a distinct anterior border furrow.

The glabella with weak glabellar furrows recalls *P. olenusorum* Orłowski (1968, p. 275) but the occipital ring in that species has a node and the free cheek appears to have a longer spine. Of the *Protopeltura* with ten(?) thoracic segments, *P. planicauda* (Brögger) has pleural spines and *P. holtedahli* Henningsmoen has an occipital node and a long cheek spine (Henningsmoen 1957*a*).

Family ELVINIIDAE Kobayashi
Genus IRVINGELLA Ulrich and Resser *in* Walcott, 1924
Irvingella nuneatonensis (Sharman, 1886)

Text-fig. 6*a*

Discussion. Irvingella nuneatonensis was described fully by Rushton (1967). The remarkable structure of the thorax has been observed also in *I. perfecta* Tchernysheva (*in* Datsenko *et al.* 1968, p. 207, pl. 22); like *I. nuneatonensis* there are twelve thoracic segments of which the last five have elongated pleural spines. *I. perfecta* differs in having wider interocular cheeks, longer pleural spines, and a more abruptly tapered pygidium. *I. perfecta* is from the Chopkinsky Suite of the north-western Siberian platform, from beds above those with *G. reticulatus*. Ergaliev (1980, p. 41) recorded it from the *Ivshinagnostus ivshini* Zone in the Maly Karatau in beds with *I. major* but above the level of *I. tropica* Öpik.

Irvingella norilica Lazarenko (*in* Datsenko *et al*. 1968, pl. 17, figs. 1, 2) differs from *I. nuneatonensis* in having narrower fixed cheeks and a shorter, less tapered, and less truncate glabella. It comes from beds overlying those with *I. perfecta* (op. cit., p. 90).

Family CERATOPYGIDAE Linnarsson
Genus PROCERATOPYGE Wallerius, 1895

Discussion. Proceratopyge is widely distributed, especially in the early Merioneth Series, and if, following Henderson (1976, p. 332) *Kogenium* Kobayashi and *Lopnorites* Troedsson are included as synonyms, together with the subgenus *Sinoproceratopyge* Lu and Lin (1980, p. 128), there are at least forty-three named species. Lisogor (*in* Zhuravleva and Rozova 1977, p. 255) listed the twenty-seven species asterisked (*) below with bibliographic references; the full list is as follows, but this excludes at least eight *nomina nuda*:

    P. ajguliensis Kraskov, 1977, pl. 14, fig. 1.
    P. [Anomocare] angusta (Whitehouse, 1939, pl. 23, fig. 21).
**P. asiatica* Ivshin, 1956.
    P? brevirhachis Zhou (*in* Zhou *et al*. 1977, pl. 70, fig. 7).
**P. captiosa* Lazarenko, 1966.
**P. chuhsiensis* Lu, 1956.
**P. conifrons* Wallerius, 1895.
    P. constricus (recte *constricta*) Lu, 1964 (see Lu *et al*. 1965, pl. 115, fig. 1).
    P. corrugis Romanenko (*in* Zhuravleva and Rozova, 1977, pl. 24, figs. 18–21).
    P. cryptica Henderson, 1976, pl. 47, figs. 19–24, pl. 48, figs. 1–3.
**P. cylindrica* Chien, 1961.
**P. fenghwangensis* Hsiang, 1962.
**P. [Lopnorites] fragilis* Troedsson, 1937.
**P. [L.] grabaui* Troedsson, 1937.
**P. gracilis* Lermontova, 1940 [not seen].
    P. incondita Harrington and Leanza, 1957, p. 185, fig. 94, 10.
    P. inexspectata Harrington, 1938, pl. 5, fig. 17.
    P. kiangshenensis Lu, 1964 (see Lu *et al*. 1965, pl. 115, fig. 5).
**P. lata* Whitehouse, 1939.
    P. latilimbatus (recte *latilimbata*) Zhou (*in* Zhou *et al*. 1977, pl. 70, figs. 11–13).
    P. latilimbata Lee *in* Yin and Lee 1978, pl. 179, fig. 11.
    P. latirhachis Zhou (*in* Zhou *et al*. 1977, pl. 70, figs. 14–16).
**P. liaotungensis* Kobayashi and Ichikawa, 1955.
**P. longispina* Ivshin, 1962.
**P. magnicauda* Westergård, 1947.
**P. nathorsti* Westergård, 1922.
**P. nectans* Whitehouse, 1939.
    P. [Lopnorites] orthogonialis Yang, 1978, pl. 3, figs. 1–4; pl. 7, fig. 21.
**P. polita* Whitehouse, 1939 [referred to *Mapania* by Henderson 1976, p. 333].
**P. portentosa* Lazarenko, 1966.
**P. [L.] rectispinata* Troedsson, 1937.
**P. [L.?] robusta* Kobayashi, 1962.
**P [Kogenium] rotunda* Kobayashi, 1935.
**P. rotunda* Kryskov, 1960.
**P. rutellum* Whitehouse, 1939 [referred to *Aplotaspis* by Henderson 1976, p. 333].
**P. schaganica* Hajrullina, 1962.
**P. similis* Westergård, 1947.

P. *taojiangensis* Zhou (*in* Zhou *et al.* 1977, pl. 70, fig. 17).
*P. *tenuita* Lazarenko, 1966.
*P. *triangula* Ivshin, 1962.
P. [*Kogenium*] *triangularis* Kobayashi, 1935.
P. *truncatum* (recte *truncata*) Yang (*in* Zhou *et al.* 1977, pl. 70, figs. 18, 19).
*P. *tullbergi* Westergård, 1922.

With so many species to consider, determination of material is difficult, especially as the variation and ontogenetic changes are insufficiently known in most species. Yang (1978, p. 66) has shown that in

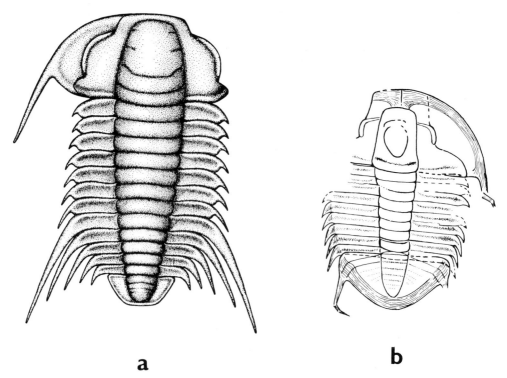

a   **b**

TEXT-FIG. 6. *a*, *Irvingella nuneatonensis* (Sharman), reconstruction × 4, from Rushton 1967, text-fig. 1a. *b*, *Proceratopyge* cf. *rectispinata* (Troedsson), partial reconstruction (× 5) to suggest an interpretation of the specimen in Plate 19, fig. 12. The cephalic doublure shows a median suture.

P. *fenghwangensis* the cylindrical glabella seen in small individuals becomes more conical in later growth-stages, and that the eyes become proportionately smaller; and Henderson (1976, p. 388) has shown variation in adult cranidia of *P. lata*. A thorough review of *Proceratopyge* will probably show that too many species have been proposed. A further difficulty is that *Proceratopyge* had a thin exoskeleton (cf. Troedsson 1937, p. 37) so that material in shale is apt to be poorly preserved; therefore determination of specimens from the Outwoods Shales is tentative.

The median suture inferred in *Proceratopyge* by Westergård (1947, p. 10) is confirmed by two Outwoods Shales specimens. The specimen in Plate 19, fig. 12 and text-fig. 6b has a backward median expansion of the doublure, comparable with that of *P. nathorsti* (Westergård 1947, pl. 2, fig. 3), which is presumably related to the plectral lines observable on the dorsal surface of many *Proceratopyge* cranidia. Kobayashi (1962, pl. 9, fig. 21) illustrated conjoined free cheeks of *P.* (*Kogenium*) *rotundum*. If the cheeks are correctly referred to that species they demonstrate an 'advanced' character in the fusion of the median suture in the type species of *Kogenium*.

Proceratopyge cf. *rectispinata* (Troedsson, 1937)

Plate 19, figs. 12, 13; text-fig. 6*b*

cf. 1937   *Lopnorites rectispinatus* sp. nov.; Troedsson, p. 32, pl. 2, figs. 1, 2.
cf. 1962   *Proceratopyge* (*Lopnorites*) *rectispicatus* [*sic*] Troedsson; Kobayashi, p. 120, pl. 6, figs. 11, 12.
cf. 1968   *Proceratopyge* (*Lopnorites*) *rectispinatus* (Troedsson); Palmer, p. B54, pl. 10, figs. 1–6.
     1972   *Proceratopyge* cf. *rectispinata* (Troedsson); Taylor and Rushton, pp. 14, 20, pl. 1 [recorded].
? 1980   *Proceratopyge rectispinatus* (Troed.); Ergaliev, p. 41 [recorded].

Material. A dorsal shield and a fragment, from the *cataractes* Subzone at loc. 1 in Oldbury Quarry. Fragments including a thorax and pygidium from the M1 Borehole, also *cataractes* Subzone. A poorly preserved dorsal shield from Chilvers Coton railway cutting (IGS JR 3006) (horizon uncertain) may be referable here.

Remarks. The figured dorsal shield resembles Troedsson's (1937, pl. 2, fig. 1) holotype closely but the preocular sutures converge slightly forwards whereas they diverge slightly in the holotype and are practically parallel in the cranidium from Alaska figured by Palmer (1968, pl. 10, fig. 2). Little significance is accorded to this feature, however, because comparable variation is known in other *Proceratopyge* species including that described below. *P. rectispinata* has been recorded from China (Tien Shan) and south Korea, apparently associated with *Irvingella* in both places, and from the approximately coeval early 'Franconian 1' fauna in Alaska (Palmer 1968, p. 13). Ergaliev (1980) recorded it from a slightly higher horizon, the *Pseudagnostus pseudangustilobus* Zone, in the Maly Karatau, Kazakhstan.

Proceratopyge tullbergi Westergård, 1922

Plate 19, figs. 10, 11

1922   *Proceratopyge tullbergi* n. sp.; Westergård, p. 121, pl. 2, figs. 6, 7.
1947   *Proceratopyge tullbergi* Westergård; Westergård p. 11, pl. 2, figs. 8–10.
1972   *Proceratopyge tullbergi* Westergård; Taylor and Rushton, pp. 15, 20 [recorded].

Material. A number of fragments on a loose block of silty pyritous shale at loc. 1, Oldbury Quarry, among tipped material from above the sill; blocks of similar lithology yielded *Olenus cataractes* so the occurrence is referred to the *cataractes* Subzone.

Remarks. A cranidial fragment shows the rather transverse postocular suture described by Westergård (1947, p. 11) and the pygidium is similar in size and details to the small specimen in his plate 2, fig. 10. In Sweden *P. tullbergi* occurs in the base of the *P. spinulosa* Zone and probably also in the upper part of the *O. dentatus* Subzone.

Proceratopyge aff. *nathorsti* Westergård, 1922

Plate 19, figs. 14–16; text-fig. 7

1972   *Proceratopyge* cf. *fragilis*; Taylor and Rushton, pl. 1 [recorded].

Material. Numerous poorly preserved specimens from two horizons at locs. 7 and 8 in Stockingford Railway Cutting, *cataractes* Subzone. A pygidium and some fragments from the M1 Borehole, also *cataractes* Subzone.

Remarks. The cranidia have a tapered glabella with poorly marked glabellar furrows. The preocular sutures diverge forwards more or less strongly in different specimens (text-fig. 7) and the postocular sutures have an even or uneven curvature outwards and backwards. The cranidia differ from many species of *Proceratopyge* in the exsagittal length of the postocular cheeks, such that the distance from the back of the eye to the posterior margin is only a little less than the transverse width of the postocular cheek (in most species the distance is about half this width). The most comparable species in this feature are *P. corrugis*, *P. nathorsti*, *P. magnicauda*, *P. fragilis*, and *P. rectispinata*, but the preocular sutures seem generally more divergent than in any of these species. *P. tenuita* and *P. latilimbata* Lee (*non* Zhou) have unusually divergent preocular sutures but in both these species the postocular sutures are more transverse than in *P.* aff. *nathorsti*, giving pointed, slender postocular cheeks.

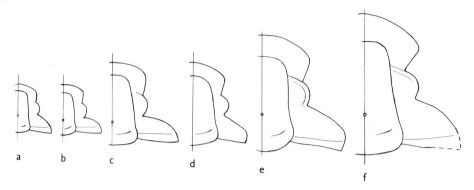

TEXT-FIG. 7. *Proceratopyge* aff. *nathorsti* Westergård. Sketches of cranidia, all × 4, to show the variation in the course of the facial suture. Based on (left to right) IGS Zl 8285, Zl 8266, Zl 8287, Zl 8283, SM A59087, A59081, all from the Stockingford Cutting, loc. 8.

The pygidium of *P*. aff. *nathorsti* resembles many *Proceratopyge* species in being twice as wide as it is long. It does not appear as 'long' or 'triangular' as many published species because the exsagittal length '*Le*' (measured from the notch at the base of the marginal spine) is comparatively great, being two-thirds of the sagittal length ('*Ls*', which excludes the articulating half-ring), whereas in many *Proceratopyge* species *Le* is only half (or less) of *Ls*, as in *P. magnicauda*, *P. rectispinata*, and *P. fragilis*. In *P. corrugis Le* is more than half of *Ls* but the figured pygidia have a narrower rhachis than in *P*. aff. *nathorsti*; in *P. nathorsti* (s.s.) *Le* is likewise more than half of *Ls* but the rhachis is proportionately wider than in *P. corrugis* and agrees more closely with the Stockingford Shales species. In other species, such as *P. captiosa*, in which *Le* approaches two-thirds of *Ls* the postocular cheeks are short and transverse in shape.

P. aff. *nathorsti* is known only in the *cataractes* Subzone in the Outwoods Shales. *P. nathorsti* (s.s.) has been recorded from the *pisiformis* Zone and possibly the base of the *Olenus* Zone in Scandinavia, Siberia, and England (Rushton 1978, p. 275).

EXPLANATION OF PLATE 19

Fig. 1. *Olenus* sp. Fragmentary cranidium, IGS RU1583, × 4; *cataractes* Subzone, M1 Borehole. The background was blackened anterolaterally to show the course of the facial suture.

Figs. 2, 3. *Protopeltura* sp. Incomplete(?) thorax and pygidium and incomplete dorsal shield, IGS RU1409 and RU1395, both × 4; *cataractes* Subzone, M1 Borehole.

Figs. 4–9. *Cliffia*? sp., all from *cataractes* Subzone, Stockingford Cutting, loc. 6. 4, 7, incomplete(?) thorax and pygidium, internal mould and latex cast of external mould, SM A59078a, b, × 4. 5, small partial thorax, SM A59077a, × 6. 6, small fragmentary cranidium, SM A59076a, × 6. 8, pygidium, SM A59079, × 4. 9, fragment of pygidium seen from the ventral side, showing a trace of a narrow doublure, SM A59080a, × 5.

Figs. 10, 11. *Proceratopyge tullbergi* Westergård. Postocular cheek and pygidium, SM A59094a and A59095b (latex cast), both × 4; *cataractes* Subzone, Oldbury Quarry, loc. 1.

Figs. 12, 13. *Proceratopyge* cf. *rectispinata* (Troedsson). 12, dorsal shield, SM A59093, × 5; *cataractes* Subzone, Oldbury Quarry, loc. 1. 13, pyritized thorax and pygidium, IGS RU1529, × 4; *cataractes* Subzone, M1 Borehole.

Figs. 14–16. *Proceratopyge* aff. *nathorsti* Westergård. 14, external mould of pygidium with pyritized doublure, IGS RU1684, × 4; *cataractes* Subzone, M1 Borehole. 15, cranidium, SM A59088, × 3, and 16, small pygidium, SM A59084, × 10; both *cataractes* Subzone, Stockingford Cutting, loc. 8.

PLATE 19

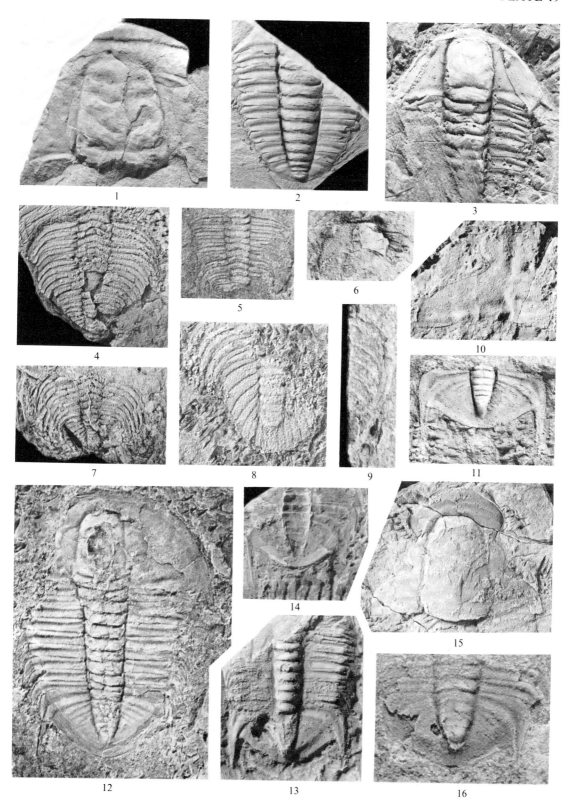

FAMILY UNCERTAIN
Genus CLIFFIA Wilson, 1951

Discussion. The type species is *Acrocephalites lataegenae* Wilson, 1949; following the redescription by Wilson (1951) it cannot be classified with *Acrocephalites*, especially as the pygidia of these genera differ markedly.

Cliffia? sp.

Plate 19, figs. 4–9

Material. A small scattered exuvia, two partial thoraces (one with a pygidial fragment attached), and two pygidia, one of which is fairly complete, all from *cataractes* Subzone, Stockingford Cutting, loc. 6.

Description. The only cephalon collected is small (probably juvenile) and fragmentary, especially anteriorly (Pl. 19, fig. 6). Glabella wider than long, rounded in front, occipital ring not preserved. S1, S2 both short, slightly oblique backwards; S3 small, pit-like. Preglabellar field is buckled or broken but appears to have been short. Eyes remote from glabella, placed opposite S3 and L3; eye-ridges not observed. Fixed cheek approximates to a quadrant. Free cheek inadequately known. Hypostome with oval middle body, margin not preserved.

Thorax of unknown length, with wide (tr.) pleurae, the posterior segments being falcate. The larger thorax (Pl. 19, fig. 4) has thirteen segments visible in front of the pygidium and the anterior segments are falcate, whereas the smaller thorax (fig. 5) shows some straight transverse segments in front of the falcate ones. The scattered exuvia shows thirteen segments but no pygidium.

Pygidium slightly wider than long, semi-elliptical. Rachis less than three-quarters of total length, composed of about six rings and a terminal part. Pleural areas wider than rhachis anteriorly, crossed by seven distinct pairs of oblique pleural furrows, and two faint pairs of furrows directed more or less straight backwards behind the rhachis. No interpleural grooves. Margin entire(?), without border but with a narrow doublure (Pl. 19, fig. 9).

The surface of all convex parts is covered densely with granules.

Remarks. This species has a distinctive pygidium, quite dissimilar from most upper Cambrian trilobites. That of *Erixanium* Öpik (1963, pl. 8, fig. 1) is similar in outline but has a shorter rhachis, a border, fainter pleural furrows, and definite interpleural grooves. The pygidium of *Cliffia lataegenae* (Wilson 1951, pl. 90, fig. 22) is rather calymenid-like; it is somewhat triangular in outline, with rounded anterolateral corners, is finely granulose, and is evidently more convex than that of the present species, as preserved; the rhachis is longer and bears median nodes. However, the pleural regions in *C. lataegenae* resemble the present species in having strong pleural furrows which reach the margin, no interpleural grooves, and no border.

The cranidium (Pl. 19, fig. 6) is small and fragmentary but although it does not show a long preglabellar field like that of *C. lataegenae* the part that is preserved resembles the posterior part of cranidia of *C. lataegenae* behind the level of the eyes, for example in the short, deep glabellar furrows and long, convex postocular suture (Wilson 1951, pl. 90, figs. 18, 21, 23).

Acknowledgements. I thank the several geologists who assisted me with collecting fossils in the Nuneaton area over many years, and who freely donated specimens. I am grateful to Sir James Stubblefield for reading the manuscript and to him and Dr. R. A. Fortey for discussion. I thank the staff at the Sedgwick Museum, Cambridge, for the loan of specimens, J. M. Pulsford, H. J. Evans, Dr A. T. Thomas, and my wife for assistance with photography, and Linda Dubé for typing the manuscript. This paper is published by permission of the Acting Director, Institute of Geological Sciences (N.E.R.C.).

REFERENCES

ALLEN, P. M., JACKSON, A. A. and RUSHTON, A. W. A. 1981. The Stratigraphy of the Mawddach Group in the Cambrian succession of North Wales. *Proc. Yorks. geol. Soc.* **43**, 295–329, pls. 16, 17.

ANGELIN, N. P. 1851. *Palaeontologia Suecica. I: Iconographia crustaceorum formationis transitionis.* Fasc. I, 1–24, pls. 1–24. Lund.

BELT, T. 1867. On some new trilobites from the Upper Cambrian rocks of North Wales. *Geol. Mag.* **4**, 294–295, pl. 12, figs. 3–5.

BRÖGGER, W. C. 1882. *Die silurischen Etagen 2 und 3* . . . viii + 376 pp., 12 pls. Kristiania.

COWIE, J. W., RUSHTON, A. W. A. and STUBBLEFIELD, C. J. 1972. A correlation of the Cambrian rocks in the British Isles. *Spec. Rep. geol. Soc. Lond.* no. 2, 42 pp., 6 pls.

DALMAN, J. W. 1827. Om Palaeaderna, eller de så kallade Trilobiterna. *K. svenska Vetensk-Akad. Handl.* [for 1826], 113–152, 226–294, pls. 1–6.

DATSENKO, V. A., ZURAVLEVA, I. T., LAZARENKO, N. P., POPOV, YU. N. and CHERNYSHEVA, N. E. 1968. [Biostratigraphy and fauna of the Cambrian deposits of the north-west part of the Siberian Platform.] *Trud. Nauchno-Issled. Inst. Geol. Arkt.* no. 155, 213 pp., 23 pls. [In Russian.]

EASTWOOD, T., GIBSON, W., CANTRILL, T. C. and WHITEHEAD, T. H. 1923. The geology of the Country around Coventry. *Mem. geol. Surv. Gt Br.* H.M.S.O. London, 149 pp., 8 pls.

ERGALIEV, G. Kh. 1980. [*Trilobites of the Middle and Upper Cambrian of the Maly Karatau.*] 211 pp., 20 pls. Akad. Nauk Kazakh. S.S.R., Alma-Ata. [In Russian.]

FORTEY, R. A. 1974. The Ordovician trilobites of Spitsbergen. I. Olenidae. *Skr. Norsk Polarinst.* no. 160, 129 pp., 24 pls.

—— 1980. The Ordovician trilobites of Spitsbergen. III. Remaining trilobites of the Valhallfonna Formation. Ibid. no. 171, 163 pp., 25 pls.

HARRINGTON, H. J. 1938. Sobre las faunas del Ordoviciano Inferior del Norte Argentino. *Revta Mus. La Plata, Seccion. paleont.* N.S. 1, 109–289, pls. 1–14.

—— and LEANZA, A. 1957. Ordovician Trilobites of Argentina. *Spec. publs Univ. Kans., Dep. Geol.* no. 1, 276 pp., 140 figs.

HAWLE, I. and CORDA, A. J. C. 1847. *Prodrom einer Monographie der böhmischen Trilobiten.* 176 pp., 7 pls. Prague. (Also 1848. *Abh. K. böhm. Ges. Wiss.* **5,** 117–292, pls. 1–7.)

HENDERSON, R. A. 1976. Upper Cambrian (Idamean) trilobites from western Queensland, Australia. *Palaeontology* **19,** 325–364, pls. 47–51.

HENNINGSMOEN, G. 1957a. The trilobite family Olenidae. *Skr. norske Vidensk.-Akad.* 1 Mat.-naturv. Kl., 1957, no. 1, 303 pp., 31 pls.

—— 1957b. A trilobite with North American affinities from the Upper Cambrian of Sweden. *Bull. geol. Instn Univ. Uppsala,* **37,** 167–172, 1 pl.

—— 1958. The Upper Cambrian faunas of Norway. *Norsk geol. Tidsskr.* **38,** 179–196, pls. 1–7.

HOLLAND, C. H. (ed.) 1974. *Cambrian of the British Isles, Norden, and Spitsbergen.* J. Wiley, London, New York, Sydney, Toronto.

HOWELL, B. F. 1935a. New Middle Cambrian agnostian trilobites from Vermont. *J. Palaeont.* **9,** 218–221, pl. 22.

—— 1935b. Some New Brunswick Cambrian Agnostians. *Bull. Wagner Free Inst. Sci.* **10,** 13–15, 1 pl.

ILLING, V. C. 1913. Recent discoveries in the Stockingford Shales near Nuneaton. *Geol. Mag.* (5) **10,** 452–453.

—— 1916. The Paradoxidian Fauna of a Part of the Stockingford Shales. *Quart. Jl geol. Soc. Lond.* **71** (for 1915), 386–448, pls. 28–38.

IVSHIN, N. K. 1956. [*Upper Cambrian trilobites of Kazakhstan,* part 1.] 119 pp., 9 pls. Akad Nauk Kazakh. S.S.R., Alma-Ata. [IN RUSSIAN.]

—— 1962. [*Upper Cambrian trilobites of Kazakhstan,* part 2.] 412 pp., 21 pls. Akad. Nauk Kazakh. S.S.R., Alma-Ata. [In Russian.]

—— and POKROVSKAYA, N. V. 1968. Stage and zonal subdivision of the Upper Cambrian. *Int. geol. Congr.* XXIII (Prague). 1968. Proc. Sec. 9, 97–108.

JAEKEL, O. 1909. Über die Agnostiden. *Z. dt. geol. Gesell.* **61,** 380–401.

KAUFMANN, R. 1933. Variationsstatistische Untersuchungen über die 'Artabwandlung' und 'Artumbildung' an der Oberkambrischen Trilobitengattung *Olenus* Dalm. *Abhandl. geolog.-pal. Institut Griefswald,* no. 10, 54 pp.

KOBAYASHI, T. 1935. The Cambro-Ordovician formations and faunas of South Chosen. Palaeontology. Part III. The Cambrian faunas of South Chosen with a special study on the Cambrian trilobite genera and families. *J. Fac. Sci. Univ. Tokyo* sect. II, **4,** 49–344, 24 pls.

—— 1939. On the Agnostids (part 1). Ibid. sect. II, **5,** 69–198.

—— 1962. The Cambro-Ordovician formations and faunas of South Korea, part IX, Palaeontology VIII. Ibid. sect. II, **14,** 1–152, pls. 1–12.

—— and ICHIKAWA, T. 1955. Discovery of *Proceratopyge* in the *Chuangia* Zone in Manchuria, with a note on the Ceratopygidae. *Trans. Proc. palaeont. Soc. Japan* N.S. **19,** 65–72, pl. 11.

KRASKOV, L. N. 1977. [New representatives of Ceratopygidae and Olenidae from the late Cambrian of southern Kazakhstan], pp. 56–60. *In* STUKALINA, G. A. (ed.). [*New species of early plants and invertebrates from the U.S.S.R.*], vol. 4. Paleont. Inst., Akad. Nauk S.S.S.R., 'Nauka', Moscow. [In Russian.]

LAKE, P. 1906. British Cambrian trilobites, part 1, pp. 1–28, pls. 1, 2. *Palaeontogr. Soc.* [*Monogr.*], part of vol. for 1906.

—— 1908. British Cambrian trilobites, part 3, pp. 49–64, pls. 5, 6. Ibid., part of vol. for 1908.

LAZARENKO, N. P. 1966. [Biostratigraphy and some new trilobites from the Upper Cambrian of the Olenenk Uplift and Kharaulakh Mountain.] *Uchen Zap. Paleont. Biostrat. Nauchno-Issled. Inst. Geol. Arkt.* **11**, 33–78, 8 pls. [In Russian.]

LERMONTOVA, E. V. 1951. [*Upper Cambrian trilobites and brachiopods from Boshche Kulya (north-east Kazakhstan).*] 49 pp., 6 pls. Vses. Nauchno-Issled. geol. Inst. (VSEGEI). Moscow. [In Russian.]

LU YANHAO and LIN HUANLING. 1980. Cambro-Ordovician boundary in western Zhejiang and the trilobites contained therein. *Acta Pal. Sinica*, **19**, 118–134, 3 pls.

——— 1981. Zonation of Cambrian faunas in western Zhejiang and their correlation with faunas in North China, Australia, and Sweden. Pp. 118–120. *In* TAYLOR, M. (ed.). Short papers for the second international symposium on the Cambrian System, 1981. *U.S. geol. Surv. Open-file Rep.* no. 81–743.

—— ZHU ZHAOLING, CHIEN YI-YUAN and HSIANG LEEWEN. 1965. [*Chinese fossils of all Groups. Trilobita.*] Vol. 1, 1–362, pls. 1–66, vol. 2, 363–766, pls. 67–135. Science Press, Beijing. [In Chinese].

—— *et al.* 1974. Bio-environmental control hypothesis and its application to the Cambrian biostratigraphy and palaeozoogeography. *Mem. Nanking Inst. Geol. Palaeont.* **5**, 27–116, 4 pls. [In Chinese.]

MARTINSSON, A. 1974. The Cambrian of Norden, pp. 185–283. *In* HOLLAND, C. H. (ed.), q.v.

ÖPIK, A. A. 1963. Early Upper Cambrian fossils from Queensland. *Bull. Bur. Miner. Resour. Geol. Geophys. Aust.* **64**, 133 pp., 9 pls.

—— 1967. The Mindyallan fauna of north-western Queensland. *Ibid.* **74**; vol. 1, 404 pp., vol. 2, 167 pp., 67 pls.

ORŁOWSKI, S. 1968. Upper Cambrian fauna of the Holy Cross Mts. *Acta geol. pol.* **18**, 257–291, 8 pls.

PALMER, A. R. 1960. Trilobites of the Upper Cambrian Dunderberg Shale, Eureka District, Nevada. *Prof. Pap. U.S. geol. Surv.* no. 334-C, 53–109, pls. 4–11.

—— 1962. *Glyptagnostus* and associated trilobites in the United States. Ibid. no. 374-F, 49 pp., 6 pls.

—— 1965. Trilobites of the Late Cambrian Pterocephaliid Biomere in the Great Basin, United States. Ibid. no. 493, 105 pp., 23 pls.

—— 1968. Cambrian trilobites of east-central Alaska. Ibid. no. 559-B, 115 pp., 15 pls.

RASETTI, F. 1944. Upper Cambrian trilobites from the Lévis Conglomerate. *J. Paleont.* **18**, 229–258, pls. 36–39.

RUSHTON, A. W. A. 1966. The Cambrian trilobites from the Purley Shales of Warwickshire. 55 pp., 6 pls. *Palaeont. Soc.* [*Monogr.*], publ. 511, part of vol. 120, for 1966.

—— 1967. The Upper Cambrian trilobite *Irvingella nuneatonensis* (Sharman). *Palaeontology*, **10**, 339–348, pl. 52.

—— 1968. Revision of two Upper Cambrian trilobites. Ibid. **11**, 410–420, pls. 77, 78.

—— 1969. *Cyclotron*, a new name for *Polyphyma* Groom. *Geol. Mag.* **106**, 216–217.

—— 1974. The Cambrian of Wales and England, pp. 43–121. *In* HOLLAND, C. H. (ed.), q.v.

—— 1978. Fossils from the Middle–Upper Cambrian transition in the Nuneaton district. *Palaeontology*, **21**, 245–283, pls. 24–26.

—— 1979. A review of the Middle Cambrian Agnostida from the Abbey Shales, England. *Alcheringa*, **3**, 43–61.

SHARMAN, G. 1886. On the new species *Olenus nuneatonensis* and *Obolella granulata*, from the Lower Silurian ('Cambrian', Lapworth), near Nuneaton. *Geol. Mag.* (3) **3**, 565–566.

SHERGOLD, J. H. 1977. Classification of the trilobite *Pseudagnostus*. *Palaeontology*, **20**, 69–100, pls. 15, 16.

—— 1980. Late Cambrian trilobites from the Chatsworth Limestone, western Queensland. *Bull. Bur. Miner. Resour. Geol. Geophys. Aust.* no. 186, 111 pp., 35 pls.

—— 1982. Idamean (Late Cambrian) trilobites, Burke River Structural Belt, western Queensland. Ibid. no. 187, 69 pp., 17 pls.

STUBBLEFIELD, C. J. 1930. A new Upper Cambrian section in south Shropshire. *Sum. Progr. geol. Surv. G.B.* for 1929, part 2, 55–62.

TAYLOR, K. and RUSHTON, A. W. A. 1972. The pre-Westphalian geology of the Warwickshire Coalfield. *Bull. geol. Surv. Gt Br.* no. 35, 150 pp., 12 pls. [dated 1971].

TROEDSSON, G. T. 1937. On the Cambro-Ordovician faunas of Western Qurugtagh, eastern T'ien-Shan. *Palaeont. Sinica*, N.S. B, no. 2; pp. 1–74, pls. 1–10 (whole series no. 106).

WALCOTT, C. D. 1924. Cambrian geology and palaeontology V, no. 2, Cambrian and Lower Ozarkian trilobites. *Smithson. misc. Collns*, **75**, 53–60, pls. 9–14.

WALLERIUS, I. D. 1895. *Undersökningar öfver zonen med* Agnostus laevigatus *i Vestergötland.* iii + 73 pp., 1 pl. Lund.

WESTERGÅRD, A. H. 1922. Sveriges olenidskiffer. *Sver. geol. Unders. Avh.* Ser. Ca, no. 18, 205 pp., 16 pls.
—— 1944. Borrningar genom Skånes alunskiffer 1941–42. Ibid. Ser. C., no. 459, 45 pp., 6 pls.
—— 1947. Supplementary notes on the Upper Cambrian trilobites of Sweden. Ibid. Ser. C, no. 489, 34 pp., 3 pls.
—— 1949. On the geological age of *Irvingella suecica* Westergård. *Geol. För. Stockh. Förh.* **71**, p. 606.
WHITEHOUSE, F. W. 1936. The Cambrian faunas of north-eastern Australia. Parts 1 and 2. *Mem. Queensland Mus.* **11**, 59–112, pls. 8–10.
—— 1939. The Cambrian faunas of north-eastern Australia. Part 3. Ibid. 179–282, pls. 19–25.
WILSON, J. L. 1949. The trilobite fauna of the *Elvinia* Zone in the basal Wilberns Limestone of Texas. *J. Paleont.* **23**, 25–44, pls. 9–11.
—— 1951. Franconian trilobites of the Central Appalachians. Ibid. **25**, 617–654, pls. 89–95.
YANG JIALU, 1978. [Middle and Upper Cambrian trilobites of western Hunan and eastern Guizhou.] *Prof. Pap. Stratigraphy Palaeontology* (Geological Publishing House, Peking), No. 4, 1–82, pls. 1–13. [In Chinese.]
YIN GONGSHENG and LEE SHANJI, 1978. Trilobita, pp. 385–594, pls. 144–192. *In* [*Fossils of south-west China: Guizhou. Volume 1, Cambrian to Devonian periods*], 843 pp., 214 pls. Earth Sci. Press, Beijing (Peking). [In Chinese.]
ZHOU TIEN-MEI, *et al.* 1977. Trilobita, pp. 104–266, pls. 33–81. *In* WANG XIAO-FENG and JIN YU-QIN (eds.). [*Palaeontological handbook of central and southern China. Part 1, early Palaeozoic Era*], 470 pp., 116 pls. Earth Sci. press, Beijing (Peking). [In Chinese.]
ZHURAVLEVA, I. T. and ROZOVA, A. V. (eds.) 1977. [Biostratigraphy and fauna of the Upper Cambrian and boundary strata. (New data from Asiatic part of U.S.S.R.).] *Trudy Inst. Geol. Geofiz., Akad. Nauk S.S.S.R.,* *Sib. Otdel.* no. 313, 355 pp., 31 pls.

A. W. A. RUSHTON,
Institute of Geological Sciences
Exhibition Road
London, SW7 2DE

A REVIEW OF THE
TRILOBITE SUBORDER SCUTELLUINA

by P. D. LANE and A. T. THOMAS

ABSTRACT. An assessment of the morphological characters of the styginid, scutelluid, illaenid, and bumastid trilobites and *Phillipsinella* leads to the acceptance of Hupé's concept of the superfamily Scutelloidea, here elevated to subordinal rank. *Panderia*, which is here assigned to its own family, may also belong to the suborder. The affinities of the Scutelluina are with the Corynexochida rather than the Ptychopariida.

A PECULIAR problem of the trilobites dealt with here lies in the difficult systematics and nomenclature of the effaced forms assigned to the suborder. This difficulty is highlighted both by the variety of opinions held on the systematic position of these effaced trilobites, and by the large number of species erected within some of the included genera (especially *Illaenus* itself). In this paper we review the history of classification of the styginid, scutelluid, illaenid, and bumastid trilobites, *Phillipsinella* and *Panderia*, then discuss effacement and those characters used in current classifications. We lastly offer a revised classification of the suborder based on an inferred phylogeny and discuss relationships with other major trilobite groups.

We discuss a large number of suprageneric groupings within the suborder, some of which have been considered by different authors to be of different formal taxonomic status, or groupings of no formal status. To minimize confusion in the following discussion we use the nomenclature set out below.

> Suborder Scutelluina Hupé, 1953: [*emend.* Hupé, 1955]
>   [*nom. transl.* herein *ex* superfamily Scutelloidea].
> Suborder Illaenina (*sensu* Jaanusson 1959).
> Superfamily Illaenacea (*sensu* Jaanusson 1959) = illaenacean.
> Family Illaenidae (*sensu* Hupé 1953) = illaenid.
> Family Scutelluidae (*sensu* Richter and Richter 1955) = scutelluid.
> Family Styginidae (*sensu* Skjeseth 1955) = styginid.
> Family Phillipsinellidae (*sensu* Whittington 1950) = phillipsinellid.

The use of the suffix *-ine*, for example bumastine, illaenine, indicates the nominate subfamily, where such a subfamily has been erected, or an informal group of genera close in morphology to the nominate genus.

COMPOSITION OF THE SCUTELLUINA

Hupé (1953, p. 224) proposed the superfamily Scutelloidea to include the Styginidae, Theamataspidae, Phillipsinellidae, Scutellidae, and Illaenidae (Illaeninae and Bumastinae), although without mention of *Panderia*; in 1955, however, he included *Panderia* in the Illaeninae. Except for *Phillipsinella* all these groups were included by Jaanusson (1959, p. O365) in his superfamily Illaenacea. Jaanusson placed the Illaenacea, together with the Bathyuracea, Proetacea, and the monotypic Holotrachelacea in his new ptychopariid suborder Illaenina. We follow Fortey and Owens (1975) in assigning the bathyurids and proetids to a separate order (Proetida) and *Holotrachelus* may be a late bathyurid derivative (pers. comm. R. A. Fortey). Bruton (1976, p. 704), in a revision of *Phillipsinella*, supported Whittington's (1950, p. 561) allocation of the genus to a distinct family and considered it to be derived from the same root stock as the early scutelluids. In agreement with Přibyl and Vaněk (1971), therefore, he considered the Phillipsinellidae to belong to the Illaenina.

We consequently believe the Scutelluina comprises the illaenid, styginid, scutelluid, and phillipsinellid trilobites. *Panderia* may also belong there. For reasons discussed later we consider the Scutelluina may be more closely related to the Corynexochida than to any ptychopariid.

Many groupings have been proposed within the Scutelluina, about which there has been considerable discussion or confusion as to taxonomic position and validity. These are discussed below.

Styginidae Vogdes, 1890. This taxon is one of those in the Scutelluina with the confusing distinction of having been independently erected three times (Vogdes 1890; Raymond 1920; Warburg 1925). The history of its taxonomic status may conveniently be divided into three. Authors up to Jaanusson 1959 (e.g. Reed 1935; Whittington 1950; Henningsmoen 1951; Hupé 1953, 1955) followed Vogdes in considering the group to warrant familial status. The generic composition varied slightly in these different works (Skjeseth 1955), and in Hupé's revised classification (1955, p. 206 n.) he implied the monophyly of this family with the 'Scutellidae'. In the second period, with the increased knowledge concerning genera with what was considered a combination of 'styginid' and 'scutelluid' characters, it was preferred to merge these two groups into the Scutelluidae (Whittington 1963). This gradual shift in nomenclature has caused a certain amount of confusion in the literature. Lane and Thomas (1978a, p. 352), for instance, regarded the Scutelluidae as including *Stygina* and its allies. In their reply to this paper, however, Ludvigsen and Chatterton (1980, p. 474) considered two separate families to be involved, and therefore misinterpreted some of Lane and Thomas's comments concerning the Scutelluidae. More recently, authors have preferred to consider the group as a subfamily of the Scutelluidae (e.g. Fortey 1980; see also discussion in Owen and Bruton 1980, p. 12).

Bumastinae Raymond, 1916. Of all taxa assigned to the Scutelluina, Bumastinae has had the most turbulent career. It was erected to include those 'illaenids' with 'more or less flattened shields', large eyes, wide axis, and pygidium at least as long as the cephalon, often with a concave border (= 'holcos' of Lane and Helbert 1982), Raymond originally included with *Bumastus* the genera *Actinolobus* (later considered by Jaanusson (1954) a genus embracing typical *Illaenus* species) and *Illaenoides*. Warburg (1925, p. 99) questioned the value of the characters used by Raymond to separate the Illaeninae from the Bumastinae and preferred to treat all the genera as members of an undivided Illaenidae. Jaanusson (1954, 1959) accepted the subfamily as a natural group of the Illaenidae containing *Bumastus*, *Illaenoides*, *Goldillaenus*, *Bumastus* (*Bumastoides*), *Dysplanus*, *Platillaenus*, and *Thomastus*. Lane and Thomas (1978b, p. 9), however, argued that *Bumastus* and its allies belonged to the Scutelluidae—being no more than highly effaced and convex members of that family which do not otherwise differ from 'typical' members in basic morphology. This view was challenged by Ludvigsen and Chatterton (1980) who considered the Bumastinae a monophyletic subfamily of the Illaenidae which itself gave rise to the Scutelluidae.

Panderiinae Bruton, 1968. Bruton (1968, p. 4) argued that the overall cephalic and pygidial morphology, structure of the thoracic pleurae, and presence of a rostral flange suggest that *Panderia* should be assigned to the Illaenidae. He also noted, however, a number of significant differences from other members of the family, which led him to erect the subfamily Panderiinae. The nature of these differences and the systematic position of *Panderia* are discussed below.

Goldillaeninae Balashova, 1959. Balashova considered the type genus, together with her new genus *Goldillaenoides*, and a species now referred to *Failleana* (*F. indeterminata* Walcott) to form a natural group within the Scutelluidae. Lane (1972, p. 346) accepted the subfamily with some doubt since little was known of the goldillaenine rostral plate and hypostome, but considered it useful in delimiting a group of 'smooth' scutelluids; *Goldillaenoides* Balashova 1959, however, he considered to be a scutelluine. Lane and Thomas (1978b, p. 8) merged this subfamily with the Scutelluidae. Goldillaeninae has not been generally used in the literature.

Theamataspidae Hupé, 1953. Fortey (1980, p. 57) discussed this subfamily and concluded that it should be regarded as belonging to the Scutelluidae although its members show an amalgam of scutelluid and illaenid characters. Bruton (*in* Bruton and Harper 1981, p. 170) considered that *Theamataspis* was poorly founded and preferred to regard it as a scutelluid.

Brontocephalidae Kolobova, 1977; Dulanaspinae Přibyl and Vaněk, 1971. Brontocephalidae was erected with the newly described genus *Brontocephalus* as its type and otherwise only containing *Dulanaspis* (Chugaeva 1958).

These genera are of low convexity and share a wide, concave-sided glabella, narrow, well-marked anterior cranidial border, and short pygidium with a short axis which is only faintly marked in the type genus.

Přibyl and Vaněk (1971) erected the Dulanaspinae, a taxon apparently unknown to Kolobova. This 'subfamily' was discussed by Fortey (1980, p. 57) who concluded that it was possibly to be synonymized with the Theamataspidinae. In view of the general similarity between genera referred to that subfamily, *Brontocephalus* and *Dulanaspis*, we prefer to regard Brontocephalidae and Dulanaspinae as synonyms of Theamataspidinae should this group prove to be monophyletic.

HISTORY OF CLASSIFICATION

Jaanusson (1954) reviewed the history of classification of the illaenid trilobites and drew a distinction between authors who had considered the group to consist of at most two genera (Barrande 1846, 1852, 1872; Hall 1847, 1868; Billings 1859, 1865, 1866; Holm 1882, 1886; Reed 1904, 1906, 1914, 1935; Warburg 1925, etc.), and those who had made an attempt at further subdivision.

Burmeister (1843) used the number of thoracic segments to distinguish *Dysplanus* (with nine) from *Illaenus* (with ten). *Illaenus* was further divided into the subgenera *Illaenus* and *Bumastus* on the width and distinctness of the axis. These characters—number of thoracic segments and relative width of axis—have been almost universally employed by later workers, whether attempting to erect formal subdivisions of the group, or informal species-groups within genera. Thus Angelin distinguished four genera in the Illaenidae: '*Rhodope*' (= *Panderia*), *Dysplanus*, *Bumastus*, and *Illaenus*. Salter (1867), again using these features as prime characters, divided *Illaenus* into eight subgenera. He, however, stated his unease at using the number of thoracic segments to define 'natural' groups, noting that Barrande had described eleven cases in which the number of thoracic segments varied between congeneric species. Raymond (1916, p. 3) discussed seventeen taxa which had been referred to the Illaenidae as genera or subgenera. He concluded that only seven were valid: *Illaenus*, *Thaleops*, *Dysplanus*, *Octillaenus*, *Bumastus*, *Actinolobus*, and *Illaenoides*. The first four of these along with his new genus *Wossekia* constituted his Illaeninae, and the last three the Bumastinae. These two subfamilies were principally characterized by the possession (in the latter) and lack of (in the former) a concave border in cephalon and/or pygidium.

Richter and Richter (1956) used the number of paired pleural ribs to delimit genera, of which they recognized with certainty only three, in dealing with the Scutelluidae. *Eobronteus* with six pairs was considered the primitive member of the family which gave rise to *Stoermeraspis* (and possibly *Octobronteus*) with eight on the one hand, and *Scutellum* with its six subgenera (*Scutellum*, *Planiscutellum*, *Thysanopeltis*, *Kolihapeltis*, *Scabriscutellum*, and *Paralejurus*) on the other. The number of paired pygidial ribs has been employed by nearly all subsequent workers to distinguish groups within the Scutelluidae, while the shape, relative size, and position of the lateral glabellar muscle impressions were used to delimit subgenera.

Šnajdr (1957) discussed the characters he had found useful in delimiting the thirteen illaenid genera from the Ordovician and Silurian of Bohemia. Of these characters the number of thoracic segments was taken to be of prime importance; additional diagnostic characters included the relative sizes of cephalon and pygidium, the course of the axial furrows, the size and position of the eyes, the shape and structure of the rostral plate, hypostome, and pygidial doublure, and the shape, position, and arrangement of the muscle scars on the cranidial axis. From consideration of these characters, Šnajdr preferred not to recognize the Bumastinae, but merged all the genera in an undivided Illaenidae. It is obvious from his figure 5, however, that he considered there to be three phyletic lines within the family, one including *Bumastus*, one *Illaenus*, and one *Stenopareia*.

Jaanusson (1959) included the Styginidae, Thysanopeltidae (= Scutelluidae), and Illaenidae in the superfamily Illaenacea; these, together with the superfamilies Bathyuracea (Bathyuridae and Lecanopygidae), the monotypic Holotrachelacea, and the Proetacea constituted his Illaenina. The Styginidae and Thysanopeltidae were not divided into subfamilies and were distinguished by the presence in the former of seven to nine thoracic segments (ten in the latter), a long pygidial axis often with axial rings developed, a post-axial ridge, and generally poorly developed pygidial pleural ribs and furrows. Additionally, the outline of the hypostome in the two groups was stated to be 'shield-shaped' in the styginines and 'subrectangular' in the thysanopeltids. This latter group was considered by Jaanusson to contain only four genera, one of these (*Scutellum*) with six subgenera. The Illaenidae comprised the Illaeninae, Bumastinae, Ectillaeninae, and questionably the Theamataspidinae. The Bumastinae were distinguished from other Illaenidae in lacking a rostral flange and in the pattern of the cranidial axial furrows which (when present) diverged forwards. The Illaeninae (which included *Panderia* and Ectillaeninae) had a 'club-shaped' glabella and were distinguished from the latter by the presence of rostral flange, eight to ten as opposed to nine to ten thoracic segments, and by the occurrence in the latter of a transverse or concave-backwards posterior hypostomal margin.

Šnajdr (1960), although only dealing with the Scutelluidae, presented a full discussion of the characters he had used in delimiting the twenty genera he considered comprised the group. Of prime importance he found the shape, relative size, and relative position of the lateral glabellar (muscle) impressions and the outline of the pygidium. Other characters used in his generic diagnoses were convexity of the exoskeleton, length of the preglabellar part of the cranidium, shape of the fixed cheek, position and size of the eyes, outline of the rostral plate and hypostome, relative width of the thoracic axis, shape of the pygidial axis, number and morphology of the paired and median ribs on the pygidium, presence of border spines on the pygidium, and the width of the cephalic and pygidial doublures. He also discussed the development of these characters in phylogeny, thereby indicating the 'primitive' condition of some of the characters (e.g. low convexity exoskeleton, short preglabellar area, weakly impressed and well-separated lateral glabellar impressions, narrow doublures, six or eight (but not seven) paired pygidial pleural ribs, non-bifurcate median pygidial rib). Although he recognized no subfamilial groupings, the genera with seven pairs of pygidial pleural ribs he maintained were an 'independent evolutionary branch beginning with the Silurian genus *Planiscutellum* R. and E. Richter and ending at the beginning of the Upper Devonian' (pp. 45, 238). From his figure 15 it seems that he also considered the genera with six pairs of pygidial ribs (*Eobronteus* and *Protobronteus*), and *Octobronteus* with eight, to be 'evolutionary branches'; from this figure the phylogenetic position of *Protoscutellum* is moot.

Přibyl and Vaněk (1971) considered the Scutelluidae to be divisible in the subfamilies Scutelluinae, Stygininae, Eobronteinae, Octobronteinae (Maksimova 1968), and their new Dulanaspidinae. Long diagnoses of these subfamilies were presented but these contain few distinctive characters. The Scutelluinae and Eobronteinae were both diagnosed as having ten thoracic segments but differed in the relative widths of the articulated portion of the pleurae (narrow in former, wide in latter), the number of paired pleural ribs (seven as against six), and the relative length of the pygidial axis (short in former, long and with rings indicated in latter). The Stygininae was mainly distinguished on thoracic and pygidial characters—having seven to nine thoracic segments and a post-axial ridge in the pygidium. The Octobronteinae (containing only two described genera) was characterized by the eight paired pygidial pleural ribs and lateral expansion of the glabella, the other characters given being common to many members of the family. In essence, therefore, the Přibyl and Vaněk subfamilial classification depends on the number of thoracic segments, the relative width of the thoracic axis and, in the pygidium the relative length of the axis and the number of pleural paired ribs.

Pillet (1972) considered that the Illaenidae, Styginidae, and Scutelluidae comprised the Illaenacea. Within the last family he recognized the Scutelluinae, Thysanopeltinae, Eobronteinae, and Paralejurinae; the last two of these stated to be newly erected. Although he was mistaken in considering the Eobronteinae as new, his concept of the taxon agrees with that of the original (Sinclair 1949). The convex and effaced morphology of *Paralejurus* was considered by Pillet to be worthy of subfamilial recognition. The Scutelluinae was further subdivided into four 'species-groups'.

Kobayashi and Hamada (1974) produced classifications of both Illaenidae and Scutelluidae in which numerous subfamilies were utilized. In the Illaenidae they recognized all subfamilies proposed to that time (Illaeninae, Panderiinae, Ectillaeninae, Bumastinae, Goldillaeninae, and Theamataspidinae). Within the Scutelluidae they accepted Scutelluinae, Eobronteinae, Octobronteinae, Thysanopeltinae, and Paralejurinae and proposed the new Planiscutellinae and Meroperixinae. The former was characterized as having three pairs of simple, nearly equidistant glabellar furrows and a simple median pygidial rib, the latter by anteriorly strongly expanded glabella and effaced lateral glabellar furrows, or these reduced to rudimentary pits near the axial furrow. In our view the genera included in the latter subfamily are not closely related.

EFFACEMENT

Progressive effacement of the dorsal exoskeleton occurs in many groups of trilobites—for example, in the Olenacea (*Svalbardites*; Fortey 1974, pls. 19, 20), Asaphacea (*Neoasaphus*; Jaanusson 1959, fig. 248, 2), and Proetida (*Parvigena*; Owens 1973, figs. 14K–N), as well as in the Scutelluina. In at least some cases the effaced condition was apparently independently derived in different lineages, as in the Agnostida: *Lejopyge* in the Ptychagnostinae (e.g. Westergård 1946, p. 87, pl. 13, figs. 18–26), *Rhaptagnostus* in the Pseudagnostinae (e.g. Shergold 1977, p. 87, pl. 16, figs. 3, 4; see also discussion of effacement on p. 77), and *Pagetiellus* in the pagetiids (e.g. Savitski 1972, pl. 2, figs. 1–6). Because effacement tends to render obscure many of the exoskeletal features which might otherwise be of taxonomic value, difficulties are encountered in diagnosing genera and higher taxonomic groups, and in inferring phylogenetic relationships. In the case of the Scutelluina this has lead to differences of opinion concerning the relationships of certain effaced forms. We, for instance (Lane and Thomas 1978b and below), regard such genera as *Bumastus* and *Litotix* as independently derived effaced scutelluids while others regard such forms as illaenids (Kobayashi and Hamada 1974; Chatterton 1980).

The meaning of the term 'effacement' has not previously been fully discussed and we regard it as a complex character-state, which may involve three processes which often operate together:

(i)   reduced distinctness of axial and/or pleural furrows of the dorsal surface;
(ii) increased exoskeletal convexity;
(iii) increase in relative width of the axis.

The tendency to reduced distinctness of dorsal furrows is seen in many Scutelluina. When it is not accompanied by processes (ii) and (iii) however, though it may cause difficulties in the discrimination of genera and species, no difficulties arise at a higher taxonomic level. *Stygina* (Whittington 1950, pl. 72, figs. 1–6, 9) and *Protostygina* (Přibyl and Vaněk 1971, pl. 1, fig. 3) are good examples of this category. All the dorsal furrows are weak, but the basic morphology of these genera is clearly styginine and their taxonomic position is not controversial. Only rarely does increased exoskeletal convexity occur independently of a marked reduction in distinctness of dorsal furrows (but see *Paralejurus brongniarti* Šnajdr 1960, pl. 22). Most commonly the tendency to increased convexity is accompanied by an increasingly 'smooth' appearance, until only the axial furrows are at all distinct. This condition is best seen in such unequivocally illaenid genera as *Illaenus* (Whittington 1965, pls. 45–54), *Ectillaenus* (Šnajdr 1957, pls. 2, 3) and *Cekovia* (Šnajdr 1957, pl. 9) and in such forms as *Panderia* (Bruton 1968, pls. 1–11) and *Stenopareia* (Howells 1982, pls. 3, 4).

Increase in relative width of the axis is always accompanied by increased exoskeletal convexity and typically by suppression of the dorsal furrows also. Such highly effaced trilobites are those about which there is most taxonomic controversy. Good examples are *Bumastoides* (Chatterton and Ludvigsen 1976, pl. 5), and *Bumastus* and *Cybantyx* (Lane and Thomas 1978b, pls. 1, 2; pl. 5).

As these changes are the results of processes rather than simple presence-or-absence characters it is possible for various degrees (of furrow suppression or axial widening, for example) to occur. Moreover, the effect of one process may be more extremely developed than another or may be more completely developed in a particular part of the exoskeleton. In *Paralejurus campanifer* (see Šnajdr 1960, pl. 23, figs. 2–14), for instance, there is great exoskeletal convexity and the thoracic axis is wider than the pleurae. While such axial features as glabellar furrows are highly subdued, however, more distal structures, such as pygidial pleural ribs, are still clearly defined.

MORPHOLOGICAL CHARACTERS USED IN CURRENT CLASSIFICATIONS

Glabellar morphology (text-fig. 1). The glabella expands (tr.) forwards in all Scutelluina. In some early genera, e.g. *Perischoclonus* (Whittington 1963, pl. 22, figs. 3, 4) the axial furrows are almost straight and the degree of anterior expansion slight. An occipital ring of conventional construction is present. *Bronteopsis* and *Raymondaspis* (Whittington 1965, pl. 55, figs. 1, 4; pl. 56, figs. 1, 6, 7; pl. 59, fig. 1) show a broadly similar morphology except that the anterior expansion is greater and the axial furrows describe a wide outwardly concave course. Such a pattern is characteristic of genera usually referred to the styginids and scutelluids as well as *Phillipsinella*.

The cephalic axial furrows of effaced Scutelluina tend only to be present posteriorly and the occipital furrow is lost. Because of this latter character the extent of the occipital segment is not easily determined. The forward position of the occipital tubercle (not the median glabellar tubercle of *Panderia*) in *Bumastus* for example (Lane and Thomas 1978b, pl. 2, fig. 1), and the anteriorly placed lunettes, however, suggest that the occipital ring is elongated (sag. and exsag.) in at least some of these genera.

In *Panderia* the axial furrows diverge anteriorly but, on internal moulds (Bruton 1968, pl. 6, figs. 2, 4), there is a faint indication that the glabella is barrel-shaped which is a unique condition for the suborder. Also unique is the median glabellar tubercle which is placed between the eyes in a position much farther forward than the presumed occipital tubercle of genera like *Bumastus*. *Ottenbyaspis*, which may be related to *Panderia*, shows a weak occipital furrow anterior to which lies a median glabellar tubercle (Bruton 1968, pl. 12, figs. 9, 11). A

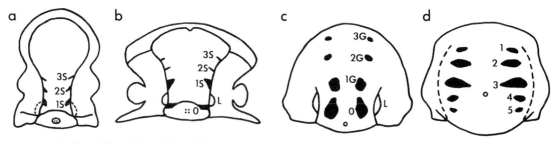

TEXT-FIG. 1. Cranidia of Scutelluina illustrating musculature and glabellar shape. *a, Phillipsinella preclara* Bruton, ×8 (based on Bruton 1976, pl. 104, fig. 1). *b, Raymondaspis reticulata* Whittington, ×5 (based on Whittington 1965, pl. 56, fig. 6). *c, Stenopareia oviformis* (Warburg), ×2·5 (based on Jaanusson 1954, pl. 2, fig. 7). *d, Panderia migratoria* Bruton, ×5·5 (based on Bruton 1968, pl. 6, fig. 4; text-fig. 3); muscle scars numbered following Bruton's nomenclature.

preoccipital tubercle is seen in some other groups of trilobites—many Asaphacea and some Cyclopygacea (*sensu* Fortey 1981) for instance. These groups contrast with *Panderia* in other respects, however, especially in the presence of a median ventral suture (although this becomes fused in advanced cyclopygaceans). In other effaced genera the axial furrows are either roughly parallel (*Nanillaenus*; Chatterton and Ludvigsen 1976, pl. 4, fig. 11) or more commonly posteriorly divergent (*Paralejurus*; Šnajdr 1960, pl. 23, figs. 2, 6) presumably depending upon the width of the thoracic axis (= distance between the points of articulation). Only rarely is there any sign of the axial furrows in the anterior part of the cephalon. The glabella of *Thaleops longispina* Shaw (1968, pl. 21, figs. 1, 2, 4, 6, 8, 9) is bluntly rounded anteriorly and not greatly widened (tr.). In such genera as *Bumastus* and *Cybantyx* (Lane and Thomas 1978*b*, pl. 1, fig. 5a; pl. 5, fig. 1c), by contrast, the anterior part of the axial furrows diverge strongly before dying out. Assuming that the anterior pit (or its internal homologue) is placed at the anterolateral extremity of the glabella (see below) then the glabella is extremely wide anteriorly in these genera.

The anterior pit is the dorsal reflection of a ventral process positioned at the anterolateral corner of the glabella close to where the 3S glabellar furrow meets the axial furrow, where this can be seen. Wherever this structure has been described in detail it forms a point of articulation with the anterior wing of the hypostome. The structure is present in many Scutelluina (e.g. *Stygina*; Skjeseth 1955, pl. 3, fig. 1: *Raymondaspis*; Whittington 1965, pl. 55, fig. 9; pl. 59, fig. 7). Bruton (1976, p. 701) described a structure in a homologous position in *Phillipsinella* as the 'posterior pit'. His 'anterior pit' lies above the cranidial doublure and it cannot, therefore, be homologous to the anterior pit of other Scutelluina. In other genera (e.g. *Failleana*; Chatterton and Ludvigsen 1976, pl. 6, figs. 1, 3, 6, 8, 39; *Bumastus*; Lane and Thomas 1978*b*, p. 12 for example) no pit is seen externally, but a presumably homologous process is seen on the ventral surface of the exoskeleton.

Unless silicified material, or specimens preserved as counterpart internal and external moulds is available, these structures may easily be missed. The anterior pit is not seen in Silurian and Devonian scutelluids for instance, but ventrally its internal homologue may be seen for example in *Scutellum calvum* Chatterton (1971, pl. 4, fig. 20).

No anterior pit or its internal equivalent is developed in any unequivocal illaenid (e.g. *Nanillaenus*; Chatterton and Ludvigsen 1976, pl. 3, figs. 41, 42; pl. 4, figs. 9, 11, 13: *Bumastoides*; ibid. pl. 5, figs. 1, 4, 14) or in *Panderia* (Bruton 1968, pl. 9, fig. 7). The occurrence of this structure in some effaced forms, and its undoubted absence in others, suggests that this might be a character of value in deciding the relationships of such forms.

Cranidial musculature

(*a*) Extra-axial muscle impressions. Most Scutelluina have a pair of extra-axial muscle impressions developed adjacent to where the glabella is most constricted. In the Illaenidae these are termed lunettes, and in this family they often mark the anterior limit of the axial furrows. Fortey (1980, p. 59) discussed the homology of the lunettes with the lateral muscle impressions (Šnajdr 1960) of scutelluids. No similar structures occur in *Panderia* species (Bruton 1968) and none is described in *Phillipsinella*. In that genus, however, well-preserved material (Bruton 1976, pl. 105, figs. 3, 7; pl. 106, fig. 2) shows that the axial furrow is widened abaxial of the occipital furrow and 1L. This may indicate the presence of a lateral muscle impression, albeit in rudimentary form.

(*b*) Glabellar muscle scars and furrows. Fortey (1980, p. 59) discussed the homology of early illaenid and scutelluid glabellar musculature which is a significant point of similarity between these two large groups. In *Illaenus* itself (Jaanusson 1954, pl. 1, fig. 6) four pairs of muscle impressions may be present though sometimes only three are seen (Whittington 1965, pl. 54, fig. 1). Where present the most posterior pair are presumably occipital since they lie behind the lunettes. All pairs may be roughly similar in size, or the occipital, or this and the 1G pair may be enlarged (e.g. *Illaenus* and *Stenopareia* respectively; Jaanusson 1954, pl. 1, fig. 6; pl. 2, fig. 7). In all these forms the muscle areas are remote from the axial furrow and no lateral glabellar furrows are present. Thus the basic pattern seen in Illaenidae is of two parallel or slightly diverging rows of scars. This suggests that the glabella is not greatly expanded (tr.) anteriorly in these genera (also see discussion of *Thaleops longispina* above and Jaanusson 1954, pl. 1, fig. 2).

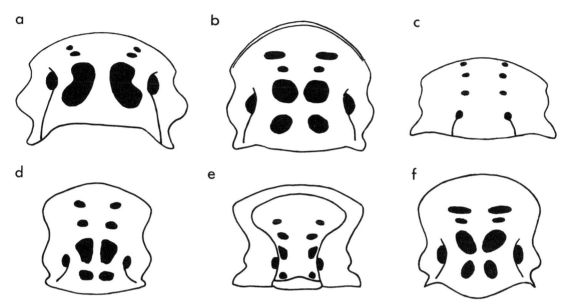

TEXT-FIG. 2. Cranidial muscle-scar patterns of Scutelluina. *a, Bumastus barriensis* Murchison, × 0·5 (based on Lane and Thomas 1978*b*, pl. 1, fig. 5a). *b, Cybantyx anaglyptos* Lane and Thomas, × 0·6 (based on Lane and Thomas 1978*b*, pl. 5, fig. 1a). *c, Illaenus gelasinus* Whittington, × 1 (based on Fortey 1980, fig. 6, p. 59). *d, B.? phrix* Lane and Thomas, × 2 (based on Lane and Thomas 1978*b*, pl. 3, fig. 1). *e, Planiscutellum planum* (Hawle and Corda), × 1 (based on Šnajdr 1960, fig. 18, p. 60). *f, Litotix armata* (Hall), × 2 (based on Lane and Thomas 1978*b*, pl. 4, fig. 8a).

This contrasts with the condition in some other effaced genera such as *Bumastus* and *Cybantyx* (text-fig. 2). Although four pairs of muscle impressions are present (occipital and 1G conjoined in *Bumastus*), the more anterior pairs of scars are placed increasingly farther from the sagittal line. This attests to the great anterior expansion of the glabella in these forms. *Litotix* is another effaced genus, superficially similar to those just discussed. This genus differs in having 3G enlarged and placed adaxial to 2G (Lane and Thomas 1978*b*, pl. 4, fig. 8). We take this contrast in the pattern of muscles to suggest that *Litotix* was derived independently of *Bumastus*. The glabellar musculature of *B.? phrix* Lane and Thomas (1978*b*, pl. 3, fig. 1a) is more similar to that of *Planiscutellum planum* (Šnajdr 1960, pl. 1, figs. 5, 6) than to that of any illaenid and this suggests an independent origin for at least this effaced form.

The cranidial musculature of *Panderia* (Bruton 1968, fig. 3, p. 21) differs from that of all other genera of the Scutelluina in that five pairs of glabellar impressions are present. Bruton suggested that the posterior two pairs (numbers 4 and 5) might be parts of a larger single scar homologous with the occipital scar of genera like *Illaenus*. Certainly those numbered 1, 2, and 3 are very similar in arrangement to the anterior three pairs of such forms.

In the early scutelluid *Perischoclonus*, lateral glabellar furrows occur adjacent to the axial furrow (Whittington 1963, pl. 22, figs. 1–6). 1S is bifurcate and placed next to the lateral muscle impression; 2S is almost transverse and isolated from the axial furrow while 3S is pit-like. In other early forms (e.g. *Raymondaspis* Whittington 1965, pl. 59, fig. 7; *Bronteopsis* Whittington 1965, pl. 55, figs. 1, 4; *Theamataspis* Fortey 1980, pl. 9, fig. 5) the furrows have become muscle scars and have begun to migrate adaxially. The tendency for furrows to efface and be replaced by muscle scars is common in the Scutelluidae, and Šnajdr (1960, p. 236) and Richter and Richter (1956, p. 81) found the pattern of musculature to be a reliable generic character within this group.

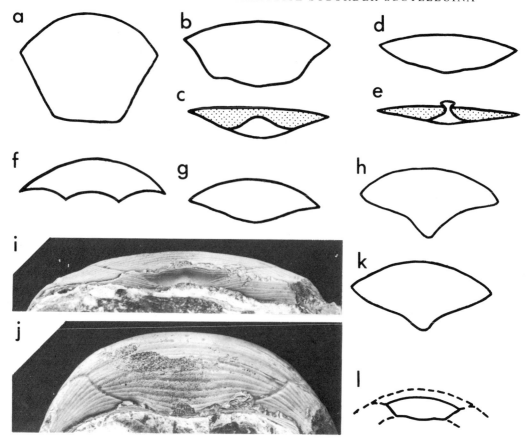

TEXT-FIG. 3. Rostral plates of Scutelluina. *a, Phillipsinella parabola* (Barrande), ×9 (based on Whittington 1950, pl. 75, figs. 3, 4). *b, c, Stenopareia glaber* (Kjerulf), ventral and posterior views to show rostral flange, ×4 (based on Owen and Bruton 1980, pl. 4, figs. 8, 9). *d, e, Illaenus marginalis* Raymond, ventral and posterior views to show rostral flange, ×2 (based on Whittington 1965, pl. 47, figs. 8, 10). *f, Dysplanus centrotus* (Dalman), ×2 (based on Jaanusson 1954, pl. 3, fig. 1). *g, Failleana calva* Chatterton and Ludvigsen, ×3 (based on Chatterton and Ludvigsen 1976, pl. 6, fig. 13). *h, Cybantyx anaglyptos* Lane and Thomas, ×1 (based on Lane and Thomas 1978b, pl. 5, fig. 7). *i, j, Bumastus barriensis* Murchison, ×1 (GSM 36090). *k, Bumastus? phrix* Lane and Thomas, ×3 (based on Lane and Thomas 1978b, pl. 3, fig. 8). *l, Panderia lerkakensis* Bruton, ×8 (based on Bruton 1968, pl. 4, figs. 8, 10).

The muscle scar pattern of *Phillipsinella* species (Bruton 1976, p. 701, pl. 104, figs. 1–6; pl. 105, figs. 1, 3, 7; pl. 106, fig. 2) is like that of early scutelluids except that the scars are concentrated more closely together towards the back of the glabella.

We believe the pattern of glabellar musculature to be of fundamental importance in tracing relationships within large groups of trilobites as first noted by Stubblefield (1936) and supported by later workers (e.g. Fortey and Owens 1979, p. 220).

Rostral plate (text-fig. 3). In all the genera here assigned to the Scutelluina the rostral plate, where known, is rather large, typically wider than long, and laterally defined by backwardly converging connective sutures. The variation which is found in this basic plan is mainly caused by differences in the angle of convergence of the connective sutures, whether their paths are straight or curved, the presence or absence of a rostral flange and (in some late *Panderia* species only) the ankylosis of the connective and rostral sutures.

The rostral plate of *Phillipsinella* is distinctive both in its large size (approaching half the cephalic length and width) and its subpentagonal outline (Whittington 1950, fig. 8, pl. 75, figs. 3, 4, 6; Ingham 1970, pl. 5, figs. 14, 17, 20). The hypostomal suture is straight and the connective sutures diverge forwards at about 30° to an exsagittal line. The rostral suture is strongly curved and marginal. Whittington (1950, p. 500) referred to the rostral plate as 'bent dorsally', which could be taken to imply the presence of an upturned rostral flange.

In *Stenopareia* the rostral suture is ventral and broadly convex forwards. The maximum width of the plate is about twice its sagittal length and a triangular, upwardly turned rostral flange is present posteriorly (Jaanusson 1954, pl. 2, figs. 4, 5; Owen and Bruton 1980, pl. 4, figs. 4, 8, 9, 13). The rostral plate of *Nanillaenus* is similar to that of *Stenopareia* except that the connective sutures converge at a more obtuse angle, and the upwardly and forwardly turned rostral flange is apically truncated (Chatterton and Ludvigsen 1976, pl. 4, figs. 23, 24). The rostral plate of *Bumastoides* is also of this type (Chatterton and Ludvigsen 1976, pl. 5, figs. 23, 24).

In ventral view (e.g. Jaanusson 1954, pl. 1, fig. 3) the rostral plate of *Illaenus* is not dissimilar from those of the genera just discussed, except that the rostral suture is farther from the cephalic margin. Posteriorly, however, a large and distinctive upwardly turned, axe-shaped rostral flange is present (Jaanusson 1954, pl. 1, fig. 7; Whittington 1965, pl. 47, fig. 10). The hypostome of *Illaenus* rises vertically behind the front of the cephalon (Jaanusson 1954, pl. 2, figs. 1, 2) and its anterior margin possesses a median embayment where it abuts against the tip of the rostral flange (Whittington 1965, p. 385).

In *Dysplanus* and *Platillaenus* (Jaanusson 1954, pl. 3, figs. 1, 4, 5) the hypostomal suture runs parallel to the curved marginal rostral suture. The connective sutures diverge at a high angle and no rostral flange is present. The rostral plates of *Bronteopsis* and *Stygina* also seem to be of this type (Whittington 1950, p. 545, fig. 4, pl. 72, figs. 7, 10). Many other Scutelluina have a rostral plate broadly of this type except, like *Stenopareia*, etc., discussed above, the hypostomal suture is roughly transverse or slightly convex backwards (e.g. *Failleana*; Chatterton and Ludvigsen 1976, pl. 6, figs. 9, 13). The variation seen is mainly due to the orientation of the connective sutures and the degree of inflation (Šnajdr 1960, p. 237). In *Cybantyx* and *Bumastus*? species (Lane and Thomas 1978b, pl. 3, figs. 4, 7, 8; pl. 4, fig. 4c; pl. 5, figs. 6, 7) the connective sutures approach the mid-line posteriorly so that the rostral plate becomes almost triangular in plan view. In *Bumastus* by contrast, the posterior ends of the connective sutures are at some distance apart so that the plate retains the quadrate outline more typical of the suborder (Lane and Thomas 1978b, pl. 1, fig. 2c; pl. 2, fig. 1b, c). The rostral plate of *Bumastus* is additionally distinguished by the possession of a forwardly convex, upturned, wide rostral flange.

The rostral plate is variable in development between different *Panderia* species (Bruton 1968, text-fig. 2, p. 8) and is sometimes fused with the cephalic doublure (Bruton 1968, p. 7). In stratigraphically older species a wide rostral plate is present adjacent to the hypostome and a strip of cranidial doublure is situated between the rostral suture and the cranidial margin. In younger species the rostral plate is fused with the other ventral cephalic elements so that connective sutures are absent and the free cheeks are united by a continuous strip of doublure. In at least some *Panderia* species, a dorsally reflexed rostral flange is present (Bruton 1968, pl. 3, fig. 7). This is in the form of a truncated triangle.

It therefore seems that the Scutelluina are united by a fundamentally similar type of rostral plate. Variations in the form of the structure within the group have been widely used as a guide in illaenine systematics. We accept that the detailed structure of the rostral plate may provide a useful generic criterion, but we caution against the uncritical use of rostral morphology at the subfamily and family levels. The rostral plates of *Bumastus* and *B.*? *phrix* for instance, which are very similar forms in other important respects, are as different from each other as they are from those of almost any other member of the Scutelluina. Moreover, the sporadic occurrence of rostral flanges—in such genera as *Panderia*, *Bumastus*, *Stenopareia*, and *Illaenus*, which are otherwise not particularly similar—argues against using the mere presence of such a structure at a high taxonomic level.

Hypostome (text-fig. 4). The general form of the hypostome, and particularly the morphology of its middle body, is consistent and therefore diagnostic among genera in many trilobite families and subfamilies.

Within the Scutelluina there are four general types of hypostome; that of *Panderia* is unknown:

1. *Phillipsinella*-type (Whittington 1950, text-fig. 8) is restricted to this genus and is long and narrow overall, as is the middle body. The hypostome seemingly reaches at least as far back as the posterior margin of the cranidium.

2. *Illaenus*-type is based on *Illaenus sarsi* (see Jaanusson 1954, pl. 2, fig. 1). This type, which has only been figured *in situ* in species of *Illaenus* s.s., is wide anteriorly owing to the often massive and subquadrate anterior wings. The middle body is narrow and ovate, and the anterior margin straight and transverse but may have a median indentation into which the rostral flange fitted. We have chosen to illustrate the hypostome of *I. consobrinus* (Whittington 1963, pl. 18) since it is the best illustrated of this type. It demonstrates the unique shape of the massive anterior wings and the anterior indentation seen in this genus.

3. *Scutellum*-type (Chatterton 1971, pl. 4, figs. 14, 15) is subtriangular or 'shield-shaped' with a rounded to subangular posterior margin, and a short to extremely short posterior lobe of the middle body, prominent maculae and generally narrow borders. The hypostome of *Dysplanus* has a middle body of this sort but larger anterior wings. *Scutellum*-type hypostomes are also found in such genera as *Kosovopeltis*, *Spiniscutellum*, and *Bumastus*.

4. *Stenopareia*-type (Owen and Bruton 1980, pl. 4, figs. 4, 10–12) is relatively short and wide with transversely elongate middle body with short posterior lobe and narrow borders. *Zdicella* (Šnajdr 1957, pl. 5) also has a hypostome of this sort.

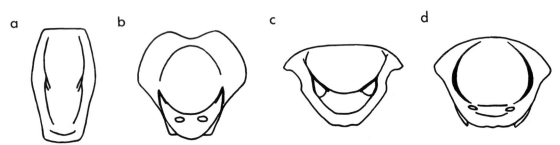

TEXT-FIG. 4. Hypostomes of Scutelluina. *a*, *Phillipsinella*-type (*P. parabola* (Barrande); based on Whittington 1950, pl. 75, fig. 5; text-fig. 6), × 7. *b*, *Illaenus*-type (*I. consimilis* Billings; based on Whittington 1965, pl. 52, figs. 6–11), × 6. *c*, *Scutellum*-type (*S. calvum* Chatterton; based on Chatterton 1971, pl. 4, fig. 14), × 8. *d*, *Stenopareia*-type (*S. glaber* (Kjerulf); based on Owen and Bruton 1980, pl. 4, fig. 11), × 3·5.

Within the last three of these types there is some variation principally in the overall proportions. The outline of the hypostome depends ultimately on the width of the cranidial axis since the processes of the anterior wings are separated by the distance between the cranidial anterior pits (although these are absent in some genera). In the Scutelluina, however, any increase in distance between these processes is largely taken up by an increase in width of the anterior wings; the middle body is relatively unaffected.

The hypostome of corynexochids, for example *Corynexochus* itself, is most similar to that of *Illaenus* and we therefore consider the morphology of the hypostome of the latter genus as primitive for the Scutelluina.

In no trilobite does the hypostome become effaced. For this reason its morphology should be useful in tracing relationships between members of the Scutelluina. Unfortunately, the hypostome is known in few genera of the suborder; in place it is known in very few.

Thoracic structure. Most holaspid illaenids and scutelluids have either nine or ten thoracic segments. *Octillaenus* like *Panderia* has eight and *Phillipsinella* six. Genera which on other grounds appear to be closely related may have different numbers of thoracic segments so that segment number does not appear to be a useful character at a high taxonomic level as it is in some other major groups of trilobites. There is, however, considerable variation in the structure of the thoracic segments of genera referred to the suborder.

In 'typical' Silurian and Devonian scutelluids the gently inflated axis constitutes 25–30% of the segment width, and there is a short half-ring (e.g. Lane 1972, pl. 60, fig. 12). No pleural furrow is developed but the horizontal part of each pleura bears a narrow (exsag.) articulating flange anteriorly and posteriorly (Šnajdr 1960, fig. 7, p. 24). At the outer ends of these, tiny articulating processes seem to be present close to the fulcrum (Chatterton 1971, pl. 4, fig. 6). The distal part of the pleura is spinose and flexed downwards, and ventrally the doublure extends inwards as far as the fulcrum (Chatterton 1971, pl. 4, figs. 7, 8). The thoracic segments of such Ordovician genera as *Stygina* (Whittington 1950, pl. 72, figs. 2, 9) and *Protostygina* (Horný and Bastl 1970, pl. 8, fig. 3) also show this type of construction. Přibyl and Vaněk's drawing (1971, pl. 1, fig. 6) suggests that the thoracic segments of *Styginella* differ in having pleural furrows. Such furrows are also present in the meraspid of *Perischoclonus* (Whittington 1963, pl. 22, fig. 10), while in *Bronteopsis* although there is no pleural furrow there is a change in slope along a transverse line (Whittington 1950, pl. 71, figs. 10–12) in a similar position.

In many illaenid genera, for instance *Illaenus* (Whittington 1963, pl. 18, figs. 2, 3), *Nanillaenus* (Chatterton and Ludvigsen 1976, pl. 4, figs. 7, 8, 12), *Thaleops* (Shaw 1968, pl. 20, fig. 23), and *Ectillaenus* (Šnajdr 1957, pl. 3, fig. 1) the axis is about one-third of the thoracic width and the half-ring is very short (Chatterton and Ludvigsen

1976, pl. 4, fig. 14). Inflation of the axis is variable and the axial furrows may sometimes be marked only by a change in slope. No pleural furrow is present but the anterior margin of each segment bears a narrow (exsag.) articulating flange which extends along the width of the (horizontal) adaxial half of each pleura. These flanges apparently articulate with a narrow section of doublure on the next anterior segment (Chatterton and Ludvigsen 1976, pl. 4, fig. 12). The distal portion of the pleura is flexed downwards and the typically rounded pleural tip bears a forwardly inclined articulating facet. In some forms the pleural tips are isolated and spinose rather than rounded; in these, no articulating facet is present. The thoracic segments of *Panderia* are similar to those of illaenids. In this genus, however, the axis becomes proportionally wider anteriorly (Bruton 1968, pl. 7, figs. 1, 3) and the horizontal part of the pleurae correspondingly narrower.

The above type of segment structure contrasts with that found in such genera as *Rhaxeros* (Lane and Thomas 1978*a*, pl. 1, figs. a, c), *Bumastus* (Lane and Thomas 1978*b*, pl. 1, fig. 5a, b; pl. 2, fig. 8a, b), and *Cybantyx* (Lane and Thomas 1978*b*, pl. 5, figs. 1a, 8a, b). In these forms the axis is extremely wide and convex, the segments only articulate at the fulcra, and the pleurae continue the general convexity of the axis over the whole width of the segment. No discrete half-ring is present. We formerly implied (Lane and Thomas 1978*b*, p. 10) that this type of thoracic structure was characteristic of what we regard as effaced scutelluids, and quite different from that of illaenids. We no longer think this is so, however, for a very similar thoracic structure is seen in both *Bumastoides* which we regard as an illaenid, and *Failleana* which we regard as a styginid (Chatterton and Ludvigsen 1976, pl. 5, figs. 8, 26, 31, 33–36; pl. 6, figs. 17–28); these genera, however, have a very short (sag. and exsag.) and wide articulating half-ring. In all these cases, outward migration of the axial furrows has resulted in the progressive elimination of the horizontal inner part of the pleurae together with the associated flanges. When the axial furrows reach the fulcrum a similar morphology is produced irrespective of whether the ancestral structure was of illaenid or scutelluid type.

The thoracic segments of *Phillipsinella* (Whittington 1950, pl. 75, figs. 4, 7; Bruton 1976, pl. 108, fig. 6) are of conservative construction. The axis is strongly inflated and articulating half-rings are well developed. The horizontal inner part of each pleura is short (tr.), crossed by a shallow transverse pleural furrow, and bears narrow (exsag.) articulating flanges anteriorly. The distal portion of each pleura is longer (tr.), blunt-tipped, and has a forwardly inclined articulating facet.

In view of the range of thoracic structure in genera which would be assigned uncontroversially to suprageneric taxa within the Scutelluina, this character should not be regarded as important at a high taxonomic level.

Pygidium. The pygidium is one of the most variable features in the Scutelluina. Typically (with the exception of some Silurian and Devonian scutelluids) it is wider than long with rounded posterior margin, lacks marginal spines, is smaller and less convex than the cephalon, lacks posterior and lateral borders, and has a doublure with prominent terrace ridges.

The pygidium of *Phillipsinella* is subquadrate in outline with up to eight rings (indicated by muscle impressions on the internal mould) on the axis, which itself is less than one-third the total anterior width. Up to five pleural ribs and furrows are present which are indistinct posteriorly and distally. A short post-axial ridge is present on the internal mould of *Phillipsinella* sp. (Bruton 1976, pl. 108, fig. 12).

In most Ordovician styginids the pygidium is often much shorter than wide, and the axis is relatively narrow, about half the length of the pygidium, has fewer than ten rings indicated, and there is usually a prominent post-axial ridge. Pleural areas normally have very weak ribs and furrows. The whole is weakly convex.

A large number of Silurian and Devonian scutelluines has a weakly convex to almost planar pygidium which is about as long as wide and broadly rounded behind. In these forms the axis is short, triangular, often trilobed due to a pair of furrows which run close to the exsagittal direction, and rarely are axial rings indicated. The pleural field bears six to eight paired ribs and furrows of variable cross-section which never reach the margin, and a single posterior rib which may be bifid.

In the pygidium of many species referred to *Illaenus*, *Bumastus*, and *Stenopareia* effacement of the pleural and axial furrows is complete, or almost so. Sometimes, in *Illaenus* and *Stenopareia* for example, an indication of the width of the axis at the anterior margin is given by a more or less pronounced median arch and by shallow features affecting the whole thickness of the exoskeleton. In species of *Bumastus*, *Cybantyx*, and *Opsypharus* there is no median arch, and the anterior margin of the pygidium describes a smooth curve, sometimes weakly interrupted by the two points of articulation. In some pygidia of this last type, effacement has not proceeded to completion, and remnant pleural ribs may be seen especially anterolaterally (*Rhaxeros*, some *Paralejurus* species).

The pygidium of *Panderia* has effaced pleural fields, but there is a short and wide quadrate axis which is up to half the length of the whole, and quite well delimited laterally; axial rings are not normally seen.

The pygidial doublure of the Scutelluina is very variable in its extent and morphology; it is always characterized by the presence of well-developed terrace ridges, although the morphology of these ridges does not always conform to that defined by Miller (1975, p. 157: see Lane and Thomas 1978b, p. 8). In a large number of forms referred to the suborder the doublure is narrow (less than one-third total sagittal length), narrows somewhat towards the anterolateral corners of the pygidium, is gently convex ventrally, and has a smoothly curved anterior margin subparallel to the posterior margin of the pygidium. In some with this general morphology (*Phillipsinella* and *Panderia*) the anterior margin of the doublure medially lies almost under the posterior of the pygidial axis; in others it is remote from this position (*Bumastus*, *Cybantyx*, *Opsypharus*). The doublure in some forms is very extensive (up to three-quarters the length of the whole pygidium) though still with its anterior margin subparallel to the posterior margin of the pygidium (many Silurian and Devonian scutelluids).

In many species referred to *Illaenus* and *Stenopareia*, however, the anterior margin of the doublure medially bears a forward projection, which may be simple, bifid, or trifid; such projections reach forward to, or surround a position under, the posterior of the pygidial axis above, where this can be seen.

Although Dean (1978, p. 102) argued that the shape of such processes could vary, and quoted the presence of both bifid and trifid processes in *S. linnarssonii*, other authors (e.g. Whittington 1963) used the morphology of the process as a specific character, and Lane (1979, p. 16) showed that in *S. somnifer* the morphology of the process was constant throughout a large range of sizes of specimens referred to the species which occurred in a single sample. It is likely that Dean's material belongs to more than one species (Owen and Bruton 1980, p. 18).

Ontogeny. In very few Scutelluina is anything known of ontogenetic development. That of *Failleana* (Chatterton and Ludvigsen 1976, p. 38, pl. 6, figs. 1–42; Ludvigsen and Chatterton 1980, p. 471, fig. 1), which we previously (1978a, p. 354) misinterpreted and therefore incorrectly assessed, is perhaps the best known and illustrated. Other more-or-less completely known ontogenies include those of species of *Nanillaenus* and *Bumastoides* (Chatterton 1980), *Scutellum* and *Dentaloscutellum* (Chatterton 1971), and *Illaenus* (Hu 1976); the details of protaspid development presented by Hu have been questioned by Chatterton (1980, p. 26).

Ludvigsen and Chatterton (1980, p. 474) considered ontogeny of prime importance in adducing relationships. They stated: 'The similarities between the late meraspid or early holaspid *Failleana* and mature styginids and between metaprotaspids of *Nanillaenus* and *Bumastoides* and metaprotaspids of *Raymondaspis* suggest origins of both the subfamilies Bumastinae and Illaeninae from the family Styginidae in the Early Ordovician. The similarity between the metaprotaspids of *Failleana* and protaspids of scutelluids suggests either that the bumastines were ancestral to the scutelluids or that both these groups shared common ancestry within the styginids (the former interpretation is more likely and that is shown in Fig. 2).' It is possible to interpret their figure in three ways (text-fig. 5), and with respect to their arguments we offer the following three observations:

(i) We doubt the usefulness of the detailed timing of the attainment of a particular morphology in ontogeny in adducing phylogeny, especially in a large group in which few ontogenies are known. Although the increasing similarity in early ontogenetic stages of more closely related taxa has been shown (Whittington 1957, p. 460) and can be expected, a particular morphology presumably more immediately reflects a function for the individual and not any exact relationship between it and an individual belonging to another taxon at any particular stage in its development. Thus the similarity of young stages in trilobites may be useful in showing general, not particular, relationships within a high level taxon. Indeed, Chatterton (1980, p. 26) has noted similarities in the growth series of *Failleana*, *Nanillaenus*, *Bumastoides*, and scutelluids such as *Breviscutellum* (*Meridioscutellum*) which we take to indicate the general relationship of the groups to which these belong within the Scutelluina.

(ii) The great similarity of mature *Raymondaspis* to meraspid *Failleana* indicates, at least in this case, that effacement is a derived condition.

(iii) In the alternatives Ludvigsen and Chatterton suggest or imply (their fig. 2) the Scutelluidae are more closely related to the Bumastinae than either is to the Illaeninae. From their figure it is difficult to construe the Illaenidae as a monophyletic grouping.

INFERRED RELATIONSHIPS WITHIN THE SCUTELLUINA

This review suggests that the Scutelluina is characterized by a number of morphological features and developmental similarities which justify the recognition of the suborder. Particularly important is the presence of a large and wide rostral plate which is defined by backwardly converging connective sutures. This rostral morphology immediately serves to distinguish members of the suborder from

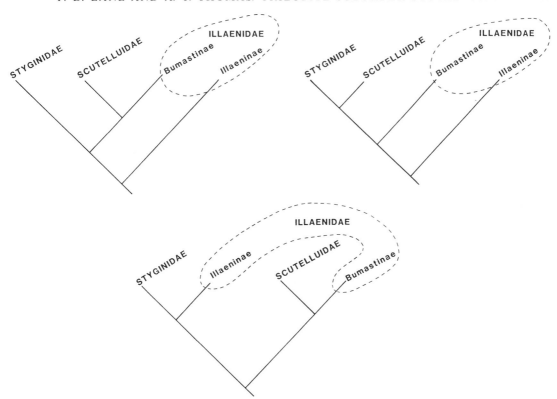

TEXT-FIG. 5. Three possible interpretations of the relationships of Scutelluina from Ludvigsen and Chatterton 1980, fig. 2.

some superficially similar groups of trilobites, like the bathyurids, which are now assigned to the Proetida.

Within the suborder we recognize the Illaenidae, Styginidae, and Phillipsinellidae; the Panderiidae is included with doubt (text-figs. 6, 7). For the reasons discussed below we consider that the trilobites previously assigned to the Styginidae and Scutelluidae should belong to the same family. Styginidae is the oldest available name for this taxon.

One of the major problems in the systematics of the suborder concerns the affinities and classification of its effaced genera. By comparison with other trilobites, and for ontogenetic reasons, effacement is clearly a derived condition. But whether it is one which evolved only once in this group, or several times, is more controversial. Effaced trilobites, especially those belonging to the same suborder, are likely to be morphologically similar whether or not they are closely related. Here, therefore, and in the diagnoses below we place most reliance on those characters (especially axial ones) least affected by effacement in order to try to distinguish mere similarity from close relationship.

Setting aside *Panderia*, which is discussed below, the effaced genera fall into two groups. One includes such genera as *Illaenus*, *Bumastoides*, *Stenopareia*, and *Thaleops* together with forms generally referred to the Ectillaeninae. These share: a glabella which does not seem to expand greatly anteriorly; the lack of an anterior pit or its internal homologue anterolaterally; two approximately parallel rows of glabellar muscle scars; a hypostome of *Illaenus*- or *Stenopareia*-type; a relatively short and wide pygidium, much less convex than the cephalon, and typically possessing doublural projections. We consider these trilobites to constitute the Illaenidae. The second group, including such genera as *Bumastus*, *Dysplanus*, and *Failleana*, we consider to be effaced styginids. These share: a greatly expanded, waisted glabella with rows of muscle scars which tend to diverge considerably anteriorly; the presence of an anterior pit, or its internal homologue; a hypostome of *Scutellum*-type; and a large pygidium, of similar size and convexity to the cephalon, and which lacks doublural projections. Comparison of the glabellar musculature both between different effaced genera, and with that of 'typical' styginids, suggests that these effaced styginids are not a monophyletic group.

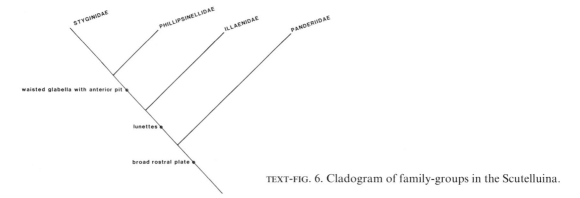

TEXT-FIG. 6. Cladogram of family-groups in the Scutelluina.

Illaenids, styginids, and *Phillipsinella* are united by their homologous cranidial musculature—especially by the possession of a lunette or its equivalent. *Phillipsinella* resembles styginids in possessing a waisted glabella and anterior pit, so that we consider (text-fig. 6) the Phillipsinellidae to be more closely related to the Styginidae than to the Illaenidae. The small size of *Phillipsinella*, the small number of thoracic segments and its generalized pygidial morphology suggest that it may be a progenetic form. Both *Phillipsinella* and styginids are described from the Arenig and a species of *Illaenus* is now known from the Tremadoc (pers. comm. R. A. Fortey), so that these groups had already differentiated early in Ordovician times.

Also known in the Arenig is *Panderia*, and the possibly related *Ottenbyaspis* is from the Tremadoc. We consider that *Panderia* possibly belongs to the Scutelluina because of the form of the rostral plate of its early species. But it contrasts with other members of the suborder in the absence of lunettes, in its glabellar shape (and possibly musculature), its convergent anterior sections of the facial sutures, and in its possession of a glabellar (not occipital) tubercle. Because of these contrasts we elevate Panderiinae to family level and refer it to the suborder with doubt; the discovery of a hypostome of the genus would help clarify the taxonomic position.

INFERRED RELATIONSHIP TO OTHER MAJOR TRILOBITE GROUPS

Many previous authors (e.g. Jaanusson 1959; Bergstrom 1973; Hahn and Hahn 1975) have regarded the trilobites here assigned to the Scutelluina as closely related to those which we would regard as Proetida (following Fortey and Owens 1975, 1979). The origins of both groups have been sought among the Ptychopariida and, for proetids, their ancestry may well lie there. The Ptychopariida, however, is a very large and heterogeneous group of trilobites (Eldredge 1977, p. 325) and some taxa are classified there mainly because they cannot be readily accommodated elsewhere.

Whittington (1966, p. 705) noted the possibility that such Cambrian genera as *Koldinia* and *Koldiniella* might be illaenid ancestors. Přibyl and Vaněk (1971, p. 362) dismissed this possibility and regarded these genera as plethopeltids. *Koldinia* has a forwardly tapering glabella (Moore 1959, fig. 411, 2a–c) which suggests that it is not closely related to the Scutelluina.

The Scutelluina show no significant similarity to any ptychopariid, contrasting particularly in having a forwardly expanding glabella and large, wide rostral plate. These are important points of similarity with certain Corynexochida, however, especially *Corynexochus* itself (see *C. plumula* Whitehouse; Shergold 1982, p. 47, pl. 14, figs. 1–7) and members of the Dolichometopidae

(Thorslund 1940; Skjeseth 1955; Přibyl and Vaněk 1971). Přibyl and Vaněk (1971, p. 362) compared certain Scutelluina with such corynexochids as *Clavaspidella*, noting their similar overall morphologies and particularly the common possession of a post-axial ridge on the pygidium. Of particular importance for implying a phylogenetic link is the glabella of corynexochids, which is anteriorly expanding and extends far forwards, the wide rostral plate, and the similarity of the hypostome (e.g. Rassetti 1947, pl. 48, fig. 5) to that of the Scutelluina. The main contrast between the two groups lies in the fusion of the corynexochid hypostome to the rostral plate. This fusion takes place early in ontogeny (Robison 1967) and is surely a derived condition, though possibly not uniquely so. If later, more detailed work confirms the significance of these similarities, and if the corynexochids are themselves monophyletic, then the Scutelluina should be regarded as the sister-group of the Corynexochida (text-fig. 7). Both then might suitably be assigned to the same Order.

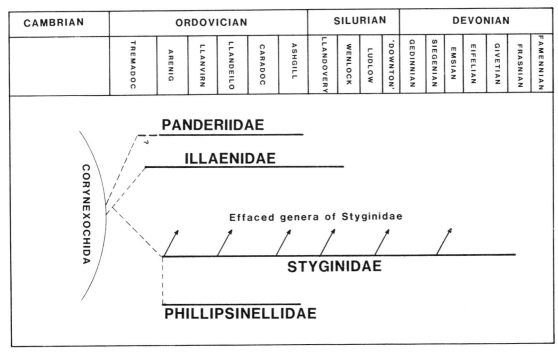

CAMBRIAN	ORDOVICIAN						SILURIAN				DEVONIAN						
	TREMADOC	ARENIG	LLANVIRN	LLANDEILO	CARADOC	ASHGILL	LLANDOVERY	WENLOCK	LUDLOW	'DOWNTON'	GEDINNIAN	SIEGENIAN	EMSIAN	EIFELIAN	GIVETIAN	FRASNIAN	FAMENNIAN

TEXT-FIG. 7. Possible relationships of family-groups in the Scutelluina with stratigraphic ranges indicated.

Ludvigsen (1982, p. 119) has argued that *Missisquoia* (= *Parakoldinioidia*; see Fortey, this volume) was the ancestor of the trilobites we refer to the Scutelluina. We have not investigated this possibility closely, but consider that the generalized morphology of the cephalon of that genus is much less like that of early genera of the Scutelluina than is that of the Corynexochida. We particularly note that the rostral plate of this genus is small and triangular, and that the pygidium does not closely resemble that of any Scutelluina (ibid. figs. 65S, 66D).

REVISED CLASSIFICATION OF THE SCUTELLUINA

Suborder SCUTELLUINA Hupé, 1953 (emended Hupé, 1955)
[*nom. transl.* herein *ex* superfamily Scutelloidae Hupé, 1953]

Diagnosis. Glabella expanding forwards (tr.), often strongly, over anterior part with four pairs of lateral glabellar muscle impressions or furrows; or barrel-shaped with five(?) pairs (*Panderia*). Opisthoparian; anterior section of facial sutures typically strongly divergent anteriorly (not so in

Panderia). Eye often placed far back on cheek and close to axial furrow. Rostral plate large, wide (tr.), and short (sag. and exsag.) except in *Phillipsinella*; laterally delimited by backwardly converging connective sutures which are ankylosed in late *Panderia* species. Terrace ridges common, especially on distal parts of dorsal exoskeleton and whole of ventral exoskeleton. Effacement of axial and pleural furrows common.

Family ILLAENIDAE Hawle and Corda, 1847

Diagnosis. Cranidial axial furrows generally effaced in front of lunettes. Glabella strongly convex (sag.) with four pairs of equally spaced lateral muscle impressions which are subequal in size and lie on arcs parallel to the axial furrow, never diverging strongly. Rostral plate trapezoid in outline, with flange. Hypostome with oval, sagittally elongate middle body; anterior wings subquadrate and massive. Pygidium shorter than wide (tr.), smaller, and less convex than cephalon, commonly highly effaced. Median arch prominent. Pygidial doublure often with median projections or cuspate (sag.).

Discussion. Because many of the genera referred to this family are incompletely known, and particularly their ventral morphology, we find it impossible to determine relationships within the taxon. At present, therefore, we recognize no subfamilial groupings.

Stratigraphic range. Tremadoc to Llandovery (?Wenlock).

Genera included. Illaenus; Bumastoides; Cekovia; Ectillaenus; Harpillaenus; Meitanillaenus; Nanillaenus; Ninglangia; Octillaenus; Ordosaspis; Parillaenus; Spinillaenus; Stenopareia; Thaleops; Thomastus; Ulugtella; Wuchuanella; Zbirovia; Zdicella.

Family PHILLIPSINELLIDAE Whittington, 1950

Diagnosis (emended from Bruton 1976, p. 700). Cephalon moderately convex; glabella waisted, defined by deep axial furrows, narrowest between palpebral lobes. Three pairs of lateral glabellar furrows are barely impressed (sometimes not visible), placed far back, and close to axial furrow. Rostral suture marginal; rostral plate long and wide, approaching half sagittal length of cephalon. Hypostome elongate (sag.) covering posterior half of cranidial length and extending a little beyond posterior margin; middle body oval in outline, anterior wings short. Free cheek with genal spines extending the length of thorax. Thorax with six segments. Pygidium short, with posteromedian notch; axis long, with three to six rings; post-axial ridge present (best seen on internal mould).

Stratigraphic range. Arenig to Ashgill.

Genera included. Phillipsinella.

Family STYGINIDAE Vogdes, 1890

Diagnosis. Early genera (e.g. *Raymondaspis*) with narrow glabella which expands gently forwards (tr.) with lateral glabellar furrows or muscle impressions not obvious; pygidium shorter than wide (tr.) with narrow axis about half the length (sag.) of the whole; doublure narrow. Later genera (e.g. *Scutellum*) with glabella expanding strongly forwards (tr.) over anterior half; pygidium as long as, or longer than wide, with very short often trilobed axis; doublure one-third or more sagittal length of pygidium. Anterior pit, or its ventral homologue present. Hypostome with subtriangular middle body with barely developed posterior lobe; maculae prominent, placed far back. Borders narrow. Genera showing varying degrees of effacement occur from Lower Ordovician to Middle Devonian; these are convex to strongly convex overall, and may have the diagnostic characters at best confined to the ventral surface of the dorsal exoskeleton.

Discussion. In preliminary work which will lead to a revision of this group for the second edition of the *Treatise on Invertebrate Paleontology*, Vol. O, we have encountered eighty genera and subgenera which we assign to this family. These have been previously assigned (with some other genera) to the Brontocephalidae, Bumastinae,

Dulanaspidinae, Eobronteinae, Goldillaeninae, Meroperixinae, Octobronteinae, Paralejurinae, Planiscutel-luinae, Scutelluinae, Stygininae, Theamataspidinae, and Thysanopeltinae. Many of these subfamilies have been based on characters which, for the reasons discussed above, we think are not significant at the subfamilial level. Unfortunately, many characters which may be important discriminators at this level (particularly ventral ones) are not known in many genera.

We are not yet sure whether the Stygininae (*sensu* Jaanusson 1959) is a monophyletic group or a stratigraphically defined group of genera which show certain primitive characters. Similarly, the Scutelluidae (*sensu* Jaanusson 1959) may comprise another group of genera showing more advanced characters; such 'scutelluids' could have evolved in more than one phyletic line from 'styginids'.

Because there are as yet no demonstrable phyletic lines of development, we list the genera in an undivided family. There may be synonyms lurking in this list in addition to those indicated.

Genera included. Scutellum (Scutellum); S. (Neoscutellum); Altaepeltis; Arctipeltis; Ancyropyge; Australo-scutellum; Bojoscutellum; Boreoscutellum; Breviscutellum (Breviscutellum); B. (Meridioscutellum); Bronteopsis (Bronteopsis); B. (Chichikaspis); Brontocephalina; Brontocephalus; Bumastus (Bumastus); B. (Bumastella); Chugaevia; Cornuscutellum; Craigheadia; Cybantyx; Decoroscutellum (Decoroscutellum); D. (Flexiscutellum); Delgadoa; Dentaloscutellum; Dulanaspis; Dysplanus; Ekwanoscutellum; Eobronteus; Failleana; Goldillaenus; Goldillaenoides; Hallanta; Hidascutellum; Hyboaspis; Illaenoides; Illaenoscutellum; Japonoscutellum; Kirk-domina; Kobayashipeltis; Kolihapeltis; Kosovopeltis [= Eokosovopeltis; = Heptabronteus]; Lamproscutellum; Leioscutellum; Ligiscus; Litotix; Meroperix; Metascutellum; Microscutellum; Octobronteus; Opoa; Opsypharus; Paralejurus; Perischoclonus; Planiscutellum; Platillaenus; Platyscutellum (Platyscutellum); P. (Izarnia); Poro-scutellum; Protobronteus; Protoscutellum; Protostygina; Ptilillaenus; Radioscutellum; Raymondaspis; Rhaxeros; Scabriscutellum; Spiniscutellum; Stoermeraspis; Stygina; Styginella; Tenuipeltis; Theamataspis; Thysanopeltella (Thysanopeltella); T. (Septimopeltis); Thysanopeltis; Tosacephalus; Turgicephalus; Unicapeltis; Uraloscutellum; Weberopeltis.

?Family PANDERIIDAE Bruton, 1968

Diagnosis (emended from Bruton 1968, p. 5). Mature exoskeleton of very small size. Axial furrows shallow, outlining barrel-shaped glabella. Anterior sections of facial sutures convergent forwards. Five pairs of lateral glabellar muscle impressions, and a median glabellar tubercle placed well forward of the posterior extent of the palpebral lobes. Thorax with eight segments. Pygidium much smaller than cephalon, axis discernible and more than half the length of the whole (sag.).

Discussion. This taxon is included with doubt in the suborder because of the atypical morphological features of *Panderia* (especially the presence of a median glabellar tubercle) and because the hypostome is unknown. The structure of the rostral plate and the possibly homologous glabellar musculature, however, suggest that it is closer to the Scutelluina than to any other trilobite group.

Stratigraphic range. (?Tremadoc) Arenig to Ashgill.

Genera included. Panderia; ?Ottenbyaspis.

Discussion. Ottenbyaspis has been included with doubt since its ventral sutures are unknown.

Acknowledgements. We thank D. L. Bruton, A. W. A. Rushton, and the referee for their comments on the manuscript. Lin Norton drew text-fig. 5.

REFERENCES

BALASHOVA, E. A. 1959. [Middle and Upper Ordovician and Lower Silurian trilobites of eastern Taimyr, and their stratigraphical significance.] *Sb. Stat. Paleont. Biostratigr.* **17**, 17–47, pls. 1–6 (relevant plates in ibid. **14**). [In Russian.]

BARRANDE, J. 1852. *Système Silurien du centre de la Bohême. 1ère partie: Recherches paléontologiques. Vol. 1. Crustacés: Trilobites.* xxx + 935 pp., 51 pls. Prague et Paris.

—— 1872. *Système Silurien du centre de la Bohême. 1 ère partie. Recherches paléontologiques. Supplément au vol. 1. Trilobites, Crustacés divers et Poissons.* xxx + 647 pp., 35 pls. Prague et Paris.

BERGSTRÖM, J. 1973. Organisation, life and systematics of trilobites. *Fossils & Strata*, **2**, 69 pp. 5 pls.

BILLINGS, E. 1859. Descriptions of some new species of trilobites from the Lower and Middle Silurian rocks of Canada. *Can. Nat. Geol.* **4**, 367–383.

BILLINGS, E. 1865. *Paleozoic fossils. Vol. 1. Containing descriptions and figures of new or little known species of organic remains from the Silurian rocks.* Pp. 169–344. Geological Survey of Canada, Montreal.

—— 1866. *Catalogues of the Silurian fossils of the island of Anticosti, with descriptions of some new genera and species.* Geological Survey of Canada, Montreal.

BRUTON, D. L. 1968. The trilobite genus *Panderia* from the Ordovician of Scandinavia and the Baltic areas. *Norsk Geol. Tidsskr.* **48**, 1–53, pls. 1–12.

—— 1976. The trilobite genus *Phillipsinella* from the Ordovician of Scandinavia and Great Britain. *Palaeontology,* **19**, 699–718, pls. 104–108.

—— and HARPER, D. A. T. 1981. Brachiopods and trilobites of the early Ordovician serpentine Otta Conglomerate, south central Norway. *Norsk Geol. Tidsskr.* **61**, 153–181, 5 pls.

BURMEISTER, H. 1843. *Die Organisation der Trilobiten, aus ihren lebenden Verwandten entwickelt; nebst einer systematischen Uebersicht aller zeither beschreibenen Arten.* 147 pp. 6 pls. Berlin.

CHATTERTON, B. D. E. 1971. Taxonomy and ontogeny of Siluro-Devonian trilobites from near Yass, New South Wales. *Palaeontographica* **(A), 137**, 1–108, pls. 1–24.

—— 1980. Ontogenetic studies of Middle Ordovician trilobites from the Esbataottine Formation, Mackenzie Mountains, Canada. Ibid. **171**, 1–74, pls. 1–19.

—— and LUDVIGSEN, R. 1976. Silicified Middle Ordovician trilobites from the South Nahanni River area, district of Mackenzie, Canada. Ibid. **154**, 1–106, pls. 1–22.

CHUGAEVA, M. N. 1958. [Trilobites of the Ordovician of the Chu-Ili Mountains] *in* [The Ordovician of Kazakhstan. III]. *Trudy geol. Inst. Akad. Nauk* SSSR **9**, 5–138, pls. 1–11. [In Russian.]

DEAN, W. T. 1978. The trilobites of the Chair of Kildare Limestone (Upper Ordovician) of eastern Ireland. Part 3. *Monogr. palaeontogr. Soc.* London: 99–129, pls. 45–52 (Publ. no. 550, part of Vol. 131 for 1977).

ELDREDGE, N. 1977. *Trilobites and evolutionary patterns.* Chap. 9 *in* HALLAM, A. (ed.). *Patterns of Evolution.* Pp. 305–332. Elsevier, Amsterdam.

FORTEY, R. A. 1974. The Ordovician trilobites of Spitsbergen. I. Olenidae. *Skr. Norsk Polarinst.* no. 160, 1–129, pls. 1–24.

—— 1980. The Ordovician trilobites of Spitsbergen. III. Remaining trilobites of the Valhallfonna Formation. Ibid. no. 171, 1–112, incl. pls. 1–25.

—— 1981. *Prospectatrix genatenta* (Stubblefield) and the trilobite superfamily Cyclopygacea. *Geol. Mag.* **118** (6), 603–614.

—— and OWENS, R. M. 1975. Proetida—a new order of trilobites. *Fossils & Strata,* **4**, 227–239.

—— —— 1979. Enrollment in the classification of trilobites. *Lethaia,* **12**, 219–226.

HAHN, G. and HAHN, R. 1975. Forschungsbericht über trilobitomorpha. *Paläont. Z.* **49**, 432–460.

HALL, J. 1847. *Descriptions of the organic remains of the lower division of the New-York System.* Geological Survey of New York, 1. xxiii + 338 pp., 87 pls.

—— 1868. Account of some new or little known fossils from rocks of the age of the Niagara Group. *Ann. Rept. Univ. St. N.Y.* **20**, 305–41, pls. 10–23.

HAWLE, I. and CORDA, A. J. C. 1847. *Prodrom einer Monographie der böhmischen Trilobiten.* 176 pp., 7 pls. Prague (also 1848, *Abh. K. böhm. Ges. Wiss.* **5**, 117–292, pls. 1–7).

HENNINGSMOEN, G. 1951. Remarks on the classification of trilobites. *Norsk Geol. Tidsskr.* **29**, 174–217.

HOLM, G. 1882. De svenska arterna af Trilobitslaget *Illaenus* Dalman. *Bih. K. svenska VetenskAkad. Handl.* **7**, xiv + 148 pp. 6 pls.

—— 1886. Revision der ostbaltischen Silurischen trilobiten. III. Illaeniden. *Mém. Acad. Imp. Sci. St.-Pétersbourg* Ser. 7, **3**, 1–173, pls. 1–12.

HORNÝ, R. and BASTL, F. 1970. *Type specimens of fossils in the National Museum of Prague. Vol. 1. Trilobita.* 354 pp. 20 pls.

HOWELLS, YVONNE. 1982. Scottish Silurian trilobites. *Monogr. palaeontogr. Soc.* London: 1–76, pls. 1–15 (Publ. No. 561, part of Vol. 135 for 1981).

HU, C.-H. 1971. Ontogeny and sexual dimorphism of Lower Paleozoic Trilobita. *Palaeontographica Americana,* **7**, 31–155, pls. 7–26.

HUPÉ, P. 1953. Classe de trilobites. *In* PIVETEAU, J. (ed.). *Traité de paleontologié,* **3**, 44–246, 140 text-figs. Paris.

—— 1955. Classification des trilobites. *Annls Paléont.* **41**, 91–317. [Part 2 of this work; part 1 published ibid. **39**.]

INGHAM, J. K. 1970. A monograph of the Upper Ordovician trilobites from the Cautley and Dent districts of Westmorland and Yorkshire. *Monogr. palaeontogr. Soc.* London: 1–58, pls. 1–9 (Publ. No. 526, part of Vol. 74 for 1970).

JAANUSSON, V. 1954. Zur morphologie und Taxonomie der Illaeniden. *Ark. Miner. Geol.* **1**, 545–583, 3 pls.

—— 1957. Unterordovizische Illaeniden aus Skandinavien. *Bull. geol. Inst. Upsala* **37**, 79–165, 10 pls.

—— 1959. Suborder Illaenina, O365–O415. *In* MOORE, R. C.

KOBAYASHI, T. 1935. The Cambro-Ordovician formations and faunas of South Chosen. Palaeontology Part III. Cambrian faunas of South Chosen with special study on the Cambrian trilobite genera and families. *Jl Fac. Sci. Imp. Univ. Tokyo*, **4**, 49–344.

—— and HAMADA, T. 1974. Silurian trilobites of Japan in comparison with Asian, Pacific and other faunas. *Spec. Pap. palaeont. Soc. Japan*, **18**, viii + 155 pp., 12 pls.

KOLOBOVA, N. M. 1977. [On the extent and systematic position of the new Ordovician family Brontocephalidae (Trilobites).] *Ezheg. Vse. Paleont. Obshch.* **19**, 84–90, 1 pl. [In Russian: volume for 1968–1971.]

LANE, P. D. 1972. New trilobites from the Silurian of north-east Greenland with a note on trilobite faunas in pure limestones. *Palaeontology*, **15**, 336–64, pls. 59–64.

—— 1979. Llandovery trilobites from Washington Land, North Greenland. *Bull. Grøn. geol. Unders.* no. 131, 5–37, 6 pls.

—— and HELBERT, G. J. 1982. Family Scutelluidae. *In* HELBERT, G. J., LANE, P. D., OWENS, R. M., SIVETER, DEREK J. and THOMAS, A. T. Lower Silurian trilobites of Norway. I.U.G.S. Subcommission on Silurian Stratigraphy. Field Meeting, Oslo Region 1982. *Palaeontological contributions from the University of Oslo*, no. 278, 129–148. D. Worsley (ed.).

—— and THOMAS, A. T. 1978a. Silurian trilobites from north-east Queensland and the classification of effaced trilobites. *Geol. Mag.* **115**, 351–358, 2 pls.

———— 1978b. Family Scutelluidae. *In* THOMAS, A. T. British Wenlock trilobites. Part 1. *Monogr. palaeontogr. Soc.* London: 1–56, pls. 1–14 (Publ. No. 552, part of Vol. 132 for 1978).

LUDVIGSEN, R. 1982. Upper Cambrian and Lower Ordovician trilobite biostratigraphy of the Rabbitkettle Formation, western District of Mackenzie. *Contr. Life Sci. Div. R. Ont. Mus.* no. 134, 1–188. 70 figs.

—— and CHATTERTON, B. D. E. 1980. The ontogeny of *Failleana* and the origin of the Bumastinae (Trilobita). *Geol. Mag.* **117**, 471–478, 1 pl.

MAKSIMOVA, Z. 1968. [Middle Palaeozoic trilobites of central Kazakhstan]. *Trudy vses. Nauchno.-issled. geol. Inst.* N.S. no. 165, 208 pp. 35 pls. [In Russian.]

MILLER, J. 1975. Structure and function of terrace lines. *Fossils & Strata*, **4**, 155–178.

MOORE, R. C. (ed.) 1959. *Treatise on Invertebrate Paleontology, Part O, Arthropoda 1*. xix + 560 pp., 415 figs. Geol. Soc. Amer. and Univ. of Kansas Press (Lawrence).

OWEN, A. W. and BRUTON, D. L. 1980. Late Caradoc–early Ashgill trilobites of the central Oslo region, Norway. *Palaeontological contributions from the University of Oslo*, no. 245, 62 pp. 10 pls.

OWENS, R. M. 1973. Ordovician Proetidae (Trilobita) from Scandinavia. *Norsk geol. Tidsskr.* **53**, 117–181.

PILLET, J. 1972. Les trilobites du Devonien inferieur et du Devonien moyen du Sud-Est du Massif Amoricain. *Mém. Soc. sci. Anjou*, **1**, 307 pp. 64 pls.

PŘIBYL, A. and VANĚK, J. 1971. Studie über die Familie Scutelluidae Richter et Richter (Trilobita) und ihre phylogenetische entwicklung. *Acta Univ. Carol. Geol.* **4**, 361–394, 1 pl.

RASETTI, F. 1948. Middle Cambrian trilobites from the conglomerates of Quebec (exclusive of the Ptychopariidae). *Jl Paleont.* **22**, 315–339, pls. 45–52.

RAYMOND, P. E. 1916. New and old Silurian trilobites from southeastern Wisconsin, with notes on the genera of the Illaenidae. *Bull. Mus. comp. Zool. Harvard*, no. 60, 1–41, 4 pls.

—— 1920. Some new Ordovician trilobites. Ibid. no. 64, 273–296.

REED, F. R. C. 1904. The Lower Palaeozoic trilobites of the Girvan district, Ayrshire. Part 2. *Monogr. palaeontogr. Soc.* London: 49–96, pls. 7–13 (Publ. No. 276, part of Vol. 58 for 1904).

—— 1906. The Lower Palaeozoic trilobites of the Girvan district Ayrshire. Part 3. Ibid. 97–186, pls. 14–20 (Publ. no. 286, part of Vol. 60 for 1906).

—— 1914. The Lower Palaeozoic trilobites of the Girvan district. Supplement. Ibid. 1–56, pls. 1–8 (Publ. no. 329, part of Vol. 67 for 1913).

—— 1931. The Lower Palaeozoic trilobites of the Girvan district, Supplement No. 2. Ibid. 1–30 (Publ. no. 382, part of Vol. 83 for 1929).

—— 1935. The Lower Palaeozoic trilobites of the Girvan district. Supplement No. 3. Ibid. 1–64, pls. 1–4 (Publ. no. 400, part of Vol. 88 for 1934).

RICHTER, R. and RICHTER, E. 1955. Scutelluidae n.n. (Tril.). *Senck. leth.* **36**, 291–293.

———— 1956. Grundlagen für die Buerteilung und Einteilung der Scutelluidae (Tril.). Ibid. **37**, 79–124, 7 pls.

ROBISON, R. A. 1967. Ontogeny of *Bathyuriscus fimbriatus* and its bearing on the affinities of corynexochid trilobites. *Jl Paleont.* **41**, 213–221, pl. 24.

SALTER, J. W. 1867. A monograph of the British trilobites from the Cambrian, Silurian and Devonian formations. Part 4. *Monogr. palaeontogr. Soc.* London: 177–214, pls. 25*–30 (Publ. no. 86, part of Vol. 20 for 1866).

SAVITSKI, V. E. 1972. [Cambrian of the Siberian platform.] *Trudy sib. Nauchno.-issled. geol. Inst.* no. 130, 200 pp., 23 pls.

SHAW, F. C. 1968. Early Middle Ordovician Chazy trilobites of New York. *Mem. N.Y. St. Mus.* no. 17, 1–163, pls. 1–24.

SHERGOLD, J. H. 1977. Classification of the trilobite *Pseudagnostus*. *Palaeontology*, **20** (1), 69–100, pls. 15, 16.

—— 1982. Idamean (Late Cambrian) trilobites, Burke River structural belt, Western Queensland. *Bull Bur. Miner. Resour. Aust.* no. 187, 1–69, pls. 1–17.

SINCLAIR, G. W. 1949. The Ordovician trilobite *Eobronteus*. *Jl Paleont.* **23**, 45–56, pls. 12–14.

SKJESETH, S. 1955. The Middle Ordovician of the Oslo region, Norway. 5. The trilobite family Styginidae. *Norsk geol. Tidsskr.* **35**, 9–28, 5 pls.

ŠNAJDR, M. 1957. [Classification of the family Illaenidae (Hawle et Corda) in the Lower Palaeozoic of Bohemia.] *Sb. ústřed. Úst. geol.* (Paleont). **23**, 125–284, 12 pls. [For 1956; in Czech.; English summary, pp. 270–284.]

—— 1960 [A study of the family Scutelluidae (Trilobitae).] *Rozpr. ústřed. Úst. geol.* **26**, 11–221, pls. 1–36. [In Czech.; English summary, pp. 227–265.]

STUBBLEFIELD, C. J. 1936. Cephalic sutures and their bearing on current classifications of trilobites. *Biol. Rev.* **11**, 407–440.

THORSLUND, P. 1940. On the *Chasmops* Series of Jemtland and Sodermannland (Tvaren). *Sver. geol. Unders. Afh.* no. C436, 1–191, pls. 1–12.

VOGDES, A. W. 1890. A bibliography of the Paleozoic crustacea from 1698 to 1889. *Bull. U.S. geol. Surv.* no. 63, 1–177.

WARBURG, E. 1925. The trilobites of the Leptaena Limestone in Dalarne. *Bull. geol. Inst. Upsala*, **17**, viii + 1–446, pls. 1–11.

WESTERGÅRD, A. H. 1946. Agnostidea of the middle Cambrian of Sweden. *Sver. geol. Unders. Afh.* no. C477 *Arsb.* **40** (1), 3–141, pls. 1–16.

WHITTINGTON, H. B. 1950. Sixteen Ordovician genotype trilobites. *Jl Paleont.* **24**, 531–565, pls. 68–75.

—— 1957. The ontogeny of trilobites. *Biol. Rev.* **32**, 421–469.

—— 1963. Middle Ordovician trilobites from Lower Head, western Newfoundland. *Bull. Mus. comp. Zool. Harvard*, no. 129, 1–118, 36 pls.

—— 1965. Trilobites of the Ordovician Table Head Formation, western Newfoundland. Ibid. no. 132, 275–442, 68 pls.

—— 1966. Phylogeny and distribution of Ordovician trilobites. *Jl Paleont.* **40**, 696–737.

P. D. LANE
Department of Geology
University of Keele
Staffordshire ST5 5BG

A. T. THOMAS
Department of Geological Sciences
University of Aston
Gosta Green
Birmingham B4 7ET

THE LOWER CARBONIFEROUS GRANTON 'SHRIMP-BED', EDINBURGH

by D. E. G. BRIGGS *and* E. N. K. CLARKSON

ABSTRACT. The sedimentology of the Granton 'shrimp-bed' (Dinantian, Lower Oil Shale Group) is described. The bed is shown to be an algal stromatolite, presumably of tidal-flat origin. The fauna, which includes lightly skeletonized and soft-bodied elements, is dominated by the crustacean *Waterstonella grantonensis* Schram, 1979, redescribed herein. Associated with it are other crustaceans, the conodont animal, rare fish, molluscs, worm-like forms, and branching organisms of uncertain affinity. The crustacean 'community' is not characteristic of an off-shore environment as recently interpreted by Schram. This algal stromatolite fauna represents an unusual example of a Konservat-Lagerstätte.

THE Granton 'shrimp-bed' was discovered in 1919 by D. Tait who described the lithology and its geological setting (1925). Tait listed the fauna and figured a slab, GrI 47498, with many crustaceans. He reported (p. 132) that B. N. Peach considered the genus *Tealliocaris* to be the common crustacean 'but that some of the species are new to science'. In fact, only four of the crustacean specimens from Granton known to us may be referred to *Tealliocaris*. In 1931 P. E. Raymond (Harvard University) spent some time collecting with Tait and prepared a description of the most abundant crustacean, which he recognized as new. This paper was never published, and it was not until 1979 that the common species was finally described and named *Waterstonella grantonensis* by Schram. Our research on the Granton fauna, which began some years ago as part of a wider project on Scottish Carboniferous Crustacea, gained added impetus with the discovery of a conodont-bearing animal with preserved soft parts in material from the locality held by the Institute of Geological Sciences, Edinburgh (IGSE) (Briggs *et al.* 1983). In our studies we have used the extensive collections in the IGSE, the Grant Institute of Geology, University of Edinburgh (GrI), the Royal Scottish Museum, Edinburgh (RSM) where our own collections are now lodged, and the Museum of Comparative Zoology, Harvard University (MCZ). Here we describe the sedimentology of the Granton beds for the first time and review the total fauna. It has also proved possible to provide a more detailed description of *Waterstonella*, which is unique to the Granton locality, than that published by Schram (1979b). The preservation of the soft tissue of the conodont animal together with abundant lightly skeletonized *Waterstonella* and other apparently soft-bodied organisms suggests that the Granton assemblage is likely to include representatives of most of the original biota, and the occurrence may therefore be termed a Konservat-Lagerstätte (Seilacher 1970). This account of an exceptionally preserved Palaeozoic fauna dominated by arthropods seems appropriate for inclusion in a volume to honour H. B. Whittington, who has made such a major contribution to our knowledge of the famous middle Cambrian Burgess Shale fauna of British Columbia.

STRATIGRAPHY AND SEDIMENTARY ENVIRONMENT

The Granton 'shrimp-bed' occurs within the Dinantian Granton Sandstones of the Lower Oil Shale Group, Calciferous Sandstone Measures. The Granton Sandstones consist of two units of sandstone, the higher Ravelston (110 m) and the lower Craigleith (30 m) Sandstones, separated by about 100 m of shale which includes the 'shrimp-bed'. The Granton Sandstones are restricted in their distribution to the Edinburgh District (IGS 1:50,000 Sheet 32E, Edinburgh). They immediately underlie the Wardie Shales from which a well-known fish fauna has been obtained to the east of the 'shrimp-bed' exposures (Wood 1975; Dick 1981; Dick and Maisey 1980). The most recent account of the stratigraphy and sedimentology of this area is given in Mitchell and Mykura (1962). These beds were formerly considered to be Asbian–Holkerian in age (George *et al.* 1976) but the

[Special Papers in Palaeontology No. 30, pp. 161–177, pls. 20–22]

evidence for this is inconclusive. The conodonts yielded by the 'shrimp-bed' suggest an earlier Dinantian age but they are shallow-water forms and may be unreliable for correlation due to a facies-controlled distribution (Briggs *et al.* 1983).

The oil shales are of largely non-marine origin and were evidently deposited in a large fresh or brackish water lake (Lake Cadell of Greensmith 1962) which covered the eastern end of the Midland Valley of Scotland. The succession consists mainly of sandstones, mudstones, and shales with some limestones and up to nine major developments of oil shale (which were formerly mined). The lake was open from time to time to marine incursions from the east and the Granton 'shrimp-bed' occurs in one of the resultant limestone-rich intervals. Limestones are more numerous in the Upper Oil Shale Group.

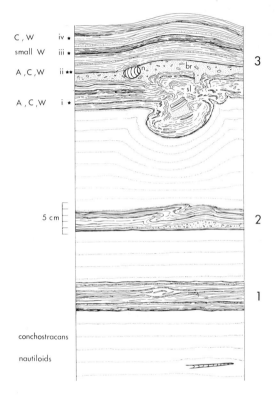

TEXT-FIG. 1. Section through the Granton 'shrimp-bed', slightly schematic, showing the three thin bands of laminated limestone within shale. Band 3 yields *Anthracophausia* (A), *Crangopsis* (C), and *Waterstonella* (W) at four main horizons (i–iv). The asterisks give an indication of relative abundance. Other symbols represent a nautiloid (n), brecciated bed (br), and slump (sl). (Cf. text-fig. 2a.)

Tait (1925, fig. 1) found the 'shrimp-bed' in two places. The more easterly exposures, in the old Granton land and sea quarries, have long been lost due to infilling during land reclamation. To the west, the more recent construction of the esplanade and breakwater along the Muirhouse shore has covered much of the original outcrop and application has been made to protect the two remaining poor small exposures (discovered by S. P. Wood in 1978 and J. G. Sharp in 1981) as a Nature Conservancy Site of Special Scientific Interest (SSSI). Within the *c.*20 m of shale exposed at Granton is an interval 45 cm thick which includes three thin bands of laminated limestones (text-fig. 1). The lower two (Bands 1 and 2), which are about 4 cm thick, are lithologically similar, but a little less limy and more fissile than the upper band (Band 3). All three bands consist of alternating dark and light laminae which in places pass laterally into microbreccias consisting of fragments of the laminated material in a mud matrix. They appear to be largely of algal origin.

Band 3 (text-figs. 1, 2*a*), which has yielded all the material collected by us, is much thicker than the lower bands, varying from about 14 to 20 cm. It was almost certainly also the source of the specimens collected by Tait and others, which share the same distinct lithology. Bands 1 and 2, on the other hand, appear unfossiliferous. The shale unit underlying Band 3 is folded in places with an amplitude of about 10 cm and a wavelength of 30 cm. Where present these folds are evident as regularly spaced ridges where the bedding surface is exposed, but they do not penetrate the underlying calcareous Band 2. The lowest part of Band 3 consists of about 4·5 cm of laminae but is thicker and highly contorted where the underlying shale is downfolded. Above this disturbed interval there

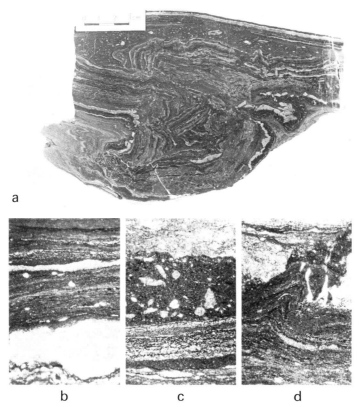

TEXT-FIG. 2. Petrography of Granton 'shrimp-bed', Band 3. *a*, polished
section through Band 3 showing slumped algal mat, GrI 47499 (cf.
text-fig. 1). *b*, thin section showing algal mat and two dolomitic bands,
GrI 47500, × 2·67. *c*, algal mat and microbrecciated horizon with dolomitic fragments, interbedded with algal laminae displaying clotted
texture, GrI 47501, × 2·67. *d*, pull-apart structure in dolomitic band
(above) infilled with contorted algal mat, GrI 47502, × 5.33.

is a breccia 2–3 cm thick, consisting of laminated fragments in a carbonate mud matrix. The top of Band 3 consists of about 7 cm of laterally continuous and largely undisturbed laminae. Abundant crustaceans are confined to four bedding-plane horizons within Band 3 (i, ii, iii, iv—text-fig. 1), but other elements of the fauna (branching organisms, worm-like forms) occur throughout the band.

All three limestone bands are predominantly dolomitic. Thin sections reveal alternating laminae of calcareous mud and dark brown or black carbonaceous material up to 1 mm thick (text-fig. 2*b*, *c*, *d*) which in some cases preserve evidence of algal filaments. Microsparitic laminae up to 3 mm thick are also common but preserve little evidence of internal structure. In places laminae up to 2 mm thick display a clotted texture (text-fig. 2*c*), possibly representing faecal pellets, the voids filled with drusy calcite. A fenestral fabric is common, resulting from the trapping of gas bubbles. Shrinkage cracks on bedding planes, breccias, and pull-apart structures (text-fig. 2*c*, *d*) suggest desiccation and, in association with discontinuity surfaces, indicate intermittent subaerial exposure. The formation of occasional algal tufts may also be the product of exposure (Mayall and Wright 1981). The well-preserved fossils are normally found within undisturbed laminae. Convolution of the laminae is presumably the result of contemporaneous slumping and overturning of the algal mats.

The features of the laminated limestones at Granton are closely similar to many Recent (e.g. Davies 1970; Gebelein 1969; Hardie 1979) and fossil (e.g. Leeder 1975; Gill 1977) tidal-flat stromatolites and suggest that the assemblage was preserved among intertidal algal mats. Schram's (1981*a*, p. 132) interpretation of the *Waterstonella*-dominated community as 'off-shore marine' was therefore clearly in error. Tait (1925, p. 132), however, in the original description of the 'shrimp-bed', noted that 'the surface markings in the limestone indicate very shallow water conditions, and that at times during deposition the beds were exposed to the air and subjected to drying and cracking'.

A REVIEW OF THE BIOTA

Vertebrates. Two poorly preserved specimens of a palaeoniscid fish from the Granton 'shrimp-bed' have been identified by S. P. Wood (pers. comm.) as *Rhadinichthys*. One of these may be *R. brevis* Traquair, 1903.

The conodont animal. The part and counterpart of a single specimen of an elongate animal bearing ray-supported fins posteriorly, and showing repeated traces which may represent segmentation, was found at Granton, probably by Tait, and lay undescribed in the collections of the IGSE until February 1982. A detailed study of this specimen revealed the presence of an *in situ* conodont apparatus at the anterior end, and a subsequent search for conodonts by R. J. Aldridge, using standard extraction techniques, has yielded many specimens including a fused cluster of hindeodelliform elements. Since the conodont animal (referred to *Clydagnathus*? sp. cf. *C. cavusformis* Rhodes, Austin and Druce, 1969) has been described elsewhere (Briggs *et al.* 1983), no further comments are made here.

Molluscs. An uncompacted specimen of a coiled nautiloid (T3020F) free of matrix, 4·5 cm in diameter, the shell replaced and infilled by calcite, was discovered in the IGSE collections. It is, however, too poorly preserved for identification. A similarly preserved orthocone 1·5 cm in diameter was found in a large block (T3339F) in the field. A thin-section indicates that the upper surface of the shell was exposed above a layer of newly sedimented breccia allowing the growth of stromatolite upon it (text-fig. 1). An extensive search of old and new collections has failed to substantiate Tait's (1925, p. 132) record of rare gastropods; thin sections of the limestones, however, revealed possible evidence of gastropod spat. The only other mollusc known is a small bivalve (T2033F) so poorly preserved as to preclude identification.

Worm-like forms. Two worm-like forms, which lack any traces of structure or segmentation, are preserved in fluorapatite. The larger of these, which averages about 0·5 mm in width and reaches lengths of up to 10·0 mm, is usually curved and tapers slightly in one direction (Pl. 20, fig. 11). The smaller consists of very thin (up to 0·2 mm wide) usually tangled filaments (Pl. 20, figs. 9, 12). The majority of specimens of both forms occur in crustacean-bearing horizons, but rare individuals have been found at levels containing only fragmentary branching organisms. Substantially larger but otherwise similar forms have been described from the Mazon Creek (Westphalian) biota of Illinois (Schram 1973) and the Bear Gulch Limestone (Namurian) of Montana (Schram 1979a). These, however, in addition preserve evidence of structures found in nematodes (fine hair-like projections of the cuticle, oral papillae) and nemerteans (proboscis) and were assigned by Schram to these worm phyla. In the absence of such diagnostic structures the specimens from the Granton 'shrimp-bed' might represent megascopic algae.

Branching organisms. At many horizons within the shrimp-bed there occur abundant branching organisms, usually fragmentary, preserved in a thin layer of fluorapatite like the crustaceans but more yellow-brown in colour and showing relatively little detail to indicate their biological affinity. They often occur in association with the crustaceans and are commonly attached to them (Pl. 21, figs. 2, 3), having presumably grown on exuviae or carcasses. These organisms are common, however, at some horizons where crustaceans are very rare or absent. The nature of the branching among these organisms is variable but the different forms appear to represent a continuum; it is, therefore, unlikely that more than one species is represented. The original cross-section is unknown, but may have been cylindrical. The commonest form consists of a cluster up to 10·0 mm long of unsegmented, longitudinally ribbed branches up to 0·5 mm wide which apparently bifurcate irregularly (Pl. 20, figs. 15, 17). The ribbing is not always distinctly preserved. Successively more distal branches are slightly narrower, but some expand slightly at their termination, which is rounded. A small nodule of a dark mineral is often present just within this termination (Pl. 20, fig. 15). Some examples show this feature to mark the distal end of a dark trace running along the middle of the branch, which may represent a lumen (Pl. 20, fig. 8). Specimens comprised of a cluster of branching straps show that the angle between branches varies from about 5° to 30° (Pl. 20, figs. 8, 14). Plate 20, fig. 14 shows two small clumps apparently arising from a thicker common 'stem'. The branches of some specimens are more widely spaced, and divide in a more regular, apparently dichotomous fashion (Pl. 20, figs. 13, 16). Other fragments have a swollen appearance and bear one or more lateral buds (Pl. 20, fig. 10). What appears to be a median longitudinal division of some branches in some specimens (Pl. 20, fig. 13) is a result of the diagenetic growth of fluorapatite within the branch. The form of these branching organisms prompts a comparison with three groups: the hydroid cnidarians, bryozoans, and algae.

The periderm of hydroids, like the exoskeleton of crustaceans, is composed of chitin and therefore might likewise be associated with diagenetic fluorapatite. The mode of branching in the fossils is similar to that in some colonial hydroids (monopodial, racemose, branching in all planes). The branches appear to have been hollow tubes (equivalent to the tubular coenosarc), the slight distal expansions possibly representing hydrothecae. The hydranths of Recent hydroids lack a chitinous cuticle, and this would explain why they are not preserved in the fossils. Recent hydroids commonly live attached to live or inert objects, thus the attachment of the fossils to crustaceans (Pl. 21, figs. 2, 3) is consistent with this interpretation.

The fossil record of hydroids is sparse. The most remarkably preserved are the chitinous hydroids which have been isolated from Ordovician limestones from Poland, Öland, and Greenland (Kozlowski 1959; Skevington 1965; Frykman 1979). Two organisms from the Mazon Creek biota have also been assigned to the Order Hydroida, the solitary hydra *Mazohydra megabertha* Nitecki and Schram, 1975, and the colonial *Drevotella proteana* Nitecki and Richardson, 1972. Apart from its much larger size and more expanded distal terminations *Drevotella* is strikingly similar to the Granton forms with which it shares irregularly dichotomous branching, a decrease in the width of the more distal branches, and rounded terminations (Nitecki and Richardson 1972; Foster 1979).

The form of branching is also reminiscent of some Bryozoa. Known Carboniferous bryozoans branching in this fashion, however, have a mineralized (calcium carbonate) skeleton, and are not usually preserved either compacted or altered to fluorapatite. The more irregularly branching Granton organisms are similar in appearance, however, to the phylactolaemate Bryozoa such as *Plumatella*. The fossil record of phylactolaemate colonies is limited to a single record from the Cretaceous of Bohemia, disputed by Mundy *et al.* (1981). The colony walls of phylactolaemates are chitinous, and might have undergone a similar diagenetic alteration to that affecting the exoskeleton of the crustaceans. *Plumatella* is similar in scale to the fossils and the individual zooids protrude from an orifice at the termination of each branch (Mundy 1980, p. 21, fig. 9a). A number of factors, however, reduces the likelihood of the branching organisms representing a phylactolaemate. Some of the fossils show a regularly dichotomous pattern (Pl. 20, figs. 13, 16) unknown in the phylactolaemates. The chitinous statoblasts which allow the phylactolaemates to survive adverse conditions first appear in the late Tertiary (Mundy *et al.* 1981). The residues from the Granton lithology have included nothing resembling a statoblast (pers. comm. R. J. Aldridge). All living phylactolaemates are freshwater whereas these Carboniferous branching organisms are associated with marine fossils.

The regular and irregular dichotomous branching is similar to that in some algae (compare the brown alga *Dictyota* for example). The branches would then correspond to the thallus and the distal nodules might represent reproductive conceptacles. The thalli of few algae are hollow, however, and longitudinal ribbing is less likely to occur in algae than in hydroids or bryozoans with their chitinous exterior. Further, as Nitecki and Richardson (1972, p. 1) observed with regard to *Drevotella*, a similar form from the Mazon Creek biota, 'if they are plants, there is nothing in the known Palaeozoic flora to which they may be compared'. Thus while the possibility of an algal or bryozoan affinity for these branching organisms cannot be ruled out entirely they are more likely to represent hydroids.

Other organisms. A small (1·5 mm in diameter) sub-circular form made up of numerous narrow radiating strands is known from two incomplete specimens (Pl. 20, fig. 7). It could represent a sponge with radiating spicules but in the absence of more material its affinities remain uncertain.

Crustaceans. All the crustaceans are preserved in the dark algal-rich layers, and specimens are usually very prominent on the bedding planes as they are preserved in a whitish mineral which contrasts with the dark matrix. It is not certain whether this mineral, which proves on XRD analysis to be fluorapatite, replaces the cuticle or is an infilling of the body. The latter is suggested by the preservation of patches of cuticle in association with this fluorapatite, particularly in specimens of *Crangopsis*.

Two forms of *Crangopsis*, which may represent separate species, are present at certain horizons in the Granton 'shrimp-bed'. The smaller form (usually 14 to 22 mm in length) occurs in association with *Waterstonella* and in approximately equal numbers at three horizons (text-fig. 1). It can be distinguished readily from *Waterstonella* by its stouter morphology and thicker cuticle. The carapace expands more ventrally and the appendages, especially the antennae, are shorter and stouter. In addition, the expanded outline of the epimeron on the second pleomere, which is characteristic of the genus, is occasionally preserved (Pl. 22, figs. 6–8).

Rare specimens of a larger form of *Crangopsis* (reaching lengths of up to 50 mm), many with preserved cuticle, also occur in Band 3 (Pl. 22, fig. 9). Schram (1979b, p. 74) identified this species as *C. socialis* (Salter, 1861), which is the common species in the Midland Valley of Scotland (Schram 1979b, p. 30, fig. 3).

Only four poorly preserved specimens of *Tealliocaris* are known from the Granton 'shrimp-bed'. The fourth crustacean genus, *Anthracophausia* (Pl. 20, figs. 2–6), is several times larger than the other crustaceans. It is also rare; only ten specimens from Granton are known to us. *Anthracophausia* was erected by Peach in 1908. With that genus Brooks (1969) synonymized *Palaemysis* Peach, 1908, which was based on isolated tail fans. In a more

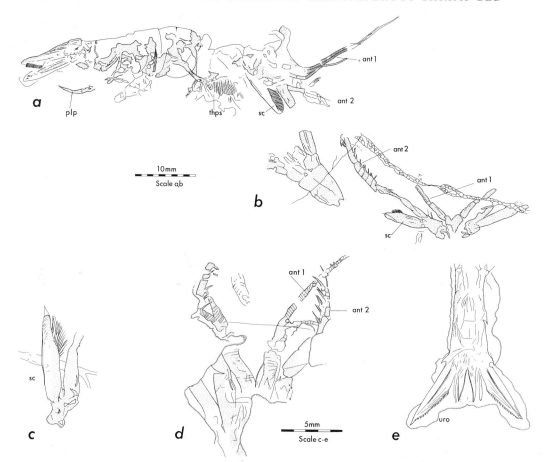

TEXT-FIG. 3. *Anthracophausia dunsiana* Peach, 1908. Lower Oil Shale Group, Granton. *a*, RSM.GY.1982.66.13 (cf. Pl. 20, fig. 2). *b*, RSM.GY.1982.66.14 (cf. Pl. 20, fig. 3). *c*, IGSE 13883 (cf. Pl. 20, fig. 4). *d*, RSM.GY.1975.11.3 (cf. Pl. 20, fig. 6). *e*, RSM.GY.1975.11.1 (cf. Pl. 20, fig. 5). Abbreviations: ant 1—first antenna, ant 2—second antenna, plp—pleopod, sc—scaphocerite, thp—thoracopod, uro—uropod.

EXPLANATION OF PLATE 20

Fig. 1. *Waterstonella grantonensis* Schram, 1979. GrI 47498, lateral aspect, figured by Tait (1925), probably from horizon ii, Band 3, × 3·75.

Figs. 2–6. *Anthracophausia dunsiana* Peach, 1908. 2, RSM.GY.1982.66.13, lateral aspect, × 1·35 (cf. text-fig. 3*a*). 3, RSM.GY.1982.66.14, part of first antennae, second antennae with scaphocerite, × 1·35 (cf. text-fig. 3*b*). 4, IGSE 13883, scaphocerite with base of second antenna, first antenna lying transversely, × 4 (cf. text-fig. 3*c*). 5, RSM.GY.1975.11.1, tail fan and posterior abdominal somites, × 2·75 (cf. text-fig. 3*e*). 6, RSM.GY.1975.11.3, first and second antennae of large specimen, × 2·5 (cf. text-fig. 3*d*).

Figs. 7–17. Worm-like forms and branching organisms (?hydroids). 7, RSM.GY.1982.66.2, unidentified organism, × 9. 8, RSM.GY.1982.66.2a, ?hydroid, × 5·25. 9, RSM.GY.1982.66.2b, vermiform structure, possibly a nemertean, × 4. 10, RSM.GY.1982.66.2, ?hydroid, × 9. 11, RSM.GY.1982.66.1g, two overlapping vermiform structures, possibly nematodes, × 5·25. 12, RSM.GY.1982.66.22, possibly small nematode or nemertean, × 5·25. 13, RSM.GY.1982.66.17, ?hydroid, × 6. 14, RSM.GY.1982.66.20, ?hydroid with two small clumps arising from a common stem, another detached and lying to the left, × 5·25. 15, RSM.GY.1982.66.18, ?hydroid, × 5·25. 16, RSM.GY.1982.66.21, ?hydroid, × 5·25. 17, RSM.GY.1982.66.19, ?hydroid, × 5·25.

PLATE 20

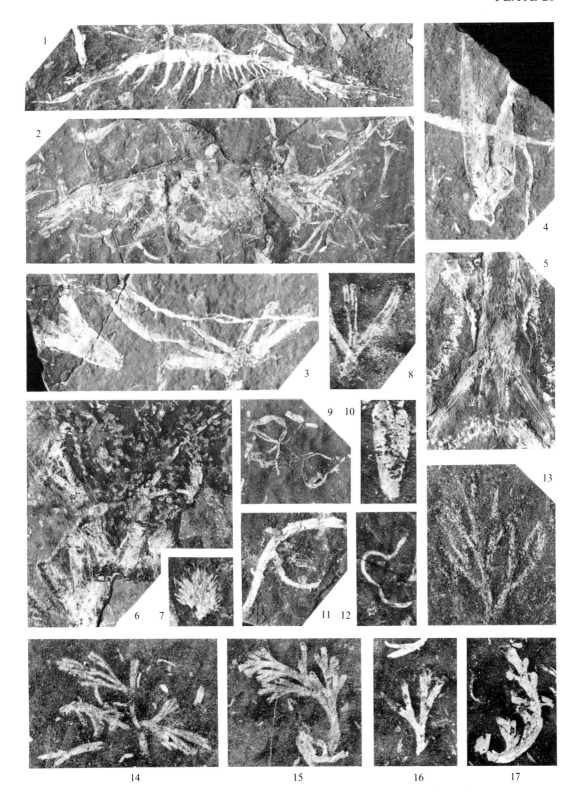

BRIGGS and CLARKSON, Granton 'Shrimp-Bed'

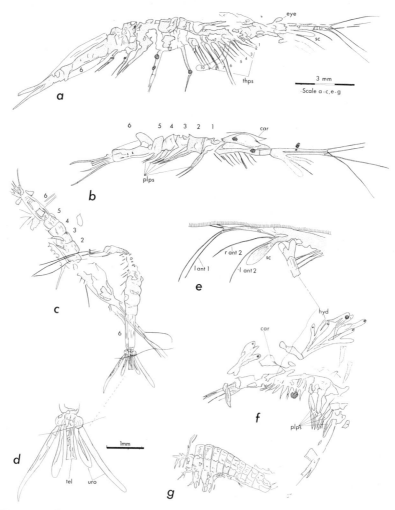

TEXT-FIG. 4a–f. *Waterstonella grantonensis* Schram, 1979. Lower Oil Shale Group, Granton. *a*, IGSE 13053, holotype, Schram 1979a, fig. 30 (cf. Pl. 21, fig. 5). *b*, IGSE 13054 (cf. Pl. 21, fig. 6). *c*, RSM.GY.1982.66.3 (cf. Pl. 21, fig. 4). *d*, tail fan of same enlarged. *e*, RSM.GY.1982.66.7 (cf. Pl. 21, fig. 9). *f*, RSM.GY.1982.66.2 (cf. Pl. 21, fig. 3). *g*, Arthropod of uncertain affinity, GrI 47498 (cf. Pl. 21, fig. 1). Abbreviations as for text-fig. 3, except: l, r—prefixes indicating left and right, car—carapace, hyd—?hydroid, tel—telson, 1, 2, 3, 4, 5, 6—abdominal somites.

EXPLANATION OF PLATE 21

Fig. 1. Arthropod of uncertain affinity, GrI 47498, × 5·25 (cf. text-fig. 4g); same slab as Plate 20, fig. 1.
Figs. 2–9. *Waterstonella grantonensis* Schram, 1979. All specimens from horizon iii, Band 3, Granton. 2, RSM. GY.1982.66.1, lateral aspect, with ?hydroid attached at rear, × 6·5. 3, RSM.GY.1982.66.2c, lateral aspect, flexed, with attached ?hydroids, × 5·5 (cf. text-fig. 4f). 4, RSM.GY.1982.66.3, two specimens, dorsoventral aspect, × 7·25 (cf. text-fig. 4c). 5, IGSE 13053, holotype (Schram 1979a, fig. 30), lateral aspect, × 4·25 (cf. text-fig. 4a). 6, IGSE 13054, lateral aspect, × 5·25 (cf. text-fig. 4b). 7, RSM.GY.1982.66.6, dorsoventral aspect, anterior part of complete specimen showing carapace (anterior to top right), × 4. 8, RSM. GY.1982.66.6, dorsoventral aspect, first and second antennae, and thorax (anterior to left), carapace absent, × 8. 9, RSM.GY.1982.66, lateral aspect, showing first and second antennae and scaphocerite, × 9·5 (cf. text-fig. 4e).

PLATE 21

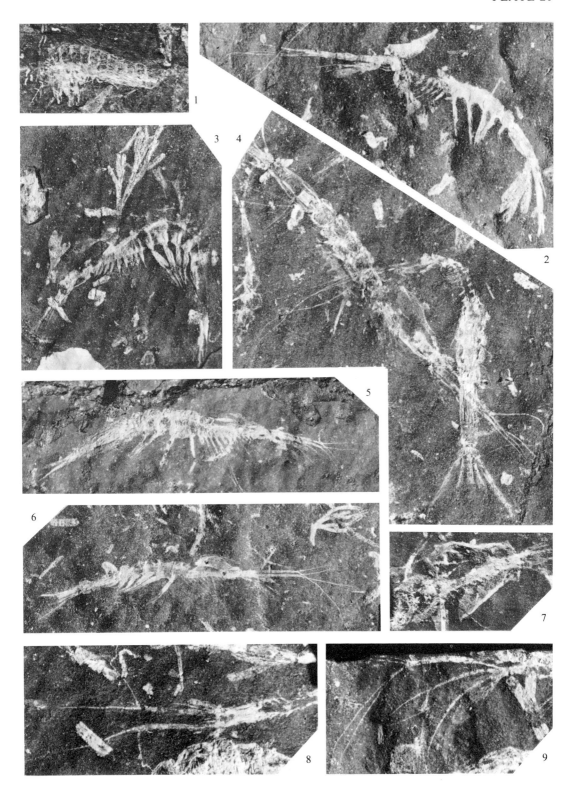

BRIGGS and CLARKSON, Carboniferous arthropods

extensive revision Schram (1979*b*) synonymized a number of additional forms with *A. dunsiana* which remains the only soundly based species referred to the genus. Apart from the uropods (Pl. 20, fig. 5), the appendages of *A. dunsiana* have been poorly known to date (Schram 1979*b*, fig. 29). Schram reported that 'despite all the preservational variants, nothing has survived that would allow the determination of the affinities of *A. dunsiana*'. An exceptionally preserved rich new fish and crustacean fauna at Bearsden, Glasgow (Wood 1982), of basal Namurian age has yielded a number of crustaceans including complete specimens of *A. dunsiana*. These show a paired spinose raptorial appendage projecting anteriorly (Wood 1982, fig. 4; cf. Pl. 20, figs. 3, 6). Rolfe (*in* Wood 1982, p. 577) identified this appendage as the second antenna although it is strikingly similar to the second thoracopod of stomatopod hoplocarids. A raptorial antenna is rare among Crustacea (it occurs in some valviferan isopods) and it is tempting to suggest, based on Wood's (1982, fig. 4) illustration of *Anthracophausia*, that the raptorial appendage is a maxilliped extending forward beneath the head. Re-examination of a fragmentary specimen of *Anthracophausia*, found in the Granton 'shrimp-bed' in 1976, clearly reveals an antennal scale (scaphocerite) attached near the base of the raptorial ramus (Pl. 20, figs. 3, 4) and thus appears to vindicate Rolfe's observation. The presence of this unusual specialization does not, however, enable us to remove *Anthracophausia* from the Eocaridacea and assign it more satisfactorily to another taxon. Further interpretation of this crustacean must be delayed until detailed work on the Bearsden crustaceans is completed.

A single specimen found at Granton appears to represent a fragment of an additional different arthropod (Pl. 21, fig. 1), but it is too incomplete to allow a determination of its nature or affinities. The specimen consists of ten short divisions of similar length, the series gradually widening in one direction. A bundle of filamentous structures at the wider end may also belong to this arthropod.

The commonest crustacean *W. grantonensis* Schram 1979 (text-fig. 6), which is redescribed below, is normally found complete and about 95% of these complete individuals are laterally compacted. This presumably reflects a laterally flattened cross-section in life. Complete dorsoventrally compacted individuals (Pl. 21, fig. 4) are very rare. On some surfaces disarticulated specimens are common. The posterior abdominal somites and tail fan are quite often preserved in isolation and are usually dorsoventrally compacted (Pl. 22, fig. 5). More rarely the carapace and cephalothorax are found isolated, sometimes in association with the tail fan. Such disarticulated fragments usually appear to be thinner and less well preserved than complete specimens; it is probable that they represent exuviae. Conversely, the greater thickness of fluorapatite in complete specimens and the occasional preservation of probable gut traces (Pl. 22, figs. 1, 2) suggest that they are carcasses. The very lightly skeletonized appendages tend to be compacted into a single plane so that it is difficult to distinguish between those of opposing sides. The long and finely preserved antennae, for example, may be superimposed to such effect that they give the spurious impression of a single antenna with supernumerary flagella (Pl. 21, fig. 9).

SYSTEMATIC PALAEONTOLOGY

Class MALACOSTRACA Latreille, 1806
Subclass EUMALACOSTRACA Grobben, 1892
Order WATERSTONELLIDEA Schram, 1981
Family WATERSTONELLIDAE Schram, 1979
Genus WATERSTONELLA Schram, 1979

Type and only species. Waterstonella grantonensis Schram, 1979.

Waterstonella grantonensis Schram, 1979

Plate 20, fig. 1; Plate 21, figs. 2–9; Plate 22, figs. 1–5; text-figs. 4–6

1925   Unnamed shrimp; Tait, p. 132, pl. 17.
1979*a* *Waterstonella grantonensis*, Schram, p. 72, figs. 30, 31.

Holotype. IGSE 13053, Plate 21, fig. 5, original of Schram 1979*b*, fig. 30.

Other material. Extensive collections in the Institute of Geological Sciences, the Royal Scottish Museum and the Grant Institute of Geology, Edinburgh, and the Museum of Comparative Zoology, Harvard University.

Description. Body very elongate in outline; long second antennae and narrow uropods which project at either end give the specimens a wispy appearance.

Head with two pairs of antennae. First antenna very long and consists of a slender peduncle, tapering distally, which bears two long flagella. Evidence for number of peduncle segments is equivocal—there are at least two

(Pl. 21, fig. 2) and probably three (the number characteristic of many eumalacostracans). There are no traces of segmentation in the flagella. Specimens in lateral aspect usually show both flagella, one directed anterodorsally, the other anteroventrally (Pl. 21, figs. 5, 6, 9), suggesting that they were attached to the peduncle one above the other. First antennae commonly superimposed in lateral aspect; peduncles may overlap almost exactly, giving the erroneous impression of a single antenna with four flagella (Pl. 21, fig. 9). Second antenna projects ventral of first and is much shorter (Pl. 21, figs. 2, 9). Sub-oval antennal scale about half the length of flagellum which extends dorsally beyond it (but generally not beyond peduncle of first antenna). At least three short segments proximal of flagellum but they are rarely evident; number of segments at base of scale unknown (Pl. 21, fig. 9).

Poorly preserved traces on a number of specimens may represent pedunculate eyes arising near base of antennae and projecting anterolaterally (Pl. 21, fig. 5; Pl. 22, fig. 3). No details are evident. Although anterior head appendages (which extend beyond carapace) are clearly evident in most specimens, posterior head appendages are usually obscured, and the distinction between them and thoracopods is not clear. There is no evidence of the mandible.

Carapace thin, rather insubstantial, attached anteriorly, and extends posteriorly to cover head and most of thorax (Pl. 21, figs. 2, 6, 7). In lateral aspect carapace subtriangular in outline, dorsal margin curving slightly posterodorsally (Pl. 21, fig. 2), carapace expanding posteroventrally to cover base of thoracopods. Small anterolateral projection evident in some specimens preserved in parallel aspect (Pl. 21, fig. 7) is presumably the result of compaction and folding through a slight anteroventral expansion of carapace (text-fig. 6). Pronounced posterior indentation, which exposes posterior thoracic somites dorsally, is apparent in lateral (Pl. 21, fig. 6) and dorsoventrally compacted aspect (Pl. 21, fig. 7; Pl. 22, fig. 3). Carapace not evident in many specimens (Pl. 21, figs. 4, 8), and may have been less substantial than the rest of the exoskeleton.

Thorax very short (about 0·25 length of body excluding antennae) and somites not clearly preserved (Pl. 22, figs. 1, 2). Appendages are difficult to use as an indication of number of thoracic somites. Distinction between posterior cephalic and anterior thoracic limbs not clear. In addition it is difficult to distinguish between left and right appendages as they may be superimposed in lateral aspect. A series of nine similar appendages posterior of antennae is evident, however, on RSM.GY.1982.66.55 (Pl. 22, fig. 2). First and second pleopods clearly preserved in this specimen, left and right limbs side by side in contact and apparently in the same plane. Proximal parts of pleopods almost overlap. Only one member of each pair of thoracopods is preserved; their very regular spacing appears to negate the possibility that any two could be the opposite members of a pair. A number of specimens including the holotype (Pl. 21, fig. 5) and RSM.GY.1982.66.7 preserve evidence of ten appendages apparently attached on same side of thorax. The absence of limbs preserved between this series of ten and antennae suggest that the anterior two may represent maxillae which are similar, in general structure at least, to the thoracopods (a suggestion also made by Raymond in his unpublished manuscript). Appendages of thorax decrease slightly in length anteriorly and are generally directed anteroventrally. Individually they consist of a basal peduncle/protopod with an unknown number of segments which bears two slender rami, one anterior of the other (Pl. 22, figs. 1, 2). Detailed structure of these rami unknown.

Abdominal somites much longer than those of thorax, roughly square in outline in both lateral (Pl. 20, fig. 1; Pl. 21, fig. 3) and parallel (Pl. 21, fig. 4) aspect. Cuticle of overlying tergites, which extend ventrally to overlap base of pleopods, rarely preserved. Individual tergites taper slightly posteriorly (Pl. 20, fig. 1); the series decreases in height both anteriorly and posteriorly from third abdominal somite. Epimera evident in both lateral (Pl. 20, fig. 1; Pl. 21, fig. 3) and parallel (Pl. 21, fig. 4) aspect. The first five somites of approximately equal length; sixth considerably longer.

Pleopods, like thoracopods, generally poorly preserved. They are attached to posterior part of somites (Pl. 21, fig. 6; Pl. 22, figs. 1, 2) and are usually preserved directed posteroventrally (as opposed to anteroventrally as are thoracopods) but some variability is evident, even within specimens. Pleopods consist of a long stout protopod (possibly a single segment) which bears two slender rami which are preserved curved posteriorly in some specimens and were probably flexible. Detailed structure of these rami not preserved.

A wide band of fluorapatite running the length of the thorax and/or abdomen of some specimens (Pl. 22, figs. 1, 2) probably represents the course of the gut. The mineral infill coalesces in some cases with that in the proximal part of the appendages, including antennae.

Sixth abdominal somite bears a telson and biramous uropods which together make a tail fan, the form of which is only evident in specimens compacted in parallel aspect (Pl. 21, fig. 4; Pl. 22, figs. 4, 5). Each uropod appears to consist of a basal segment which bears an elongate outer flap projecting posterolaterally and a shorter inner flap which projects more directly posteriorly. The uropods flank the telson which is elongate, sub-rectangular in outline, and widens slightly posteriorly (Pl. 21, fig. 4; Pl. 22, fig. 4). Position of the anus has not been observed.

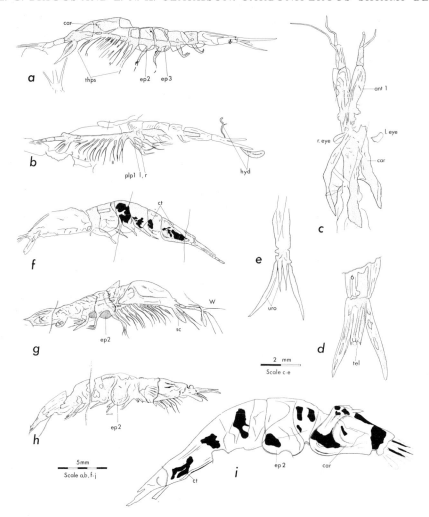

TEXT-FIG. 5a–e. *Waterstonella grantonensis* Schram, 1979. Lower Oil Shale Group, Granton. *a*, RSM. GY.1980.48.1 (cf. Pl. 22, fig. 1). *b*, RSM.GY.1982.66.5 (cf. Pl. 22, fig. 2). *c*, RSM.1980.48.1 (cf. Pl. 22, fig. 3). *d*, tail fan of same (cf. Pl. 22, fig. 4). *e*, RSM.GY.1982.66.5 (cf. Pl. 22, fig. 5). *f–i, Crangopsis* sp. *f*, IGSE 13884 (cf. Pl. 22, fig. 6). *g*, GrI 43753 (48) (cf. Pl. 22, fig. 7). *h*, GrI 43753 (48) (cf. Pl. 22, fig. 8). *i, Crangopsis socialis* Salter, 1861, IGSE 13885 (cf. Pl. 22, fig. 9). Abbreviations as for text-fig. 4 except: ct—original cuticle (black), ep—epimeron of numbered abdominal somite.

EXPLANATION OF PLATE 22

Figs. 1–5. *Waterstonella grantonensis* Schram, 1979. All specimens from horizon ii, Band 3, Granton. 1, RSM. GY.1980.48.1, × 4 (cf. text-fig. 5*a*). 2, RSM.GY.1982.66.5, lateral aspect, ?hydroids attached posteriorly, × 4·25 (cf. text-fig. 5*b*). 3, RSM.GY.1980.48.1, dorsoventral aspect, anterior end, × 6·5 (cf. text-fig. 5*c*). 4, same specimen, tail fan, × 8 (cf. text-fig. 5*d*). 5, RSM.GY.1982.66.5, tail fan, ventral view, telson absent, × 12 (cf. text-fig. 5*c*).

Figs. 6–8. *Crangopsis* sp. Granton, probably horizon i or ii, Band 3, all lateral aspect showing large second pleomere epimeron. 6, GrI 43753 (48a), × 4·75 (cf. text-fig. 5*f*). 7, IGSE 13884, × 5·5 (cf. text-fig. 5*g*). 8, GrI 43753 (48b), × 4·75 (cf. text-fig. 5*h*).

Fig. 9. *Crangopsis socialis* Salter, 1861, IGSE 13885, probably horizon i or ii, Band 3, lateral aspect, × 2·5 (cf. text-fig. 5*i*).

PLATE 22

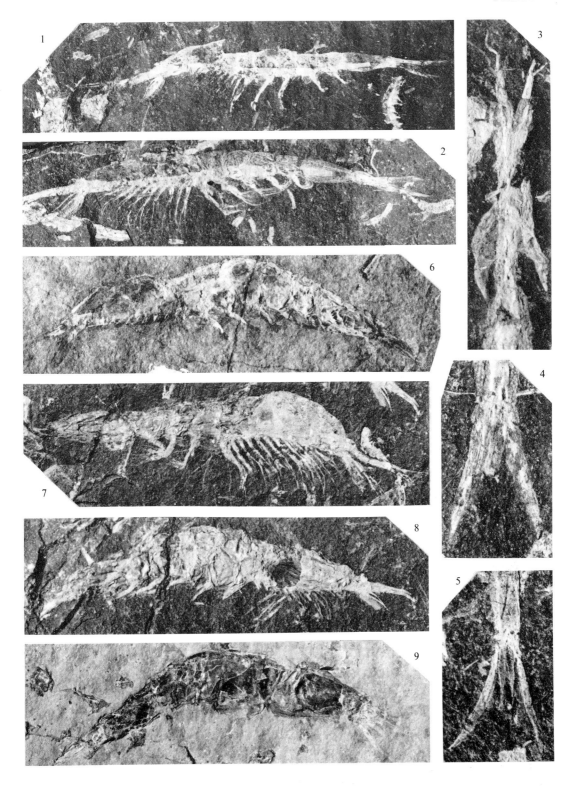

BRIGGS and CLARKSON, Carboniferous arthropods

Specimens of *Waterstonella* from horizons i, ii, and iv (text-fig. 1) are similar in size, those from horizon iii are smaller. Sagittal length (exclusive of the antennae) was measured for about eighty specimens from both horizon ii and iii (horizon ii, range 5–22 mm, mean *c*. 18 mm; horizon iii, range 2–18 mm, mean *c*. 13·5 mm).

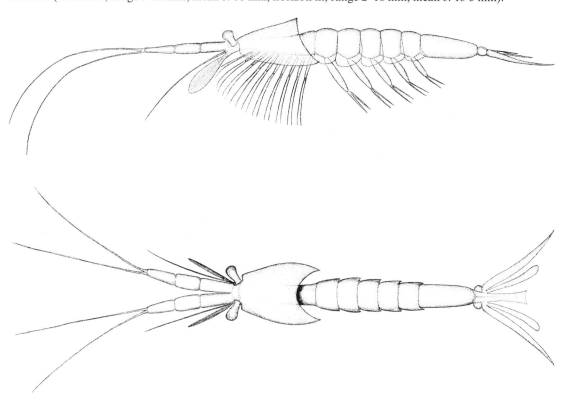

TEXT-FIG. 6. *Waterstonella grantonensis* Schram, 1979. Reconstruction in lateral and dorsal view, × 5.

Remarks. The description above, as well as amplifying that of Schram (1979*a*), differs from it in a number of details (text-fig. 6). The first antenna is shown to be much longer than reconstructed by Schram (1979*b*, fig. 31). The presence of ten pairs of similar appendages beneath the carapace, the first two of which may be maxillae, is demonstrated. In this respect *Waterstonella* may differ from other Eumalacostraca in which the maxillae are characteristically small and specialized. Schram reconstructed only eight pairs and considered them to 'carry long, terminal setae' for which we can find no evidence. We have also failed to observe the distal setae which Schram reconstructed on both the pleopods and uropods. Finally, we could find no evidence for the subdivision of the uropodal rami into two segments.

Schram (1979*b*, p. 72) erected a new family to accommodate *Waterstonella* within the Order Eocaridacea. In reviewing the Mazon Creek caridoid Crustacea (Schram 1974), he had previously noted in how few of the Eocaridacea the presence of the diagnostic characters could be verified and he had predicted that 'we shall end up reassigning these organisms as more information becomes available'. It has proved necessary to retain the Order Eocaridacea (Schram 1979*b*, p. 61; 1981*b*, p. 8), however, to accommodate 'unassignable schizopodous caridoids'. In a consideration of the classification of the Eumalacostraca as a whole, Schram (1981*b*) delineated sixteen morphotypes based on the possible combinations of three characters: the nature of the carapace, the presence or absence of a brood pouch, and whether the thoracopods are uniramous or biramous (i.e. schizopodous). *Waterstonella* is the sole representative of a distinct group within this matrix (unfused carapace, lack of a brood pouch, schizopodous) and on this basis Schram removed it from the

Eocaridacea and assigned it to a separate order Waterstonellidea. Schram considered this new order to be most closely related to the Mysidacea, which shares the schizopodous thoracopods and unfused carapace. While retaining Schram's order Waterstonellidea we would emphasize, as he did (1979*b*, p. 74), that the character separating it from the Mysidacea, the lack of a brood pouch, is impossible to verify in specimens of the size and type of preservation in question.

DISCUSSION

The algal stromatolites of the Granton 'shrimp-bed' are clearly preserved *in situ*. Many nearshore facies in the Lower Carboniferous include well-developed algal stromatolites with associated dolomitization and evidence of desiccation (Leeder 1975, 1978). The algae were tolerant of hyper-salinity and often form the only colonizers evident in Lower Carboniferous intertidal or shallow subtidal facies (Leeder 1978, p. 151, tidal-flat community A). A small number of euryhaline organisms including bivalves and serpulids is sometimes also present. A 'restricted' fauna of similarly low diversity is associated with modern stromatolites (Hardie 1979). The preservation of a crustacean-dominated fauna in association with the Granton stromatolites is therefore unusual. If the stromatolites were intertidal, as seems likely, the concentrations of crustaceans in particular areas of the fossiliferous layers may represent dried-out pools, as Tait (1925) first suggested, or alterna-tively the crustaceans were stranded on the surface of the stromatolite and subsequently covered by sediment and algal film. The abundance and delicate nature of the crustaceans and ?hydroids suggests that they inhabited intertidal or subtidal environments associated with the stromatolites and were not transported far. *Crangopsis* and *Waterstonella* were probably filter-feeders, and *Anthracophausia* with its raptorial second antenna a predator. There is no evidence of the activity of an infauna. The nektonic fish, conodont animal, and nautiloids are sufficiently rare to represent occasional or even chance introductions.

The fact that *Waterstonella* occurs only at Granton may indicate that it is restricted to the particular, possibly hyper-saline, environment represented there. The other crustaceans, especially the abundant *Crangopsis*, which also occur elsewhere, would by implication be tolerant of a wider range of salinities. Alternatively the lightly skeletonized *Waterstonella*, like the soft-bodied elements of the fauna, required the exceptional conditions for preservation which prevailed at Granton. It is likely that the algal laminae which enclose the fossils inhibited decay. Despite the unusual faunal characteristics of the Granton 'shrimp-bed' a second shrimp-bed at an approximately equivalent horizon about 30 km to the east at Gullane (Peach 1908; Traquair 1907) is also preserved in a stromatolite complex. The Gullane shrimp-bed, which we are presently studying, contains abundant specimens of the crustacean *Tealliocaris* in various stages of growth, like the Granton fauna pre-served in fluorapatite, but with more of the original cuticle remaining. The assemblage, which is even less diverse than that at Granton, also includes rare fishes and ?hydroids, but not *Waterstonella*. Although *Tealliocaris* is more robust than *Waterstonella* the presence of ?hydroids in the Gullane assemblage suggests that conditions favourable for the preservation of soft-bodied and lightly skeletonized organisms prevailed there as at Granton. The low faunal diversity of both these crustacean-dominated 'shrimp-beds' and the almost total absence of normal marine invertebrates, is therefore presumably a reflection of the composition of the original stromatolite communities. The absence of eumalacostracans (as opposed to ostracods) in association with other Lower Carboniferous stromatolites from northern Britain (Leeder 1975, 1978) may indicate a difference in their precise shallow marine environment (Leeder 1975).

Four Carboniferous crustacean communities were defined by Schram (1981*a*) on the basis of the taxa represented and their feeding habits: nearshore marine, brackish water, offshore marine, and lagoonal. These communities persisted for long periods of time even though some individual taxa changed. Schram (1981*a*, p. 127) recognized the preliminary nature of his conclusions which he presented 'as a working hypothesis to be tested by future empirical data from the Late Paleozoic record as a whole'.

Schram (1981a, pp. 131, 132) considered *Tealliocaris* to belong to the brackish water community, but noted that it is occasionally found with elements more characteristic of the near-shore marine community. *Tealliocaris* does occur in facies other than algal stromatolites. It was first described from a marine band in the largely non-marine mudstones of Belhaven Bay in East Lothian (Etheridge 1877), and also occurs in grey shales of uncertain depositional environment in Tarras Water, Dumfriesshire. The recently discovered Namurian fauna at Bearsden, near Glasgow, yields *Anthracophausia*, *Crangopsis*, *Pseudotealliocaris*, and *Minicaris*?, elements of both the near-shore and brackish water communities of Schram 'indicating a broader ecological niche than has been appreciated' (Wood 1982, p. 576) or the transitional mixing of two assemblages. The discovery that the Granton 'shrimp-bed' occurs in algal stromatolites, however, shows that Schram was in error in interpreting the *Waterstonella–C. socialis* association as an offshore community. Clearly a more complete understanding of the community structure and environmental preferences of Carboniferous crustaceans will depend not only on detailed morphological studies, but also on an analysis of the sedimentary rocks in which they occur.

It is not clear how the Granton and Gullane faunas came to be preserved nor is the origin of the diagenetic fluorapatite understood. We are not aware of any comparable soft-bodied or lightly skeletonized faunas from tidal-flat stromatolites. These shrimp-beds represent an unusual example of a Konservat-Lagerstätte (Seilacher 1970).

Acknowledgements. J. G. Sharp undertook basic mapping at Granton, supported by the Weir Fund of the University of Edinburgh. We thank S. P. Wood for helpful discussion and W. D. I. Rolfe for access to his notes on *Waterstonella* and to Raymond's unpublished manuscript. J. Miller and G. Loftus gave much-appreciated help with the sedimentology, and P. D. Taylor discussed the affinities of the branching organisms. S. Conway Morris and C. D. Waterston kindly reviewed the paper. Invaluable photographic and technical assistance was provided by D. Baty and C. Chaplin. D. E. G. B.'s research was financed by Goldsmiths' College Research Fund.

REFERENCES

BRIGGS, D. E. G., CLARKSON, E. N. K. and ALDRIDGE, R. J. 1983. The conodont animal. *Lethaia*, **16**, 1–14.

BROOKS, H. K. 1969. Eocarida. *In* MOORE, R. C. (ed.). *Treatise on Invertebrate Paleontology*, Part **R**, 332–345. Geol. Soc. Am. and Univ. of Kansas Press.

DAVIES, G. R. 1970. Algal-laminated sediments, Gladstone embayment, Shark Bay, Western Australia. *Am. Assoc. Petrol. Geologists Mem.* **13**, 169–205.

DICK, J. R. F. 1981. *Diplodoselache woodi* gen. et sp. nov., an early Carboniferous shark from the Midland Valley of Scotland. *Trans. Roy. Soc. Edin. Earth Sci.* **72**, 99–113.

—— and MAISEY, J. C. 1980. The Scottish Lower Carboniferous shark *Onychoselache traquairi*. *Palaeontology*, **23**, 363–374.

ETHERIDGE, R. 1877. On the occurrence of a Macrurous Decapod (*Anthrapalaemon*? *woodwardi* sp. nov.) in the Red Sandstone, or lowest group of the Carboniferous Formation in the South-East of Scotland. *Q. Jl geol. Soc. Lond.* **33**, 863–878.

FOSTER, M. W. 1979. Soft-bodied Coelenterates in the Pennsylvanian of Illinois. *In* NITECKI, M. H. (ed.). *Mazon Creek Fossils*, 191–268. Academic Press, New York, San Francisco, London.

FRYKMAN, P. 1979. Ordovician chitinous hydroids from Peary Land, eastern North Greenland. *Rapp. Grønlands geol. Unders.* no. 19, 25–27.

GEBELEIN, C. D. 1969. Distribution, morphology and accretion rate of Recent subtidal algal stromatolites, Bermuda. *J. Sediment. Petrol.* **39**, 46–69.

GEORGE, T. N., JOHNSON, G. A. L., MITCHELL, M., PRENTICE, I. E., RAMSBOTTOM, W. H. C., SEVASTOPULO, G. D. and WILSON, R. S. 1976. A correlation of Dinantian rocks in the British Isles. *Geol. Soc. Lond. Special Report*, no. 7, 1–87.

GILL, D. 1977. Saline A-1 Sabkha cycles and the late Silurian palaeogeography of the Michigan Basin. *J. Sediment. Petrol.* **47**, 979–1017.

GREENSMITH, T. 1962. Rhythmic deposition in the Carboniferous Oil-Shale Group of Scotland. *J. Geol.* **70**, 355–364.

HARDIE, L. A. (ed.) 1979. Sedimentation on the Modern Carbonate Tidal Flats of Northwest Andros Island, Bahamas. *Johns Hopkins Univ. Studies in Geology*, no. 22, 1–202. Johns Hopkins Univ. Press, Baltimore and London.

KOZLOWSKI, R. 1959. Les Hydroides ordoviciens à squelette chitineux. *Acta Palaeont. Polonica*, **4**, 209–271.

LEEDER, M. 1975. Lower Border Group (Tournaisian) stromatolites from the Northumberland basin. *Scott. J. Geol.* **11**, 207–226.

—— 1978. Carboniferous-Tidal-flat communities A and B. *In* MCKERROW, W. S. (ed.). *The Ecology of Fossils*, 151–152. Duckworth, London.

MAYALL, M. J. and WRIGHT, V. P. 1981. Algal tuft structures in stromatolites from the Upper Triassic of south-west England. *Palaeontology*, **24**, 655–660.

MITCHELL, C. H. and MYKURA, W. 1962. Geology of the neighbourhood of Edinburgh. *Mem. geol. Surv. Scotl.* 159 pp.

MUNDY, S. P. 1980. A key to the British and European freshwater bryozoans. *Freshwater Biol. Assn Sci.* Publn no. 41, 32 pp.

—— TAYLOR, P. D. and THORPE, J. P. 1981. A reinterpretation of phylactolaemate phylogeny. *In* LARWOOD, G. P. and NIELSEN, C. (eds.). *Recent and Fossil Bryozoa*, 185–190. Olsen Olsen, Fredensborg, Denmark.

NITECKI, M. and RICHARDSON, E. S. 1972. A new Hydrozoan from the Pennsylvanian of Illinois. *Fieldiana: Geol.* **30**, 1–7.

—— and SCHRAM, F. R. 1975. Hydra from the Illinois Pennsylvanian. *J. Paleont.* **49**, 549–551.

PEACH, B. N. 1908. Monograph on the Higher Crustacea of the Carboniferous rocks of Scotland. *Mem. Geol. Surv. G.B. Palaeontology*, 82 pp.

SCHRAM, F. R. 1973. Pseudocoelomates and a nemertine from the Illinois Pennsylvanian. *J. Paleont.* **47**, 985–989.

—— 1974. Mazon Creek caridoid crustacea. *Fieldiana: Geol.* **30**, 9–65.

—— 1979a. Worms of the Mississippian Bear Gulch Limestone of Central Montana, U.S.A. *Trans. San Diego Soc. Nat. Hist.* **19**, 107–120.

—— 1979b. British Carboniferous Malacostraca. *Fieldiana: Geol.* **40**, 1–129.

—— 1981a. Late Palaeozoic crustacean communities. *J. Paleont.* **55**, 126–137.

—— 1981b. On the classification of Eumalacostraca. *J. Crustacean Biol.* **1**, 1–10.

SEILACHER, A. 1970. Begriff und Bedeutung der Fossil-Lagerstätten. *N. Jb. Geol. Paläont. Mh.* **121**, 34–39.

SKEVINGTON, D. 1965. Chitinous Hydroids from the Ontikan Limestones (Ordovician) of Oland, Sweden. *Geol. Fören. Fört. Stockholm*, **87**, 152–162.

TAIT, D. 1925. Notice of a shrimp-bearing limestone in the Calciferous Sandstone Series at Granton, near Edinburgh. *Trans. Edin. Geol. Soc.* **11**, 131–135.

TRAQUAIR, R. H. 1907. Report on Fossil Fishes collected by the Geological Survey of Scotland from Shales exposed on the shore near Gullane, East Lothian. *Trans. Roy. Soc. Edin.* **46**, 103–117.

WOOD, S. P. 1975. Recent discoveries of Carboniferous fishes in Edinburgh. *Scott. J. Geol.* **11**, 251–258.

—— 1982. New Basal Namurian (Upper Carboniferous) fishes and crustaceans found near Glasgow. *Nature*, **297**, 574–577.

D. E. G. BRIGGS

Department of Geology
Goldsmiths' College
University of London
Creek Road
London SE8 3BU

E. N. K. CLARKSON

Grant Institute of Geology
University of Edinburgh
West Mains Road
Edinburgh EH9 3TW

CAMBRIAN-ORDOVICIAN TRILOBITES FROM THE BOUNDARY BEDS IN WESTERN NEWFOUNDLAND AND THEIR PHYLOGENETIC SIGNIFICANCE

by R. A. FORTEY

ABSTRACT. The Cambrian–Ordovician boundary may have been less drastic as an extinction event than has been supposed. Long-ranging trilobites continued with little change at shelf-edge sites, and acted as the source of subsequent platform faunas. The Cambrian–Ordovician boundary acted as a selective filter of late Cambrian taxa.

Seventeen species of trilobites are described from the critical Cambrian–Ordovician boundary sections at Broom Point, western Newfoundland; these are additional to those in Fortey *et al.* (1982). *Symphysurina* is not an asaphid, but may be an effaced lecanopygid or nileid; the latter is favoured. Isocolidae are related to the Cambrian family Catillicephalidae. *Missisquoia* Shaw is a subjective synonym of *Parakoldinioidia* Endo, and the new name *P. stitti* is proposed for the zonal form hitherto known as *Missisquoia typicalis* Shaw, 1951. The following new taxa are named: *Acheilus monile* sp. nov., *A. oryx* sp. nov., *Calculites* gen. nov., *C. stevensi* sp. nov., *Hystricurus paucituberculatus* sp. nov., *Lunacrania spicata* sp. nov., and *Symphysurina brevis* sp. nov.

PROFESSOR WHITTINGTON, together with his co-worker C. H. Kindle, provided the first detailed stratigraphic account of the Cow Head Group in western Newfoundland (Kindle and Whittington 1958, 1959; Whittington 1968; Kindle 1981, 1982). They correctly interpreted the boulder beds, so spectacularly displayed on the Cow Head Peninsula, as slide deposits derived from an adjacent shelf, and recognized that the fossils contained within the blocks were, by and large, of the same age within a single bed. All this was done at a time when transport along the Northern Peninsula was much more difficult than it is now. Their work remains the basis for all subsequent research, whether on the sedimentology of this remarkable group of rocks or on its fossil content. The Cow Head Group is now known to be one of the best-preserved continent edge sequences in the Palaeozoic, shifted on to the platform as part of the 'taconic' allochthon, and protected thereby from excessive tectonism. It spans a mid Cambrian to early Middle Ordovician interval.

One small part of this great time interval has recently become the focus for intensive research: the Cambrian–Ordovician boundary. Using Kindle and Whittington's work to locate the critical interval, sections suitable for bed-by-bed collecting were identified, the best of which proved to be at Broom Point, one of their localities. The sections were of particular importance because, almost alone at this horizon on the North American plate, they include trilobites, graptolites, and conodonts in the same site, enabling refinements in the correlation of the boundary interval between Europe and North America. A detailed account of the stratigraphy and faunal distribution has now been prepared (Fortey and Skevington 1980; Fortey *et al.* 1982). The base of the Tremadoc Series (*sensu* Rushton 1982) does *not* correlate with the base of the 'Canadian' series in North America, but lies some way up, most likely within the early *Symphysurina* Zone. In the 1982 paper I was able to give only the briefest account of the trilobites, and confined attention to those species which occur *in situ* rather than in boulders, together with those from boulders which could be compared with previously described forms. There remained a number of species which was either new, puzzling, or required more discussion than I was able to give in the stratigraphic account. These form the subject of this paper. They have some general importance apart from their relevance to the particular correlation problems, because this time interval was a critical one for the phylogeny of the trilobites. Some

[Special Papers in Palaeontology No. 30, pp. 179–211, pls. 23–27]

suggestions are made indicating that the taxonomic break between Cambrian and Ordovician trilobite families may not be as complete as has been supposed.

AGE RANGE OF THE TRILOBITES

The locality map for the fossils was given in Fortey and Skevington (1980, fig. 1; Kindle 1982, figs. 1, 2) and the detailed stratigraphic columns in Fortey *et al.* (1982, text-figs. 1, 2). The fossils described in this paper come from two horizons (Table 1).

1. *Symphysurina* conglomerates. This conglomerate horizon (B1) forms a prominent feature on sections at Broom Point North and South. In both places it lies above the appearance of the '*Dictyonema*' *flabelliforme* plexus in shales below and was emplaced, therefore, in indubitably early Ordovician (Tremadoc) times, taking the base of the system at the equivalent of the base of the Tremadoc Series. The fossils figured in this paper come from a large boulder at Broom Point South, the same boulder which has yielded *S.* cf. *spicata*, *Hystricurus millardensis*, *S. brevispicata*, and *Neoagnostus* described in Fortey *et al.* (1982). This assemblage proves that the trilobites from the same boulder described herein are early Ordovician, from the *Symphysurina* Zone. Because they include forms with a decidedly 'Cambrian' appearance this is an important point to establish.

TABLE I. Summary of the faunas obtained from boulders in conglomerates at Broom Point, western Newfoundland, spanning the basal Tremadocian, described in Fortey *et al.* (1982) and herein.

Symphysurina Zone conglomerate (B1) and equivalent at Broom Point North	Largest boulder in B1 (boulder 1)	*Symphysurina* cf. *spicata* *Symphysurina brevispicata* *Symphysurina brevis* sp. nov. *Hystricurus millardensis* *H. paucituberculatus* sp. nov. *Pseudagnostus* sp. nov. *Neoagnostus aspidoides* *Hyperbolochilus*? sp. nov. *Calculites stevensi* sp. nov.
	Olenid-bearing boulder in B1	*Parabolinella* sp. cf. *P. hecuba* *Plicatolina* sp. indet. *Symphysurina goniura*
	Second large boulder in B1	*Symphysurina* sp. nov. *Hystricurus paucituberculatus* sp. nov. *Parakoldinioidia* sp. A
	Large boulder at Broom Point North	*Hystricurus paucituberculatus* sp. nov. *Pseudohystricurus* sp. aff. *rotundus* *Neoagnostus aspidoides* *Symphysurina brevispicata*
Conglomerates immediately under the *Radiograptus* fauna: B2 at Broom Point South, and conglomerates at base of section, Broom Point North	*Acheilus monile* fauna, 3·5 m from base of section, Broom Point North ('Saukia Zone boulder' of Fortey *et al.* 1982)	*Plethometopus dubius* *Acheilus*? *falcatus* *A. monile* sp. nov. *A. oryx* sp. nov. *A.* sp. indet. *Raymondina* cf. *immarginata* *Calymenidius* sp. nov. *Platydiamesus* sp. indet. *Heterocaryon* sp. nov. aff. *tuberculatum*
	Acheilus monile fauna, boulder from base of section Broom Point North	*Acheilus monile* sp. nov. *Bynumina* sp. aff. *B. laevis*
	Probably *A. monile* fauna, boulder in B2	*Acheilus oryx* sp. nov. *Plethometopus dubius*
	Boulder with *Parakoldinioidia stitti* (= *Missisquoia typicalis*), Broom Point South, B2 (figured in Fortey *et al.* 1982)	*Parakoldinioidia stitti* *Symphysurina* cf. *cleora* *Highgatella cordilleri*
	Boulders with *Symphysurina*, basal conglomerate, Broom Point North	*Symphysurina* cf. *cleora* *Loganellus* sp. indet.

2. Boulders below *Radiograptus* fauna. Underlying '*Dictyonema*' of the *flabelliforme* group on both Broom Point sections there are several metres of bedded limestones and shales with the distinctive discoidal graptolite *Radiograptus*. *In situ* trilobites occur in the same interval. Conodonts described by Landing (1982) indicate 'high *C. proavus* Zone or lower Conodont Fauna B', and the interval is likely to correlate with the early *Symphysurina* Zone (or possibly the upper part of the *Missisquoia typicalis* subzone). Whether this is ultimately regarded as Cambrian or Ordovician depends on where the boundary is chosen, but the interval is undoubtedly early 'Canadian'—post-Cambrian in the usual usage of North American workers. Beneath this interval again there is another lenticular conglomerate (B2) on Broom Point South, and a series of smaller conglomerates on Broom Point North, forming the base of the section in Fortey *et al.* (1982). Blocks from these conglomerates have yielded the other trilobites described in this paper.

The age (or ages) of the faunas in the blocks from these conglomerates pose something of a problem. There are two distinct types of boulders present:

(*a*) Boulders with *S.* cf. *cleora* (Walcott) and *Loganellus* sp.; one has *M. typicalis*, *Highgatella cordilleri*, etc., associated, and was the source of the specimens figured in Fortey *et al.* from B2. This fauna is probably equivalent to (?upper) *Missisquoia* Zone.

(*b*) *Acheilus monile* fauna. This includes boulders with what was loosely termed '*Hungaia magnifica* faunules' in Fortey *et al.* 1982, text-fig. 2. These contain a rich assemblage of genera of Trempealeauan aspect: *Acheilus*, *Heterocaryon*, *Raymondina*, *Plethometopus*, *Stenopilus*, *Leiocoryphe*, etc., several new species being described in this paper. This is not the typical *Hungaia magnifica* fauna from boulders in Quebec (Rasetti 1944, 1945), because when examined in detail most of the species are different, although the generic assemblage is closely similar. This is likely to be because of the similarity in facies of the shelf-edge faunas throughout the latest Cambrian, with its preponderance of catillicephalids and effaced species. Furthermore, the common species from these boulders— *Acheilus monile* sp. nov. for example—are apparently not present in earlier conglomerate beds in the Cow Head Group, which include faunas more like the typical *H. magnifica* assemblages (Kindle 1981, 1982).

There are two possibilities for the age of this fauna. Either the boulders are contemporary with the *Symphysurina*-bearing ones in the same conglomerate, i.e. earliest 'Canadian', or they are derived from an upper or uppermost Trempealeauan horizon. From the generic composition alone the latter is most likely. Chippings from the boulders have not yielded conodonts. However, Dr. J. F. Miller has sampled for conodonts in some beds exposed immediately *below* the basal conglomerate in the section in Fortey *et al.* (1982, text-fig. 2) and reports (written comm. 1982) that conodonts indicating 'the upper half of the *Proconodontus* Zone' occur there. Hence the situation seems comparable with that at Broom Point South. There is a hiatus between Trempealeauan and 'Canadian' representing part (at least) of the *Missisquoia* (= *Parakoldinioidia*) Zone and the *Corbina apopsis* Subzone of the *Saukia* Zone. The conglomerates at the base of the Broom Point North section are the likely equivalent of the B2 conglomerate at Broom Point South on the seaward side of what Kindle (1982) terms 'Dictyonema Hill'. This is supported by the occurrence in B2 of a boulder containing *Acheilus oryx* sp. nov., which also occurs at Broom Point North in the *A. monile* fauna at the base of the section there. The boulders containing *A. monile* are likely to have been recruited from shelf-edge deposits at the very top of the Trempealeauan, although there is no way yet of eliminating the possibility of an earliest *Missisquoia* Zone age, other than generic composition.

Cambrian-Ordovician boundary. No international agreement has been reached on the level to draw this boundary. While it is still *sub judice* it is necessary to be careful about terminology. As far as western Newfoundland is concerned the base of the Tremadoc lies between the lower and upper faunas described here; a boundary defined at this level would relegate faunas with *Symphysurina* cf. *cleora* and *Parakoldinioidia* (= *Missisquoia*) to the Cambrian. This goes against usual North American usage, which divides the Cambrian from the Ordovician at the major faunal change at the end of the Trempealeau. In this paper I use Ordovician to refer to trilobites from the *Symphysurina* conglomerate (B1), and refer those below it either to the 'Canadian' or to the *A. monile* fauna, which is probably latest Trempealeau. 'Canadian' is used in quotes because its definition is under revision at the moment, and there seems to be some support using a new term 'Ibexian'.

PHYLOGENETIC RELATIONSHIPS OF TRILOBITES ACROSS THE CAMBRIAN-ORDOVICIAN BOUNDARY

It is a familiar fact that the Cambrian-Ordovician boundary marks a watershed in trilobite evolution. Many of the Ordovician or younger families have records extending back to the Arenig or Tremadoc, but their earlier relatives are obscure. Conversely, many Cambrian families apparently terminate below the Ordovician, and, at least as reflected in the classification, their relationship to younger families is unknown.

A number of families passes through the boundary, and apparently escapes the perturbations that affect the other trilobites. These include:

Agnostida—some genera such as *Micragnostus* and *Pseudagnostus* pass across the boundary.

Olenidae—Ludvigsen (1982) regards the appearance of *Parabolinella* as typical of Canadian strata; some genera, such as *Plicatolina* and *Bienvillia*, pass across the boundary unscathed.

Shumardiidae—see Shergold (1975), who demonstrated that *Koldinioidia* is a shumardiid, and occurs typically in pre-Canadian strata.

Asaphidae—most of the important morphological features, other than the forked hypostoma, appear to have been established before the Tremadoc.

Ceratopygidae—this group has a long late Cambrian history and continues through the Tremadoc. *Macropyge* is a related form (Owens *et al.* 1982, p. 14).

Eulominae—this group persists from the Franconian (Palmer 1968) to the Arenig.

Remopleurididae—the boundary interval was accompanied by considerable diversification in China and North America in groups allied to *Kainella* and *Apatokephalus*.

Leiostegiacea—Tremadoc and later representatives of this superfamily include both effaced (*Leiostegium*) and strongly furrowed (*Annamitella*) forms. Missisquoiids, which appear at the base of the Canadian, are probably progenetic leiostegiaceans.

Saukiidae—several late representatives of this group appear to survive the late Cambrian.

Lecanopygidae—if *Benthamaspis* is correctly assigned to this family it survived until the Arenig. See also comments on *Symphysurina* below.

Catillicephalidae—some genera appear to cross the boundary without much modification (see Fortey 1980, and herein).

It is interesting to examine why this quite long list should have survived the Cambrian-Ordovician 'event'. Some, at least, are predominantly deep-water forms, having lived in exterior sites (the first six groups listed above). If the Cambrian-Ordovician boundary represents a regressive-transgressive event one assumes that these forms would have been immune from the consequences of a regression on the platform. The Olenidae and Agnostida particularly come to mind, because the relevant genera mostly have ecological preferences which do not overlap with those of other trilobites. The same habitat preferences presumably account for the long stratigraphic ranges of these genera compared with their platform contemporaries.

The same arguments cannot be applied to the last five groups listed above (Remopleurididae, Leiostegiacea, Saukiidae, Lecanopygidae, Catillicephalidae), which are found in shelf to shelf-edge sites. Whatever the nature of the Cambrian-Ordovician boundary crisis, these groups were provided with refuges which allowed them to survive. I would suggest that such refuges included the mounds which formed near the shelf edge (often associated with the algae *Epiphyton* and *Renalcis*). During the regressive phase the platform sites were severely stressed (Ludvigsen 1982), but the shelf-edge mounds could survive in patches, and with them a range of Cambrian trilobite families. The record of such shelf-edge sites at the Trempealeauan-'Canadian' boundary is probably only in the allochthonous conglomerates, into which blocks were occasionally derived, in Quebec and Newfoundland.

Although the idea of a sub-'Canadian' regression is not invariably accepted, there seems to be a consensus that the earlier part of the 'Canadian' represents a shorewards onstep of exterior biofacies, which is most simply explained as a transgressive event. The fact that such events occur simultaneously in what were separate and dispersed continental blocks argues for a world-wide eustatic change at this time. Over North America this event is accompanied by the appearance of Olenidae and '*Geragnostus*' (actually *Micragnostus*) in cratonic sites, but these do not form the ancestral stock of subsequent Canadian diversification on the platform. If Fortey and Owens (1975) are correct in identifying *Hystricurus* and allied genera as the ancestral group from which bathyurids and dimeropygids evolved, the appearance of hystricurines early in the Canadian is of more phylogenetic significance. *Symphysurina*, *Leiostegium*, and remopleuridids are also important. The question arises whether these groups originated as part of the extracratonic faunas, or whether they were survivors from shelf-edge faunas. Hystricurids are fairly generalized ptychoparioids, and one supposes that the conventional explanation (text-fig. 1) for their appearance at the base of the Canadian is that they were derived from 'conservative' ptychoparioids purportedly present in extracratonic facies. A simpler explanation might be that they were derived from shelf-edge forms just below the 'Canadian', in the same way that catillicephalids persisted in this environment. There is very little difference between *Onchopeltis spectabilis* Rasetti, 1944 and *Hystricurus* spp. The greater width of the cranidial border of the former compared with most hystricurids can be matched by *H. paucituberculatus* sp. nov., while the glabellar furrows of *Onchopeltis* are similar to those of *H. paragenalatus* Ross and *H. millardensis* Hintze. The only difference of any importance is the flattened border on the pygidium

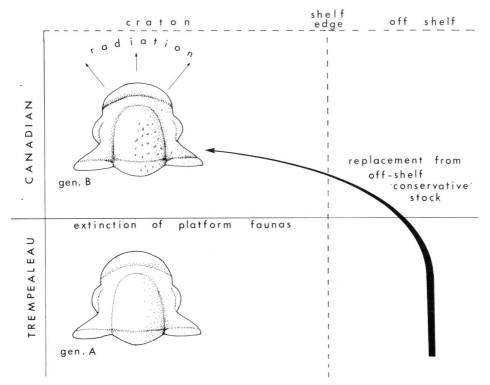

TEXT-FIG. 1. The 'major extinction' theory of trilobite faunas at the Cambrian–Ordovician boundary. Resemblances between the generalized ptychoparioid genus A below the boundary and genus B above the boundary are a matter of convergence—and separate classification is possible. 'Eventually all species of the biomere were eliminated, a new group of oceanic trilobites completed their invasion of the craton, and the whole cycle repeated itself' (Stitt 1971, p. 181).

of *Onchopeltis*. There are other similar species to *Hystricurus* in Upper Cambrian boulder faunas elsewhere; see, for example, *Meteoraspis? minuta* (Raymond) from the Rockledge conglomerate, Vermont, figured by Shaw (1952, pl. 57, fig. 9). It seems a simpler explanation to invoke persistence of a few such forms from the Cambrian in the appropriate facies, rather than to suppose that any resemblance between hystricurids and such forms is a matter of convergence, and that the Ordovician forms were 're-evolved' from a putative ancestor in the marginal facies (text-fig. 2).

Evidence given below suggests that *Symphysurina* was derived from a nileid relative, rather than a plausible candidate (*Rasettia*) from earlier shelf-edge faunas. Leiostegiids and remopleurids may have survived from the Cambrian via shelf-edge sites. It is interesting to add that certain other groups, such as isocolids, harpedids, and illaenids, make their first appearance in *in situ* or derived mound faunas later in the Tremadoc: these may have had ancestors in earlier shelf-edge faunas. Certainly, once established, some genera in these families had remarkably long ranges through one or more Systems.

Isocolids and catillicephalids. Whittington (1963) noted the resemblance between Catillicephalidae and certain isocolids, an Ordovician family of cryptogenetic origins. The discovery that catillicephalids pass beyond the Upper Cambrian into the early Ordovician (Fortey 1980, and herein), and the resemblance between such catillicephalids as *Calculites* gen. nov. and *Isocolus*, makes a phylogenetic connection between the two families likely. Whittington showed that *Isocolus* probably had a median suture, as has *Acheilus*; and there is little resemblance between other early Ordovician groups primitively with median sutures (Asaphacea, Cyclopygacea, Remopleuridacea) and either isocolids or catillicephalids. Later isocolids are secondarily blind in some genera, but early *Isocolus* have small eyes in an advanced position (e.g. *I. dysdercus* Whittington, 1963) much like

Calculites. Like catillicephalids, isocolids show much variation in pygidial structure; this may help to indicate more than one phyletic grouping within the catillicephalid–isocolid group. Finally, even within the Catillicephalidae there are some genera (*Catillicephalus*) with a rostral plate, others (*Acheilus*) with a median suture. I know of no other trilobite family with this kind of dimorphism, and it seems possible that the Catillicephalidae as at present conceived is polyphyletic.

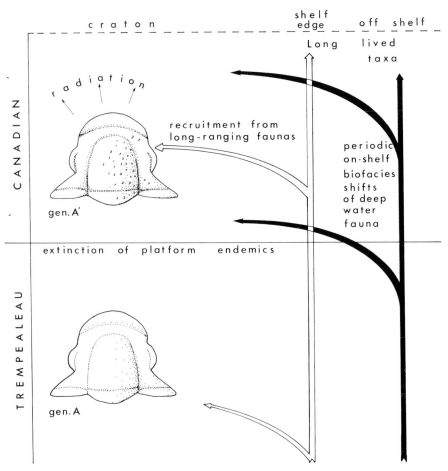

TEXT-FIG. 2. The 'partial filter' theory of trilobite extinctions at the Cambrian-Ordovician boundary favoured here. Extinctions affect platform endemics only. Recruitment from stable biofacies generates new faunas. On this model, resemblances between genera below and above the boundary (A and A') are because of common ancestry, and they should be classified together.

SYSTEMATIC PALAEONTOLOGY

The trilobites are treated family by family in the same order as in the *Treatise* (Moore 1959), and the terminology is the same as in this work, except that 'glabella' incorporates the occipital ring. Specimens are curated in the Royal Ontario Museum (ROM), British Museum (Natural History) (BM), or Smithsonian Institution, Washington (USNM). The *Acheilus monile* fauna consists of the following species, of which those marked with an asterisk are described herein: *A. monile**, *A. oryx**, *A.? falcatus* Rasetti (see Fortey *in* Fortey *et al.* 1982, pl. 2, fig. 33), *A.* sp. indet.*, *Bynumina* sp. aff. *B. laevis* Raymond*, *Calymenidius* sp. A*, *Heterocaryon* sp. nov. aff. *H. tuberculatum* Rasetti*,

Platydiamesus sp. indet.*, *Plethometopus dubius* Rasetti (Fortey *in* Fortey *et al.* 1982, pl. 2, figs. 25, 30), *Leiocoryphe* sp., *Raymondina* cf. *R. immarginata* Rasetti (Fortey *in* Fortey *et al.* 1982, pl. 2, fig. 28).

Family SOLENOPLEURIDAE Angelin, 1854
Genus HYSTRICURUS Raymond, 1913

Type species. Bathyurus conicus Billings, 1865

Hystricurus paucituberculatus sp. nov.

Plate 23, figs. 1–7

  1982   *Hystricurus* sp. nov. Fortey *in* Fortey *et al.*, p. 108, pl. 3, figs. 14, 17.
 ?1982   *Hystricurus*, sp.; Kindle, plate 1. 5, fig. 22.

Type and figured material. Holotype, cranidium with exoskeleton preserved, ROM 42292. Other cranidia: ROM 39623, 39627, 42293; free cheek: ROM 42294; pygidium: ROM 42295.

Horizon. Widely distributed in boulders from the B1 conglomerate (*Symphysurina* Zone) from both Broom Point Sections.

Diagnosis. Hystricurus with short genal spines, broad (sag.) cranidial anterior border, and lacking occipital spine. Surface sculpture of few, large, scattered tubercles, Pygidium with paired axial tubercles.

Name. From the Latin 'few-tubercled'.

Discussion. Since the original brief description of this species it has proved possible to prepare a good testate cranidium, and assign a free cheek and pygidium. Few *Hystricurus* are as completely known, and the recognition of a new species here is now justified. There is a certain amount of variation between specimens in different boulders; all the material illustrated here is from a single boulder from Broom Point South. Several good descriptions of cephalic parts of *Hystricurus* have been given (e.g. Ross 1951) and the Newfoundland species agrees in most general features, so a full description is not necessary. The following important morphological features are noted:

1. The cranidial anterior border is wide (sag.), as wide or wider than the preglabellar field as seen in dorsal view. The preglabellar field is steeply down-sloping, and in testate material is seen to bulge forwards slightly medially. This is not conspicuous on internal moulds, although the border furrow shows a tendency to become shallower in the same area.
2. Palpebral lobes are large, length (exsag.) approaching half that of pre-occipital glabella; their hind ends are well in front of the lateral parts of the occipital furrow.
3. Cephalic tuberculation is always sparse and scattered. On the free cheek there is a single row of tubercles on the genal fields. Internal moulds more weakly reflect the tuberculation, and not at all on the border, which accounts for our earlier observation that they were absent thereon, which is only true of internal moulds. Testate specimens show that these border tubercles have perforate tips (Pl. 23, fig. 3); these would have presented a marginal fringe of setae. Strictly speaking, only the latter were true tubercles; those covering the rest of the exoskeleton were pseudotubercles *sensu* Miller 1976. A few raised lines are interspersed among the tubercles on the border only.
4. The small, triangular pygidium can be confidently associated. Although only two rings are clearly defined, four pairs of tubercles along the axis reveal the minimum number of segments. The pleural fields show only two pairs of pleural furrows. Narrow border carries three or four large tubercles on each side.

The sparse, coarse tuberculation and broad cranidial border serves to distinguish *H. pauci-tuberculatus* from most other species of *Hystricurus*, including the type species. In these tuberculation is finer and denser: several lines of tubercles pass across the preglabellar field. This applies to almost all the species listed in Ross (1951, p. 40), who revised earlier work as well as introducing several new species, and other forms described and figured in Endo (1935), Hintze (1953), Gobbett (1960), Ogienko (1972), and Lu *et al.* (1976). Three species—*H. eurycephalus* Kobayashi, 1934, *H. nudus* Poulsen, 1937, and *H. politus* Ross, 1951—are smooth; none has a wide cranidial border, nor steeply sloping preglabellar field like the new species.

One of the cranidia figured previously (Fortey *et al.* 1982, pl. 3, fig. 14) is generally more transverse than the specimens figured here, and the glabella has a greater forward taper. Fragments of cuticle adhering to this specimen show that the tubercles were more subdued than on the holotype. This specimen is regarded as an extreme variant of *H. paucituberculatus*.

Very few *Hystricurus* species have pygidia assigned. Various unassigned *Hystricurus* pygidia were illustrated by Ross (1951) and Hintze (1953); all are small and transverse with up to four axial rings and narrow borders. None has the paired tubercles on the axis typical of *H. paucituberculatus*.

Family CATILLICEPHALIDAE Raymond, 1938
Genus ACHEILUS Clark, 1924

Type species. Acheilus marcoui Raymond *in* Clark, 1924, subsequently designated Raymond, 1924.

Emended diagnosis. Catillicephalids lacking anterior cranidial border. Squat glabella gently forward-expanding, with truncate anterior outline. Glabellar furrows usually defined; 1P backward-hooked at inner end. Palpebral lobes of moderate size (and smaller than *Acheilops*), curved, and medially placed. Occipital spine may be present; occipital tubercle often in forward position, adjacent to occipital furrow. Whole cranidium may become effaced anteriorly.

Nomenclatural discussion. The problems associated with this generic name have been discussed by Shaw (1952) and Rasetti (1954). The general practice since the *Treatise* has been to use *Theodenisia* Clark, 1948 (a replacement for *Denisia* Clark, 1924, preoccupied) for this group of Upper Cambrian catillicephalids. *Acheilus* Clark, 1924 has been confined to the poor holotype of *A. levisensis* Clark, 1924. The generic name was originally proposed inadvertently by Clark one month before a good description by Raymond of *A. marcoui*, a valid proposition because differences from other related genera were discussed by Clark. Rasetti (1954) states that Clark 'published *Acheilus* as a monotypic genus by describing . . . *A. levisensis*'. This is not strictly correct, because Clark also listed *A. marcoui*, and stated that it differed from *A. levisensis* in the characters: 'pustulose and has fairly strong glabellar furrows.' This should be regarded (ICZN Art. 12) as a valid proposition of *A. marcoui*. In this case Raymond's (1924) subsequent nomination of *A. marcoui* as type species also stands, a species known from well-preserved material. This would render *Theodenisia* redundant. This seems to have several advantages that outweigh the change in nomenclatural custom:

1. As thus interpreted, the type species of *Acheilus* is known from well-preserved, numerous, and relatively complete material; the type species of *Theodenisia* is a solitary cranidium.

2. *A. marcoui* is a typical member of the genus with a gently forward-expanding glabella and visible glabellar segmentation; the type species of *Theodenisia* has (Rasetti, 1954, p. 608) a subglobose glabella with poorly defined furrows.

Because the retention of the name *Acheilus* is also in accordance with the *Rules of Zoological Nomenclature*, there seems to be a good case to resuscitate it here.

EXPLANATION OF PLATE 23

Figs. 1–7. *Hystricurus paucituberculatus* sp. nov. All figures × 12. Specimens from a single boulder in conglomerate B1 at Broom Point South associated with *Symphysurina* sp. nov. and *Parakoldinioidia* sp. A. 1, 3, 4, holotype, cranidium retaining exoskeleton, in dorsal, oblique, and anterior views, ROM 42292. 2, free cheek, ROM 42294. 5, pygidium, ROM 42295. 6, 7, exoliated cranidium in lateral and dorsal views, like those figured in Fortey *et al.* (1982), ROM 42293.

Fig. 8. *Calymenidius* sp. A. Cranidium, × 12. *Acheilus monile* boulder at Broom Point North. Spec. no. ROM 42303.

PLATE 23

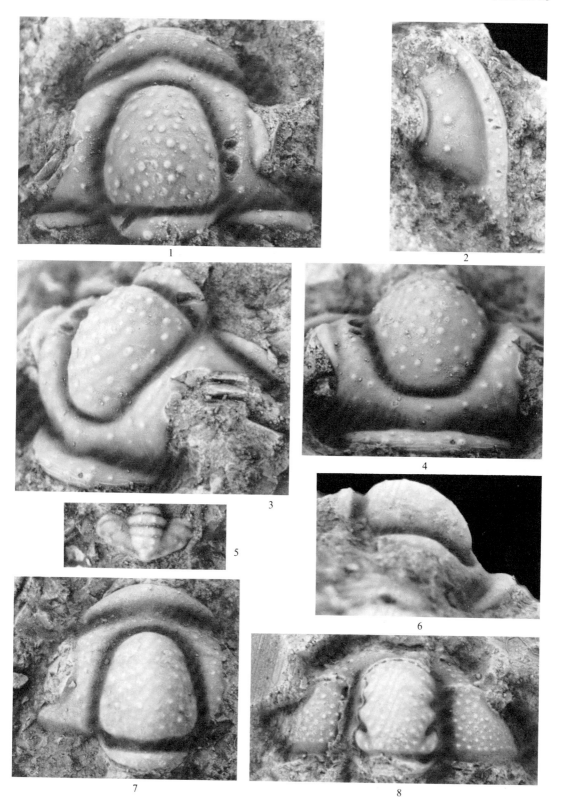

Acheilus monile sp. nov.

Plate 24, figs. 1-7, 9, 10

Type and figured material. Holotype: Cranidium, ROM 42271. Paratypes: cranidia, ROM 42272–42276; free cheeks, ROM 42277–42278, pygidium possibly associated ROM 42279.

Horizon. Common in a boulder from basal conglomerate on section at Broom Point North, and from the boulder from conglomerate 3·5 m above base of section.

Name. monile, Latin, a collar, referring to the diagnostic occipital structure.

Diagnosis. Acheilus with relatively large eyes for the genus, and surface sculpture on glabella of subdued tubercles. Glabellar furrows not deeply incised. Occipital ring runs without a break into posterior fixed cheek, thereby isolating the forward part of the glabella.

Description. Glabella transversely tumid, and downward-sloping anteriorly, expanding in width forwards to rounded anterolateral corners, but somewhat truncate at front. Maximum glabellar width at front is equal to sagittal length from occipital furrow. Glabellar furrows not incised, but always visible as smooth depressions, extending less than one-third across glabella, 1P curved quite sharply backwards, with a triangular muscle impression at its inner end. 2P gently arcuate and directed slightly forwards; 3P a faint, slit-like impression isolated within the glabella. Axial and occipital furrows deep, the latter curving forwards laterally completely isolating the preoccipital glabella. No continuation of axial furrows to define edges of occipital ring even on small cranidia. Convex occipital ring merges with postocular cheek, smooth except for prominent tubercle placed forwards adjacent to occipital furrow. Palpebral lobe with front end close to glabella, one-third to one-half length of preoccipital glabella, with strong inward bend at posterior end, midpoint only slightly in advance of outer end of 1P glabellar furrow. Palpebral rim narrow, with a few, low tubercles. Posterior border furrow very deep, adaxially sharply terminated by occipital-genal band; convex posterior border widening laterally. Preocular cheek an extremely narrow, steeply downsloping band; no anterior cranidial border. Surface sculpture on preoccipital glabella of low tubercles, densely packed, with a tendency to form low transverse ridges on frontal lobe; occipital ring and genal areas smooth.

Free cheeks narrowly triangular, extending into stout genal spine. Beneath the eye there is an elevated, convex eye socle, bounded on the outside by a deep furrow. Posterior border furrow deep, but rapidly fading towards

---

EXPLANATION OF PLATE 24

Cambrian–Ordovician Catillicephalidae, all figures × 12.

Figs. 1–7, 9, 10. *Acheilus monile* sp. nov. *Acheilus monile* boulders from conglomerates at base of section, Broom Point North. 1, 9, free cheeks; 1, dorsal view, ROM 42277; 9, anterior view of same, with second specimen to left (ROM 42278) from ventral side showing median suture. 2, 5, cranidium in dorsal and lateral view, ROM 42272. 3, 4, 7, cranidium in dorsal, lateral, and anterior views, ROM 42273. 6, incomplete cranidium, but showing well the inflated genal/occipital area, ROM 42274. 10, holotype, cranidium, ROM 42271.

Fig. 8. Dorsal view of pygidium most likely to belong to *A. monile*, ROM 42279.

Fig. 11. *Calculites depressus* (Rasetti, 1944), paratype, cranidium, BM It6535, boulder 52, Levis, Quebec, for comparison with *C. stevensi* sp. nov.

Figs. 12–15. *Acheilus oryx* sp. nov. 12, 13, 15, holotype, cranidium, ROM 42280, lateral, dorsal, and oblique anterior views. *Acheilus monile* boulder from basal conglomerate at Broom Point North. 14, cranidium, boulder from B2, Broom Point North, ROM 42281.

Figs. 16–24. *Calculites stevensi* gen. et sp. nov., largest boulder in B1 (*Symphysurina* Zone) at Broom Point South. 16, cranidium, ROM 42284. 17, 22, 24, holotype, cranidium, ROM 42283, left lateral, oblique, and dorsal views. 18, pygidium, probably belonging to this species, ROM 42288. 19, 20, pygidium, probably belonging to this species, ROM 42289, oblique lateral and dorsal views. 21, small cranidium, ROM 42286. 23, free cheek, ROM 42287.

Fig. 25. *Acheilus* sp. indet. Incomplete cranidium, ROM 42282. *Acheilus monile* fauna, Broom Point North, boulder 3·5 m from base of section.

Figs. 26, 27. Undetermined pygidium, dorsal and posterior views, the less likely candidate for the pygidium (cf. fig. 8) of *Acheilus monile* with which it is associated, ROM 42304.

PLATE 24

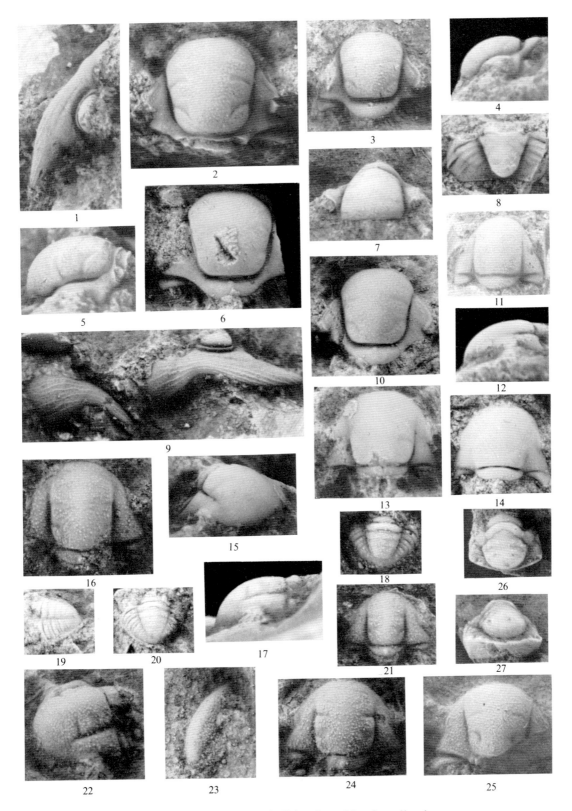

FORTEY, Catillicephalidae from Newfoundland

genal spine. There is a faint indication of the lateral border furrow, but this becomes obsolete both towards the anterior branch of the facial suture and towards the genal angle. The whole border is turned downwards in a convex roll, merging with the doublure, which carries a few, strong ridges running slightly oblique to the margin. The doublure is very broad beneath the cranidium, and terminates in a median suture; the ridges are interspersed with fine pits in this region.

Two different types of pygidium are present in the same boulders as *A. monile* cephalic fragments (Pl. 24, figs. 8, 26, 27). One is highly distinctive, with a short stubby axis, with only one ring well-defined, having curious lateral lobes, and narrow pleural areas developed as almost horizontal platforms and sharply truncated at the lateral edges; the posterior border is vertical, carries fine terrace lines, and has a strong median arch. The other is transverse (Pl. 24, fig. 8), gently emarginate, with a tapering axis extending nearly to the margin, carrying faint indications of two axial rings. The first pleural segment is clearly differentiated, with interpleural furrow particularly deep at the margin; second segment fainter, and a third only indicated by a short section of the pleural furrow. Of these two types of pygidium the latter is the more similar to those described for *Acheilus*, and the complete definition of the anterior segment is very like *A. marcoui* Raymond. It is not certain that it belongs to *A. monile* rather than the other *Acheilus* species in the boulders, but because *A. monile* is by far the commonest species it is most likely to be attributable here. The second type is recorded under open nomenclature; I cannot find any like it in the literature, but the stubby axis does suggest it might be referable to a catillicephalid. It may be referable to *A. ?falcatus* which also has a sculpture of scattered tubercles.

Discussion. This is a very distinctive and easily recognizable species of *Acheilus*, because of the fusion of the occipital ring and the posterior part of the fixed cheek to produce the collar-like structure isolating the forward part of the glabella. The type species, *A. marcoui*, is otherwise similar in glabellar form, but with more incised furrows and smaller palpebral lobes (Rasetti 1944, pl. 36, figs. 32, 33). The closest of the other described species is *A. convexus* Rasetti, 1945, which (his pl. 60, fig. 4) shows the beginnings of inflated areas at the lateral edges of the occipital ring; the axial furrows here, although clearly visible, are shallower than those defining the glabella in front. *A. convexus* is additionally distinguished from *A. marcoui* in having a pitted surface sculpture, and in the palpebral lobes being very close to the glabella. *A. latus* Rasetti, 1944, is another similar species, with a comparable surface sculpture to *A. monile*, and eyes of a similarly large size for the genus; it does not show the characteristic occipital structure, the palpebral lobes are slightly further forwards, and lack rims, and the occipital ring and fixed cheeks are also granulose.

Acheilus oryx sp. nov.

Plate 24, figs. 12–15

Type and figured material. Holotype, ROM 42280; cranidium ROM 42281.

Horizon. The boulder 3·5 m above base of section, Broom Point North; boulder from *Missisquoia*-bearing conglomerate (B2) at Broom Point South.

Name. Oryx, Greek, a mattock.

Diagnosis. Acheilus with small palpebral lobes and without surface sculpture. Axial furrows on cranidium completely effaced anteriorly.

Discussion. Although only two cranidia of this species are known, it is such a distinctive catillicephalid that it merits formal recognition as a new species. The forward-expanding glabella, position of the palpebral lobe, and structure of the posterior cranidial border show that it should be referred to *Acheilus*. Glabellar furrows show only as the faintest impressions, but the two pairs indicated are of usual form. The occipital furrow, by contrast, is deep and narrow, like the posterior border furrow. Axial furrows are distinct in front of the occipital ring, but fade out anteriorly; on one specimen they extend as far as the palpebral lobes, on the other they fade out just before. Palpebral lobes are anterior to the 1P glabellar furrow and nearly straight, lack rims, and are about one-third length of preoccipital glabella. The forward part of the cranidium is entirely smooth.

This is the first known catillicephalid to show a strong tendency towards effacement, a character widespread among contemporary trilobites. Fortunately, the effacement has not proceeded too far to be sure of the generic assignment. In general proportions the new species is most like *A. latus* Rasetti, 1944, which, however, is at once distinguished by its well-defined glabella, incised glabellar furrows, and tuberculate sculpture. Although the effacement of *A. oryx* is such a distinctive feature, there is no reason to consider a new genus for the reception of this species. Effacement seems to be capable of appearing in many different groups, and soon introduces a superficially different external appearance; the obvious polyphyly of this character shows that it should not be given high taxonomic weight unless accompanied by other changes.

Acheilus sp. indet.

Plate 24, fig. 25

Figured material. Cranidium, ROM 42282.

Horizon. The boulder 3·5 m above base of section, Broom Point North.

Discussion. A third species of *Acheilus* is represented by a single cranidium, lacking the occipital ring. It seems to be another new species, but the material is insufficient to name it. The glabellar furrows are deeply incised compared with the other two species in the Broom Point boulders, and the surface of fixed cheeks and glabella is covered with tubercles and strong, wandering ridges. The closest species is probably *A. spinosus* Rasetti, 1945 (also Rasetti 1963, pl. 129, figs. 17–20) which shows a similar sculpture. Our species shows several differences: the 1P furrow on the glabella is deeply curved rather than hooked at the inner end; the palpebral lobe does not come as far back as the outer end of the 1P glabellar furrow; and the fixed cheeks are wider. It is not known whether the Newfoundland specimen had an occipital spine.

Genus CALCULITES gen. nov.

Type species. Acheilus microps Rasetti, 1944.

Diagnosis. Catillicephalids differing from *Acheilus* in having a relatively narrow, subrectangular glabella, or very gently forward-expanding; width of glabella at frontal lobe is considerably less than sagittal glabellar length in front of occipital ring. Palpebral lobes of small to middle size, in advance of the 1P glabellar furrow, and not strongly curved, so that the facial sutures run in an almost straight course from the anterior margin to the posterior border. Anterior cranidial border absent. Occipital tubercle placed in the middle of the occipital ring, rather than adjacent to the occipital furrow.

Name. From *calculus* (Latin), a pebble, referring to its typical allochthonous occurrence in Newfoundland and Quebec, + *ites* (masculine).

Discussion. This genus is erected to include a small group of closely related species formerly referred to *Acheilus* (or *Theodenisia*). The diagnostic characters listed above distinguish the new genus from *Acheilus* as understood here; in that genus the length of the glabella measured in dorsal view is equal or nearly equal to the sagittal length of the glabella in front of the occipital ring; in *Calculites* the glabella is proportionately much longer. It shares with *Acheilus* the incorporation of the cranidial border into the front of the glabella. The free cheek and pygidium are known only in *C. stevensi* sp. nov.; the cheek is typically catillicephalid, while the pygidium carries an additional segment (perhaps two) compared with those that have been assigned to *Acheilus*.

The genus is of some phylogenetic importance because it is one of the few which cross the Trempealeauan-'Canadian' boundary; *C. stevensi* is from an undoubtedly 'Canadian' (*Symphysurina* Zone) boulder, while the remainder of the species assigned to the genus are from Trempealeauan occurrences. As stated above, *Calculites* is regarded as sharing a common ancestor with the Isocolidae, and hence forms a stratigraphic link between characteristically Cambrian and Ordovician families.

Species included. Besides the type species and *C. stevensi* sp. nov., *A. depressus* Rasetti, 1944, from Quebec, and *Theodenisia paulula* Khramova, 1977, from the Siberian platform, are to be included. *A.? marginatus* Rasetti,

1945, is closer to *Calculites* than it is to *Acheilus*, but differs from both in having a clearly defined preglabellar furrow.

Calculites stevensi sp. nov.

Plate 24, figs. 16–24

Type and figured material. Holotype, cranidium, ROM 42283; other cranidia ROM 42284–42286; free cheek ROM 42287; pygidia probably belonging to this species ROM 42288–42289.

Horizon. Early Ordovician (*Symphysurina* Zone) boulder in Conglomerate B1 at Broom Point South; this is the same large boulder that has yielded a rich early Ordovician fauna.

Name. After R. K. Stevens, Memorial University, who did much to help the author in this work.

Diagnosis. *Calculites* with glabellar furrows subdued; glabella almost parallel-sided; axial furrows faint anteriorly; surface sculpture of somewhat scattered tubercles.

Description. Cranidium slopes downwards evenly away from glabellar midline, the glabella itself not being highly vaulted transversely. Glabella subparallel-sided, slightly narrower in front of occipital ring, width three-quarters sagittal length, as measured in front of occipital ring in dorsal view. Axial furrows shallow anteriorly, in front of palpebral lobes. Glabellar furrows not deeply incised, short. 1P forms a broad (exsag.) smooth area, deeply forked at inner end, the posterior branch of the fork nearly isolating very small 1P glabellar lobe; 2P at glabellar (preoccipital) mid-length, very gently backward-curved, narrower (exsag.) than 1P and not forked. 3P and 4P are short depressions isolated within the forward part of the glabella, 3P parallel with the inner end of 2P, 4P somewhat exterior to this, and inclined inwards–forwards. The arrangement of these anterior furrows is much like the same furrows on *Triarthrus*. Occipital ring well defined, one-fifth to one-sixth length (sag.) of preoccipital glabella, and slightly depressed below level of glabella; a large median tubercle is developed, and placed at the mid-point, or slightly towards the posterior margin of the occipital ring (contrast *A. monile*). Fixed cheeks narrow-triangular, width at posterior margin about half width of occipital ring and half length (exsag.). Palpebral lobes very gently curved, palpebral furrows shallow; small palpebral lobes are symmetrically disposed about 2P glabellar furrow (slightly more posterior on smallest cranidium, Pl. 24, fig. 21). Posterior border well defined, narrow, convex. Preocular cheek wider (tr.) than in most species of *Acheilus*. Surface sculpture of discrete, small tubercles, somewhat less dense on the mid-line; smallest cranidium shows smooth mid-line and pitting anteriorly.

Small and narrow free cheek apparently lacks a genal spine and a defined border; as on most catillicephalid cheeks parallel ridges are developed on the exterior edge. Pygidia associated with the cranidia are from the same boulder; they are of the right size and form to belong to *C. stevensi*, nor are there other species present to which they might belong (although there is always a possibility with this kind of material that there is an allochthonous mixture of cranidia of one species with pygidia of another). The association can be made with some, but not complete, confidence. The broad axis tapers to a rounded extremity and does not reach the border; up to six axial rings are visible, but of these only the first two are at all clear; between these two rings an incipient half-ring is developed. On the steeply downward-curving pleural regions the first two segments are very well defined with interpleural furrows almost reaching the margin, and long, deep pleural furrows. Two posterior pleural segments are relatively poorly expressed fading before border. Scattered tubercles are compatible with the cephalic surface sculpture.

Discussion. Compared with the type species *C. microps* (Rasetti, 1944, pl. 36, fig. 39; 1963, pl. 129, figs. 6–9) *C. stevensi* is generally more effaced, with shallower glabellar furrows and more scattered tuberculation; the glabella on *C. microps* expands slightly in width forwards. *C. paulula* Khramova, 1977 (pl. 14, figs. 11, 12) is very like *C. stevensi*, but is apparently not tuberculate, and its glabella is well defined in front of the palpebral lobes. Examination of paratype specimens of *C. depressus* Rasetti, 1944 from Rasetti's boulder 52 in the Levis conglomerate, which are in the collections of the British Museum (Natural History) (Pl. 24, fig. 11), shows that it resembles *C. stevensi* in the relative effacement of the glabellar furrows, but which are even more subdued on the Levis specimens. *C. depressus* has a different surface sculpture; very dense granules over the rear part of the glabella, but with raised lines on the frontal lobe. All four species are closely related, and for this group at least the Cambrian–Ordovician boundary was of little moment.

Family KINGSTONIIDAE Kobayashi, 1933
Genus BYNUMINA Resser, 1942

Type species. Bynumina caelata Resser, 1942.

Bynumina sp. aff. *B. laevis* (Raymond, 1924)

Plate 25, figs. 2, 3

Figured material. Cranidium, ROM 42296.

Horizon. From a boulder in conglomerate at base of measured section in Fortey *et al.* (1982), associated with *A. monile* sp. nov.

Discussion. A single well-preserved cranidium is close to *B. laevis* (Raymond), but shows several small differences which may indicate that a new species is present in western Newfoundland. *B. laevis* was placed in *Plethometopus* by Rasetti (1945), and there is a general resemblance to this effaced genus. However, examination of *B. laevis* specimens in the Rasetti collection of the British Museum (Natural History) reveals that the glabella is faintly defined; it tapers forwards to a slightly truncate front at about three-quarters cranidial length (Pl. 25, fig. 1). This is quite different from the subrectangular glabella of *Plethopeltis* and *Plethometopus*, which occupies a large proportion of the cranidial area. The cranidium from Newfoundland shows smooth areas interrupting the general pitting which clearly correspond with glabellar furrows. Three pairs are present, 1P deeply hooked backwards at its inner end, 2P and 3P more weakly arcuate-backwards. This glabellar form and arrangement of furrows is identical to that of the type species of *Bynumina* (*see* Kurtz 1975, pl. 2, figs. 1–4), and has also been figured by Palmer (1982, pl. 1, fig. 22). The form of the postocular fixed cheeks and border furrow is also very like that of *B. caelata*. The only feature of *B. laevis* which might be described as plethometopid is the laterally tapering occipital ring. Because *Bynumina* is generally found much lower in the Cambrian in the *Elvinia* Zone, and below the pterocephaliid–ptychaspid biomere boundary, the question arises whether the resemblance between the species under discussion and *B. caelata* could be a matter of convergence. Such characters as there are on *B. laevis* agree so well with *Bynumina*, that it seems preferable to regard the convergence as being with the plethometopids, because of general effacement. Furthermore, Stitt (1971) has already described a *Bynumina* from the *Saukia* Zone of Oklahoma.

The Newfoundland cranidium differs from *B. laevis* in the following details: as seen in dorsal view the preocular part of the cranidium is shorter, because more steeply downturned; dorsal pitting is denser; and the axial furrows cannot be traced so far forwards.

Family UNCERTAIN
Genus HYPERBOLOCHILUS Ross, 1951

Type species. Hyperbolochilus marginaustum Ross, 1951.

Hyperbolochilus? sp. nov.

Plate 25, figs. 10, 11

Figured material. Cranidium, ROM 42297.

Horizon. Large boulder in B1 conglomerate, *Symphysurina* Zone, Broom Point South.

Discussion. A single, well-preserved cranidium of undoubted early Ordovician age is a particular conundrum. It has a short, tapering glabella without glabellar furrows, but a well-defined occipital ring; weak eye ridges; palpebral lobes of moderate size close to the glabella and quite far forwards; broad, downsloping preglabellar field with faint median depression; divergent preocular sutures; very wide cranidial anterior border; and lacks surface sculpture. I cannot find any species of a similar age with a cranidium resembling this one. In general appearance it has a 'Cambrian' aspect—it is similar to such asaphiscids as *Liaoyangaspis* Chang, 1957 (see Lu *et al.* 1965, pl. 57) but with the palpebral lobes further forwards, so that their anterior limits are nearly opposite the front of the glabella; there is also some comparison with *Coosia*. There are no forms in the latest Cambrian like it. The closest early Ordovician genus is the little-known *Hyperbolochilus* Ross, 1951, the type species of which differs from the Newfoundland cranidium in its less tapering glabella, smaller palpebral lobes which do not extend so far forwards, and in its convex, steeply downsloping preglabellar field. Ross noted that the palpebral furrows of *Hyperbolochilus* are unusual in being concave-outwards, and that is true also of our cranidium. This is a small feature, but one which serves to distinguish *Hyperbolochilus* from some other early Ordovician ptychoparioids, for example, smooth *Hystricurus*. The best provisional determination of the Newfoundland cranidium is to assign it with question to *Hyperbolochilus*; it is apparently a new species, but the material is inadequate to name it. In the *Treatise*, *Hyperbolochilus* was not assigned to a family; Kobayashi (1955) placed it in the Asaphiscidae, pointing out resemblances to the early Upper Cambrian asaphiscid *Homodictya imitatrix* Raymond. Compared with our form *Homodictya* has a broad and less-tapering glabella, and a distinctively concave border without a discrete border furrow. The more similar asaphiscids, like *Liaoyangaspis*, are of Middle Cambrian age. Kobayashi also pointed out resemblances between the type species of *Hyperbolochilus* and early bathyurids of *Licnocephala* type, but there is little similarity between the Newfoundland cranidium and such bathyurids.

Regardless of its familial placing this small cranidium is of some interest as another example of the reappearance of a morphological type at the early Canadian. Is it a 'parallel' form, having evolved from some (unknown) off-shelf ancestor, or is it a late asaphiscid, surviving perhaps in a shelf-edge habitat? Without the free cheek and pygidium it is not possible to resolve this problem.

Family MISSISQUOIIDAE Hupé, 1953
Genus LUNACRANIA Kobayashi, 1955

1973   *Paranumia* Hu, p. 86.

Type species. Lunacrania trisecta Kobayashi, 1955.

Discussion. Dean (1977) argued that *Lunacrania* was a subjective synonym of *Missisquoia*. I observed (*in* Fortey *et al.* 1982) that *Lunacrania* could be distinguished from *Missisquoia* on the basis of its minute palpebral lobes well removed from the glabella. The discovery of another species of the genus suggests that this view is correct; *Lunacrania* is a rare component of the Cambrian-Ordovician

---

EXPLANATION OF PLATE 25

Figures × 12, except fig. 12.

Fig. 1. *Bynumina laevis* (Raymond, 1924), cranidium, dorsal view showing outline of glabella. Rasetti Collection, BM It5671, Levis conglomerate, Quebec.

Figs. 2, 3. *Bynumina* sp. aff. *B. laevis* (Raymond, 1924). Cranidium, ROM 42296, dorsal and anterior views; note outline of glabellar furrows. *Acheilus monile* boulder 3·5 m above base of Broom Point North section.

Fig. 4. *Parakoldinioidia* cf. *P.* sp. A., incomplete cranidium, ROM 42299. Solitary in small boulder from B1 Broom Point South.

Figs. 5, 6. *Parakoldinioidia* sp. A., cranidium, ROM 42298, dorsal and lateral views. Boulder in B1 associated with *Hystricurus paucituberculatus* and *Symphysurina* sp. nov.

Figs. 7-9. *Lunacrania spicata* sp. nov. Large boulder in B1 associated with *Symphysurina* Zone fauna. 7, very small cranidium, ROM 42291. 8, 9, holotype, cranidium in anterolateral and dorsal views, ROM 42290.

Figs. 10, 11. *Hyperbolochilus?* sp. nov. Cranidium, ROM 42297, dorsal and oblique views. Early Ordovician (*Symphysurina* Zone) boulder in B1, Broom Point South.

Fig. 12. *Loganellus?* sp. nov. Cranidium, ROM 42300, × 8, associated with a cheek of *Symphysurina* cf. *cleora* (Walcott) from a small boulder in the conglomerate at the base of the Broom Point North section.

Fig. 13. *Platydiamesus* sp. indet. Pygidium, ROM 42302. *Acheilus monile* fauna, boulder 3·5 m above base of section, Broom Point North.

PLATE 25

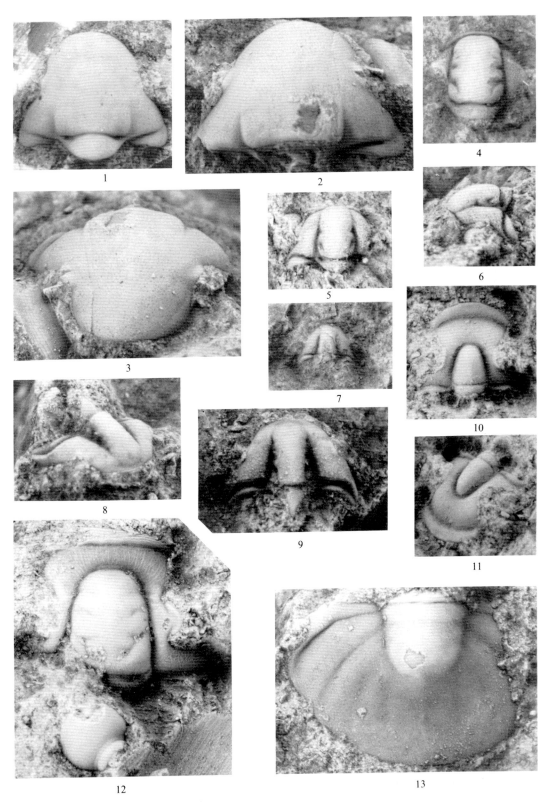

boundary faunas, so far known from western Canada and Newfoundland, and Dakota (and possibly the western United States). Hu (1973) described *Paranumia triangularia* n. gen. et sp. from the earliest Ordovician of the Deadwood Formation, South Dakota. Apart from having narrower (tr.) fixed cheeks it is closely similar to *Lunacrania*, and *Paranumia* is regarded as a junior subjective synonym of this genus. Pygidia associated with his cranidia by Hu are clearly missisquoiid.

Lunacrania spicata sp. nov.

Plate 25, figs. 7–9

Type and figured material. Holotype, cranidium ROM 42290; cranidium 42291.

Horizon. From the main *Symphysurina* Zone boulder in conglomerate B1 at Broom Point South, early Ordovician (Tremadoc).

Name. Spicata, Latin, spine-bearing.

Diagnosis. Lunacrania with prominent occipital spine; cranidial anterior border present only in front of glabella.

Description. I had doubts about formally naming this species from so little material. The larger cranidium is well preserved and obviously distinct from the poorly preserved type species; it does much to establish the validity of the genus. For these reasons it was considered preferable to name it.

Glabella slightly concave-sided, highly vaulted posteriorly, standing well above the cheeks, becoming progressively lower anteriorly to almost merge with the cheeks anterolaterally. Four pairs of glabellar furrows of usual missisquoiid type: 1P sharply backward-directed, almost cutting off narrow (tr.) basal lobe; 2P wide and very short, transverse; 3P almost immediately in front of 2P, narrow, opposite palpebral lobes; 4P far forward, faint, almost pit-like. Frontal lobe of glabella carries a median dimple. Axial and preglabellar furrows very faint around the frontal glabellar lobe. Palpebral lobes small and ill-defined, well removed from glabella, such that the transverse width of the interocular cheek is only slightly less than that of the glabella at the same place. Eye ridges weak. Posterior border furrow remarkably deep, narrowing towards, and reaching, the posterolateral corner of the cheek; posterior border narrow (exsag.). All the lateral part of the fixed cheek is steeply inclined, and is swept backwards behind the rest of the cranidium. The most conspicuous feature of the species is the extension of the occipital ring into a broad, upturned, triangular spine. A short (sag., tr.) anterior cranidial border is present only in front of the glabella, gently concave, and downward sloping; laterally the cheeks slope downwards directly to the margin. Oblique view (Pl. 25, fig. 8) shows that the free cheek would have been very narrow (tr.) and elongate-triangular. Because of the backward curve of the fixed cheek the facial sutures appear 'proparian', especially in dorsal view, but there is no reason to suppose that the sutures were other than opisthoparian. Surface of cranidium carries minute, scattered tubercles, interspersed with even finer granules.

Discussion. The holotype is the best-preserved cranidium of this little-known genus. The holotype of *L. trisecta* was illustrated by Dean (1976, pl. 1, fig. 8). General structure of the glabella and cheeks are much like the new species, but it differs notably in lacking the large occipital spine, in the tips of the fixed cheeks not being swept strongly backwards, and in its well-defined anterior cranidial border. *L.* cf. *trisecta* from the *Missisquoia typicalis/Radiograptus* autochthonous fauna in Broom Point (Fortey *et al.* 1982, pl. 2, fig. 1) is more like *L. trisecta* than *L. spicata*. Dean suggested that *Macroculites enigmaticus* Kobayashi, 1955, was conspecific with *L. trisecta*, in which case the specific name of the former would have page priority. Unfortunately, the type and only specimen of *M. enigmaticus* is incomplete, and it is not possible to see whether it had the small palpebral lobes typical of *Lunacrania*; *L. trisecta* is retained here in view of this uncertainty. *L. triangularia* (Hu, 1973) also lacks an occipital spine, has narrower fixed cheeks (tr.) than *L. spicata*, and a nearly conical glabella, well-defined anteriorly.

Ross (1951, pl. 35, figs. 1, 2) illustrated a small cranidium from Zone B (early Canadian) in Utah as 'Undetermined genus and species D'. It has an occipital ring produced into a spine and a glabella that becomes poorly defined anteriorly. Ross described this specimen as proparian, but such an apparent appearance is given by *L. spicata*. The Utah cranidium probably represents a *Lunacrania* species close to *L. spicata*, but it does differ in the fixed cheeks not being strongly backwardly curved posterolaterally.

In summary, there are certainly three, and possibly four, species of *Lunacrania*, which differ from the contemporary missisquoiids in consistent characters.

Genus PARAKOLDINIOIDIA Endo, 1937

Synonym. Missisquoia, Shaw 1951, p. 108.

Type species. Parakoldinioidia typicalis Endo, *in* Endo and Resser 1937.

Discussion. Shergold (1975) first pointed out the close similarity between *Parakoldinioidia* Endo and *Missisquoia* Shaw, 1951. Publication of good photographs of the type material of *P. typicalis* (Lu *et al.* 1965, pl. 131, figs. 20–22 *non* fig. 23) shows that the resemblance is such that *Missisquoia* should be regarded as a junior subjective synonym of *Parakoldinioidia*. The pygidium attributed to *P. typicalis* (Lu *et al.* 1965, pl. 131, fig. 23) by Endo is clearly wrongly assigned, both with regard to its large size and broad, flat borders. The only feature in which the cranidium of *Parakoldinioidia* differs from species assigned to *Missisquoia* is the relatively wide cranidial anterior border. This is inadequate as a generic discriminant, especially because *Missisquoia* also shows a good deal of variation in this character (see, for example, *M. nasuta* Winston and Nicholls, 1967). The loss of the name *Missisquoia* is unfortunate, notably because it has become familiar as a zonal index. Even more unfortunate is the fact that the type species of *Missisquoia*, *M. typicalis* Shaw, 1951, would require a new specific name if transferred to *Parakoldinioidia* (*typicalis* being preoccupied by the type species of that genus). An alternative procedure would be to restrict the use of *Parakoldinioidia* to the type species, but this would only be justifiable if the type specimens were poorly preserved or unrecognizable, which they are not. Furthermore, Ludvigsen (1982) has already noted that '*M.*' *depressa* Stitt (a senior synonym of *Tangshanaspis zhaogezhuangensis* Zhou and Zhang) occurs both in China and North America, which makes it likely that other species of the same genus are also to be found in both countries. Accordingly, the new name *P. stitti* nom. nov. (after J. B. Stitt) is proposed here to replace *P. typicalis* (Shaw 1951) *non* Endo, preoccupied.

Parakoldinioidia sp. A

Plate 25, figs. 5, 6

Figured material. Cranidium, ROM 42298.

Horizon. From an Early Canadian boulder with *Hystricurus paucituberculata* sp. nov. and *Symphysurina* cf. *brevispicata* in conglomerate B1, Broom Point South.

Discussion. This is a new species of the genus, but, as with so many of the Newfoundland forms, it is known from a single example, and cannot be named formally. It is of interest as it may derive from an interval above the range zone of *Parakoldinioidia stitti* (= *M. typicalis*). It most resembles *P. stitti* of described species, and differs from it in the following features:

1. The glabella is relatively short, and the 1P glabellar furrow is almost isolated from the axial furrow.
2. The 3P and 4P glabellar furrows are hardly developed.
3. The cranidial anterior border is extremely narrow (sag., exsag.), narrower than in any other species of the genus.
4. The transverse width of the fixed cheeks behind the eyes is less than in *P. stitti*.

A second, larger *Parakoldinioidia* cranidium (Pl. 25, fig. 4) was recovered as the solitary specimen from a small boulder in B1 at Broom Point South. Like *P.* sp. A. the glabella is shorter and squatter than in *P. stitti*, and it has a very narrow border. The glabellar furrows are of more usual missisquoiid form, however, although much less deep and pit-like than *P. stitti*. Because of the lack of associated fauna we cannot be certain of its age, although it is very probably of *Symphysurina* Zone age like the rest of the boulders in B1. It could simply be a more adult growth stage of *P.* sp. A., but perhaps the differences in glabellar structure make this unlikely. It is recorded here as *Parakoldinioidia* cf. *P.* sp. A.

AN ARTHROPOD FRAGMENT FROM THE SCOTTISH NAMURIAN AND ITS REMARKABLY PRESERVED CUTICULAR ULTRASTRUCTURE

by J. E. DALINGWATER *and* C. D. WATERSTON

ABSTRACT. A fragment of what appears to be the distal podomere of one of the posterior prosomal appendages of a large eurypterid from the basal Namurian of Lanarkshire, Scotland, is described. It is characterized by the presence of large adpressed spine-like fulcra. The cuticle between the fulcra and on the fulcral bases is ornamented by lunules of varying shapes which are associated with much smaller lunules. Many of the lunules are folliculated and many have one or more tumid areas which form rounded pustules. Follicles are present also on the spines or fulcra, where they occupy depressions of varying shapes. The size of the podomere and the nature of the ornament may be compared with the podomeres of the Carboniferous eurypterid *Dunsopterus*. Under the scanning electron microscope the specimen reveals the best-preserved cuticular ultrastructure of any eurypterid so far examined. The fibre patterns of the lamina units, fine perpendicular elements resembling the pore canals of modern arthropod cuticles, and wider perpendicular canals, probably tegumental ducts, are described. The problems and the value of cuticle research are discussed.

THE specimen, RSM GY1887.25.1074, is held in the Royal Scottish Museum. It was collected at Limekilns Quarry, East Kilbride, Lanarkshire by Andrew Patton whose collection later passed to James Coutts after whose death it was acquired by the Museum. Patton was manager of the Limekilns Quarry, as well as of other quarries in the East Kilbride district. Coutts, in Patton 1885, gave a list of species obtained from Limekilns and Patton stated (p. 322) that: 'different beds were worked at various times and, consequently, the fossils were from different horizons. The differences are not, however, great geologically, as the fossils are all from strata connected with the Calderwood limestone and the Calderwood cementstone, the position of the latter being 25 feet or so higher of the two.' The specimen to be described is probably that identified by B. N. Peach as '*Glyptoscorpius* sp.' and listed with *Eoscorpius* sp. by Coutts. Neilson (1895, p. 71) was of the opinion that these specimens came from the cementstone. The Calderwood Cementstone and associated shales are the regional equivalent of the Top Hosie Limestone, the base of which marks the base of the local Limestone Coal Group and the E_1 goniatite zone as well as the base of the Namurian (Currie 1954, p. 534).

Description of the specimen. The specimen is a curved, asymmetrical arthropod fragment broken off at the proximal end and narrowing distally to a natural termination (Pl. 29, fig. 1). The preserved length is 56 mm and the maximum breadth 17 mm. It is of flattened cylindrical form and appears to be part of a distal podomere of one of the posterior prosomal appendages of a large eurypterid.

The most striking feature is the presence of large adpressed spine-like scales or fulcra which occur throughout its length (text-fig. 1A). There are three rows of fulcra within the breadth of the podomere in the proximal half and two in the distal half while a single fulcral scale forms the terminal spine. The fulcra become less closely spaced distally. They measure from 4 to 6 mm in breadth and some 10 mm in length but there is a tendency for the proximal ones to be slightly shorter. Each fulcral scale becomes increasingly inflated distally and, although now flattened and in many cases broken, each appears to have terminated in an approximately circular aperture. The distal end of the terminal scale, however, is not preserved. Very fine anastomosing striae are visible on the inner lip of the aperture on the adpressed side. The fragment appears to be a moult since the hollow interiors of the fulcral scales have become filled with matrix. It is possible that the distal apertures formed the points of attachment of movable spines. The paired movable spines on the podomeres of *Baltoeurypterus* are set in thin cuticle, there being no other form of articulation (Holm 1898, p. 14; Selden 1981, p. 14). It is probable that in the present case the spines, being similarly articulated, were lost after ecdysis through the failure of the delicate attaching cuticles.

[Special Papers in Palaeontology No. 30, pp. 221-228, pls. 29-30]

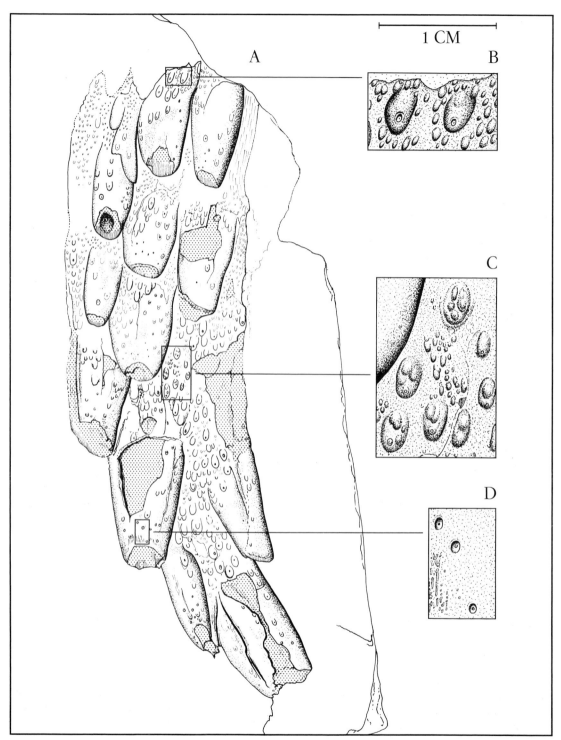

TEXT-FIG. 1. A, line drawing of specimen RSM GY1887.25.1074 from the Namurian of Limekilns Quarry, East Kilbride, Lanarkshire. Size indicated by 1 cm scale. B–C, enlarged drawings of areas indicated to show the nature of the ornament. B, lunules of two orders of size, the larger being folliculated with raised circular outer rim within which there is an annular depression with a central boss. C, larger lunules exhibiting convex swellings or pustules. D, surface of the distal end of the fulcra showing follicles in circular depressions.

The cuticle between the fulcra, and on the fulcral bases, is ornamented by lunules which are associated with much smaller lunules (text-fig. 1B, C). The larger lunules are narrow lingulate or elliptical, some becoming wide and sub-circular. Many are folliculate with a raised circular outer rim within which there is an annular depression with a central boss. In the distal half of the specimen most of the larger lunules have convex swellings or pustules (text-fig. 1C). Each lunule may have one or more of these swellings which occur irregularly on the lunules but usually in the proximal half. The smaller lunules tend to be less than quarter the size of the larger ones. They are tumid, narrow, and many are linguloid although in some areas they widen to an elliptical form. Follicles also occur on the fulcral scales but here they occur in depressions which may be lingulate or elliptical. Towards the distal end of the fulcra, however, the follicles occur in circular depressions (text-fig. 1D). The areas of cuticle between the smaller lunules have a finely granular texture.

Discussion. The size of the podomere and its general ornamentation may be compared with the Carboniferous eurypterid *Dunsopterus* which has been described from the Calciferous Sandstone Measures of Berwickshire (Waterston 1968, pl. 2, figs. 1–3, 5–6). The podomeres associated with *Dunsopterus* are more proximal but they show large scales of similar form to the fulcra of the Limekilns specimen together with smaller lunules and an ornament of much finer lunules. The cuticle of the Berwickshire podomeres is not so well preserved, but if follicles are present they are much less common than on the Limekilns specimen. This would be expected since the Limekilns specimen represents the tactile tip of a distal podomere. The ornament of the Limekilns specimen is more exaggerated than that of the Berwickshire podomeres, on which the large scales are not developed as fulcra nor terminated by the circular apertures or articulations so characteristic of the Lanarkshire specimen. These differences, however, might well be accounted for by the more distal position of the Limekilns specimen or they may be of a specific nature, the Limekilns specimen occurring considerably higher in the Scottish Carboniferous succession than the Berwickshire examples. It is unfortunate that no direct comparison can be made between *Dunsopterus* and the described specimen since the distal podomeres are unknown in *Dunsopterus*.

The finer ornament and the associated sensory follicles of the present specimen are similar to those described from *Cyrtoctenus* (Størmer and Waterston 1968, pl. 5, figs. 6–7, text-figs. 4f, 5, 6g, 13f–i) and *Thurandina* (Størmer 1974, pls. 15–16; text-figs. 39, 41–44). Similar follicles on a smaller scale have been described from *Baltoeurypterus* (Eisenack 1956, p. 120; Dalingwater 1975, p. 274, pl. 3, fig. 6; Selden 1981, p. 11, text-fig. 23a–c, 27a). Selden (1981, p. 11) has discussed the differing interpretations of these structures. Eisenack believed them to represent a form of sensillum having a thin cuticular covering, a view with which Størmer (1974, p. 416) was sympathetic in regard to structures which Størmer and Waterston (1968) had previously termed setal sockets in *Cyrtoctenus*. Selden and the present writers, however, consider that many, if not all, of the follicles may once have contained setae (Selden 1981, p. 11, text-fig. 23a).

ULTRASTRUCTURE OF THE CUTICLE

Methods. Small fragments of cuticle were removed from four areas of the specimen: (i) the middle of the most distal fulcrum (stubs A–C); (ii) the middle of a median fulcrum (stubs D–G, I); (iii) the middle of a proximal fulcrum (stub J); (iv) the posterior part of a proximal fulcrum (stub H). Each fragment was sharply rebroken, mounted using 'Durofix' with the broken side uppermost (the effective 'section' therefore parallel with the surface of the stub), gold sputter-coated for 2 minutes, and examined with a Cambridge S140 scanning electron microscope (SEM) with the accelerating voltage at 20 kV. Some of the preparations which were not satisfactorily broken the first time were rebroken on the stub and remounted, so that some stubs carry two or three preparations from the same fragment. All preparations are registered with the specimen in the R.S.M. A short working distance, medium aperture, and small spot size were used for high-resolution photography.

Description. The thickness of the cuticle ranges from a minimum of 30 μm in preparations from the posterior part of a fulcrum (on stub H) to around 100 μm in most other preparations. It is composed of lamina units throughout its thickness (Pl. 29, fig. 2), except in a few preparations where relics of what might be part of a non-laminate zone, traversed by fine perpendicular elements, persists below the laminate zone (Pl. 29, fig. 3). The lamina units characteristically are narrowest (around 0·5 μm) at the inner edge of the cuticle (Pl. 29, fig. 4) but rapidly increase to their maximum thickness of about 1·5 μm; thence, there is a gradual decrease in thickness towards the outer edge of the cuticle where the lamina units are again less than 1 μm. There may possibly be a very thin non-laminate layer at the outermost edge of the cuticle, but 'flaring' as the beam strikes the edge of the specimen, and also fracturing in this region, make it difficult to decide if this layer exists.

Numerous very fine perpendicular elements are seen in most preparations: at higher magnifications these resolve as fine channels, of the order of 0.1 μm in diameter, and also as tubular structures 0·2–0·5 μm in diameter (Pl. 29, fig. 5). Wider perpendicular canals are present in a few preparations: the first type tapers rapidly from 6 μm at the inner edge of the cuticle to around 2 μm towards the centre. These canals were found singly in a few preparations, flanking a much broader canal in one preparation (Pl. 30, fig. 1). This very broad canal measures 55 μm at its base, tapering to 10 μm towards the outer surface of the cuticle. Whilst this very broad canal is simply infilled with matrix, parts of the linings of the walls of the narrower canals are preserved (Pl. 30, fig. 2). Particularly in regions of cuticle alongside these narrower canals, perpendicular sheets with fibres of less than 0·1 μm in diameter arranged in fan-like arrays transcend the normal laminate structure of the cuticle; this aspect of the cuticle is occasionally seen in other areas of preparations.

Fortuitous breaks at about 45° to the horizontal plane 'open out' the lamina units and provide more information about their ultrastructure. Plate 30, fig. 3 gives an overall view of such a region and Plate 30, fig. 4 details of the lamina units in this region. Plate 30, fig. 5 shows lamina units from the counterpart of the break illustrated in Plate 30, fig. 3 and therefore at about 45° to the perpendicular, so that it might be said that the units are viewed from 'underneath'. Plate 30, fig. 6 illustrates lamina units from a near-perpendicular break, but viewed slightly from above. In all these preparations sheets of apparently fibrous material arc across the inter-laminae. In places these arcs appear to be pierced by fine tubular structures or by fine channels.

Although no analyses were carried out, the fracture patterns of the cuticle and the similarity of the preservation to that of previously analysed material (Dalingwater 1973, p. 181) strongly suggest that it is now composed mainly of silica.

Discussion. In its overall aspect, but most especially in the dimensions of the lamina units this cuticle closely resembles those of some previously described eurypterids. In one respect it is more like the cuticles of *Pterygotus* and *Baltoeurypterus* than that of *Mycterops*: it lacks the very extensive non-laminate zones of the latter (see Dalingwater 1973, 1975). The presence of only fragments of a possible inner non-laminate layer might indicate that the cuticle described here is part of an exuvium, but the shape of the wide canal (Pl. 30, fig. 1), expanded markedly towards its base, makes it almost certain that, exuvium or not, the great majority of the original thickness of the cuticle is preserved and that any non-laminate layer(s) would have been relatively very thin. Both this wide canal and the narrower (6 → 2 μm) canals are considered to be tegumental ducts rather than setal ducts, since in both cases it is difficult to find any opening corresponding to the position where they should reach the outer surface of the cuticle.

The numerous very fine perpendicular elements seen in all preparations are considered to be comparable to the pore canals of extant arthropod cuticles. In extant cuticles, pore canals have a discrete lining: it is possible that the perpendicular tubular elements represent lined canals and the channels the lumina of these canals, though some apparent channels may simply be spaces between adjacent lined canals.

The patterns in the lamina units in angled breaks suggest sheets of fibres arising from laminae and arcing at low angles across inter-laminae. However, although the arced nature of these sheets is

---

EXPLANATION OF PLATE 29

?*Dunsopterus* sp. (RSM GY1887.25.1074).
Fig. 1. The entire specimen. ×2.
Fig. 2. Clean perpendicular break, complete thickness of the cuticle. Stub J. ×600.
Fig. 3. Part of a possible non-laminate zone, between the laminate zone and matrix. Stub B. ×2300.
Fig. 4. Narrowest lamina units at the inner edge of the cuticle. Stub C. ×2800.
Fig. 5. Tubular structures and fine channels. Stub C. ×4000.

PLATE 29

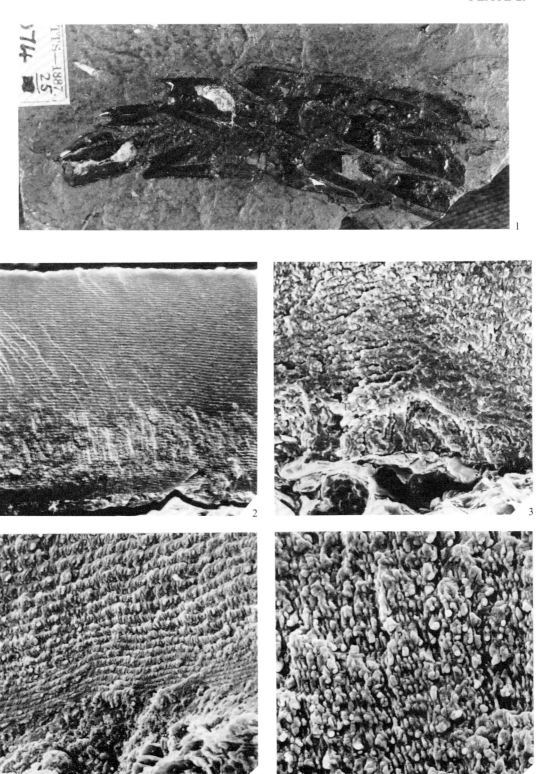

DALINGWATER and WATERSTON, Arthropod cuticle

considered to be original, their fibrous nature at the finest levels visible may represent the micro-fabric of the replacement silica. The fan-like arrays of fibrous material close to a narrower perpendicular duct (Pl. 30, fig. 2) are considered to reflect the original structure of the cuticle and strongly resemble the sub-unit structure of the 'vertical lamellae' described by Mutvei (1977) from non-laminate regions of *Limulus* cuticle: the dimensions of the fibres are similar and even branching canals can sometimes be perceived.

The composition of the cuticle, considered to be now mainly silica, raises some problems. It is difficult to envisage the processes which have resulted in the preservation of such beautiful detail and yet it seems unlikely that the cuticle was heavily impregnated with silica in life. Certainly, further work seems necessary, perhaps including a comparison of the preservation of eurypterid and xiphosuran cuticles. Professor L. J. Wills (pers. comm. 1973) remarked that the cuticles of the latter dissolve in concentrated hydrochloric acid, unlike those of the former, and Eldredge (1974, p. 21) has described the striking differences in preservation between eurypterids and xiphosurans on the same bedding plane in the Wenlockian limestones of Saaremaa, Estonia.

THE PROBLEMS AND THE VALUE OF CUTICLE STUDIES

Previous studies (Dalingwater 1973, 1975) described ultrastructural details from the cuticles of eurypterids belonging to three different families and attempted to compare these cuticles with those of other chelicerates, a comparison subsequently somewhat modified (Dalingwater 1980, p. 287) in the light of Mutvei's (1977) work on the cuticles of *Limulus* and the scorpion *Heterometrus*. The next stage in the study of eurypterid cuticles could involve examining a wider range of specimens to determine whether any general patterns exist; a problem with this approach might be in the selection of material for study. A less-biased treatment could involve examining and describing as many eurypterid cuticles as possible, drawing general conclusions only on completion of the task. An important side-effect of this strategy would be to furnish a catalogue for the identification of fragmentary remains of eurypterids; it might also provide a better basis for assessing the cuticles of both enigmatic early chelicerates and some of the rather bizarre Carboniferous animals at present assigned to the Subclass Eurypterida. On a more general note, Briggs and Fortey (1982, p. 28) have commented that there is no clear picture of the taxonomic significance of different arrangements of layers of arthropod cuticles. This is certainly true at present, and it may well be that differences of finer details will prove more helpful in interpreting fossil cuticles, in the absence of the ability to stain differentially the major subdivisions which has proved so useful in analysing extant material.

Some of the problems of interpreting fossil cuticles result from the lack of detailed information about a broad range of extant cuticles. The emphasis of cuticle work in recent years has centred around refinement of the Bouligand–Neville helicoidal architecture model and attempts to analyse all cuticle structure in terms of the 'two-system' approach (see Neville 1975). However, even before this period, workers on extant cuticles tended to describe the cuticles of a few species from each arthropodan group and to infer that the rest of the members of that particular group were similar. More than this, they tended to concentrate on the cuticle of convenient, obvious areas of the animals' exoskeletons. Mutvei (1977) highlighted the limitations of this latter approach when he showed that the cuticle of the scorpion pedipalp chela and of thickened portions of leg podomeres has features absent from other exoskeletal areas. Both these biases are, sometimes unavoidably, present in eurypterid cuticle studies: suitably preserved specimens available for study are limited in number, only part of the animal may be

---

EXPLANATION OF PLATE 30

?*Dunsopterus* sp. (RSM GY1887.25.1074).
Fig. 1. Broad canal flanked by narrower canals. Stub G. × 550.
Fig. 2. Detail of narrower canal, showing lining; perpendicular sheets with fan-like arrays of fibres to each side. Stub A. × 9500.
Fig. 3. Angled break *c*. 45° to horizontal. Stub D. × 1800.
Fig. 4. Detail of lamina units from preparation shown in fig. 3. × 7500.
Fig. 5. Detail of lamina units from break *c*. 45° to perpendicular. Stub D. × 5500.
Fig. 6. Detail of lamina units, near-perpendicular break, viewed slightly from above. Stub I. × 12000.

PLATE 30

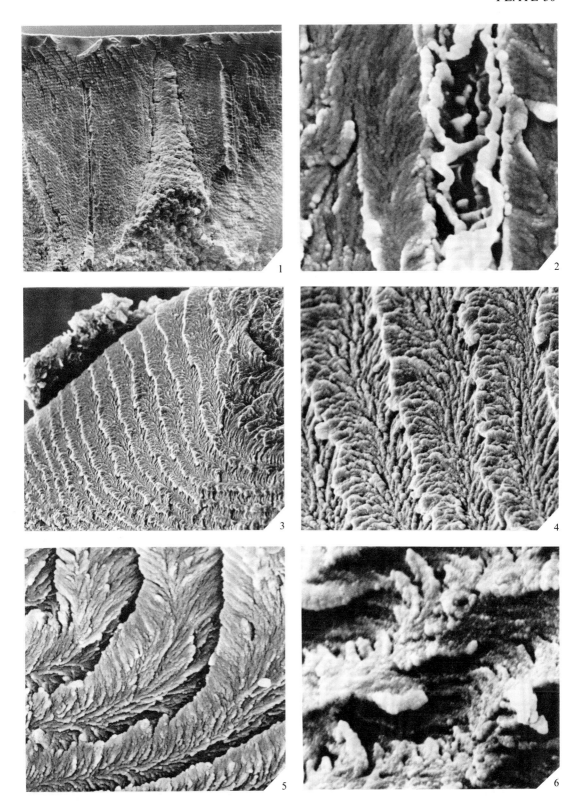

DALINGWATER and WATERSTON, Arthropod cuticle

preserved and, because of the method of preparation for SEM work, most attention is focused on animals with thicker cuticles and on the thicker regions (e.g. carinae) of otherwise relatively thin exoskeletons.

Thus, problems of cuticle research arise from an imbalance between the inductive and deductive approaches, between in-depth studies and broader treatments and, in the case of fossil cuticles, because of the limited and limiting nature of the material available for study.

Where does the present work fit in to this broader framework? Only marginally: it is presented here as an account of the best-preserved ultrastructure of any fossil eurypterid so far examined. It illustrates once again, however, the complexity of arthropod cuticle ultrastucture; indeed, it is extremely difficult to produce a reconstruction of the three-dimensional structure of this cuticle which satisfactorily incorporates all the information obtained from micrographs. The helicoidal model is no real help for there are, to paraphrase T. H. Huxley (1873, p. 229), a number of ugly facts which, if not slaying that beautiful hypothesis, at least cast further doubts as to its general validity.

Acknowledgements. We are most grateful to Dr. P. C. Macdonald for drawing the text-figures, to Dr. P. A. Selden for helpful criticism of the manuscript, and to Mr. L. Lockey for photographic work.

REFERENCES

BRIGGS, D. E. G. and FORTEY, R. A. 1982. The cuticle of the aglaspidid arthropods, a red-herring in the early history of the vertebrates. *Lethaia*, **15**, 25–29.

CURRIE, E. D. 1954. Scottish Carboniferous goniatites. *Trans. R. Soc. Edinb.* **62**, 527–602.

DALINGWATER, J. E. 1973. The cuticle of a eurypterid. *Lethaia*, **6**, 179–186.

—— 1975. Further observations on eurypterid cuticles. *Fossils & Strata*, **4**, 271–279.

—— 1980. SEM observations on the cuticles of some chelicerates. *Int. Congr. Arachnol.* **8**, 285–289.

EISENACK, A. 1956. Beobachtungen an Fragmenten von Eurypteriden-Panzern. *Neues Jb. Geol. Paläont. Abh.* **104**, 119–128.

ELDREDGE, N. 1974. Revision of the Suborder Synziphosurina (Chelicerata, Merostomata), with remarks on merostome phylogeny. *Am. Mus. Novit.* no. 2543, 1–41.

HOLM, G. 1898. Über die Organisation des *Eurypterus fischeri* Eichw. *Zap. Imp. Akad. Nauk*, **8**, 1–57.

HUXLEY, T. H. 1873. *Genesis and Abiogenesis.* In: *Critiques and Addresses.* Macmillan, London, 350 pp.

MUTVEI, H. 1977. SEM studies on arthropod exoskeletons. 2. Horseshoe crab *Limulus polyphemus* (L.) in comparison with extinct eurypterids and recent scorpions. *Zool. Scr.* **6**, 203–213.

NEILSON, J. 1895. On the Calderwood Limestone and Cementstones, with their associated shales. *Trans. geol. Soc. Glasg.* **10**, 61–79.

NEVILLE, A. C. 1975. *Biology of the Arthropod Cuticle.* Springer-Verlag, Berlin, Heidelberg, New York, 448 pp.

PATTON, A. 1885. Geological observations in the parish of East Kilbride, Lanarkshire. *Trans. geol. Soc. Glasg.* **7**, 309–333.

SELDEN, P. A. 1981. Functional morphology of the prosoma of *Baltoeurypterus tetragonophthalmus* (Fisher) (Chelicerata: Eurypterida). *Trans. R. Soc. Edinb.* **72**, 9–48.

STØRMER, L. 1974. Arthropods from the Lower Devonian (Lower Emsian) of Alken an der Mosel, Germany. Part 4: Eurypterida, Drepanopteridae, and other groups. *Senckenberg. leth.* **54**, 359–451.

—— and WATERSTON, C. D. 1968. *Cyrtoctenus* gen. nov., a large late Palaeozoic arthropod with pectinate appendages. *Trans. R. Soc. Edinb.* **68**, 63–104.

WATERSTON, C. D. 1968. Further observations on the Scottish Carboniferous eurypterids. Ibid. 1–20.

J. E. DALINGWATER
Department of Zoology
The University
Manchester M13 9PL

C. D. WATERSTON
Department of Geology
Royal Scottish Museum
Edinburgh EH1 1JF

SILICIFIED TRILOBITES OF THE GENUS *DIMEROPYGE* FROM THE MIDDLE ORDOVICIAN OF VIRGINIA

by R. P. TRIPP *and* W. R. EVITT

ABSTRACT. Three main stages in the development of the protaspis of *Dimeropyge virginiensis* Whittington and Evitt, from the Edinburg Formation, Virginia, are marked by increases from 0 to 2 in the numbers of rings in the rachis of the protopygidium. Early and late moults within stages 2 and 3 (2A, 2B, 3A, 3B) are marked by morphological changes and by progressive increases in the number of tubercles. The development of the meraspis is described. A new species, *D. dorothyae*, from the Oranda Formation is erected.

WHITTINGTON AND EVITT described *Dimeropyge virginiensis* in 1954, but at that time the protaspis was unknown and the hypostome newly discovered. Hu (1976) described the protaspis. A more detailed interpretation of the development is given in this paper. In addition, a new species from the Oranda Formation, referred to by Whittington and Evitt (1954, p. 36), is described.

Mode of occurrence, localities, preservation, and techniques employed are as described by Whittington and Evitt (1954, p. 5). Specimens studied are housed in the National Museum of Natural History, Smithsonian Institution, Washington, D.C. (USNM); the British Museum (Natural History) (BM); and the Hunterian Museum, Glasgow University (A). A list of protaspis specimens, sagittal lengths, stages of development, and localities has been deposited with, and may be obtained from, The British Library, Boston Spa, Wetherby, West Yorkshire, LS23 7BQ, U.K., as Supplementary Publication No. SUP 14020 (5 pages). All measurements were made under a binocular microscope, normal to the sagittal length.

GENERAL REMARKS ON *DIMEROPYGE VIRGINIENSIS* PROTASPIS STAGES

Three main stages in the development of the protaspis are marked by increases in the number of rings in the rachis of the protopygidium (none in Stage 1, one in Stage 2, two in Stage 3), and by increases in the number of tubercles (Table 1). The change from Stage 1 to Stage 2 is abrupt at length 0·38 mm (text-fig. 1). There is an overlap between Stages 2 and 3, but there are marked peaks in the size/frequency graph near the mean lengths of 0·44 mm and 0·57 mm respectively.

Early and late substages within Stages 2 and 3 (referred to as 2A, 2B, 3A, 3B) are marked by changes in shield outline (described under Systematic Descriptions), and by increases in the numbers of tubercles in the outer series (see next section). The size/frequency graphs for the substages are skewed towards the peak frequency of the stage. It is evident from the graph that the substage moults are less significant in the development than the moults between stages; possibly the substage moult was superimposed on the basic moulting pattern and might not have involved the normal addition of a pair of appendages.

Protaspis tuberculation. A conspicuous feature is the orderly arrangement of the tubercles. These are placed in three main series—on the rachis, on the inner part of the pleural area, and on the outer part of the pleural area. In addition, a tubercle may be present between the posterior tubercles in the inner and outer series of the cranidium, marking the posterior border (bt, text-fig. 2, 3B). The occipital tubercle may be flanked by a pair of large granules or small tubercles.

The numbers and arrangement of tubercles, numbers of specimens measured, and size ranges of substages are summarized on Table 1. The size/frequency distribution of specimens from locality 2 is graphed in text-fig. 1.

[Special Papers in Palaeontology No. 30, pp. 229–240, pls. 31–33]

Protaspis doublure. Throughout the three protaspis stages the doublure consists of a fixed pygidial part and a free anterior part, composed of the fused free cheeks, and carrying their dorsal surfaces. In Stage 1 the two parts appear to be fused although the anterior part (and free cheeks) is not always present. In stage 2 the posterior doublure extends forwards to opposite a point 35 % sagittal length from back of protaspis, considerably in advance of the posterior tubercle of the outer cranidial series which later marks the posterior extremity of the

TABLE 1. Protaspis stages of *Dimeropyge virginiensis*, localities 2, 3, and 4, Edinburg Formation, Virginia; sizes and numbers of tubercles in the three series.

| | | | | Number of tubercles (total) | | | | | |
| | Number of specimens | Mean size mm | Size range mm | Cranidium | | | Protopygidium | | |
Stages				Rachis	Inner	Outer	Rachis	Outer	Total
1	21	0·35	0·32–0·37	1	6	6	—	—	13
2A	32	0·42	0·38–0·46	5	8	8	2	2	25
2B	26	0·47	0·44–0·52	5	8	10	2	2	27
3A	18	0·54	0·49–0·56	5	8	10	4	4	31
3B	62	0·58	0·54–0·63	5	8	12	4	6	35

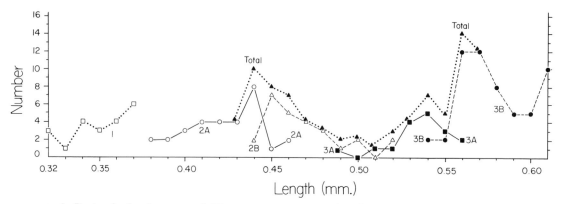

TEXT-FIG. 1. Protaspis development of *Dimeropyge virginiensis*, locality 4, Edinburg Formation, Virginia, U.S.A. Size/frequency distribution of 150 specimens by total and substages. *Ordinate*—number of specimens. *Abscissa*—length (sag.) of protaspis in mm.

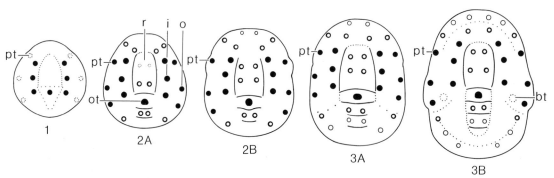

TEXT-FIG. 2. Protaspis development of *Dimeropyge virginiensis*, Edinburg Formation, Virginia, U.S.A. Diagrams of growth stages: 1, Stage 1. 2A, Stage 2 (early form). 2B, Stage 2 (late form). 3A, Stage 3 (early form). 3B, Stage 3 (late form). ot, occipital tubercle; pt, palpebral tubercle; bt, posterior cranidial border tubercle; r, i, o, mark rachial, inner, and outer series of tubercles, × 50.
Tubercles present in Stage 1 are filled in; tubercles which appear in Stage 2A are emphasized.

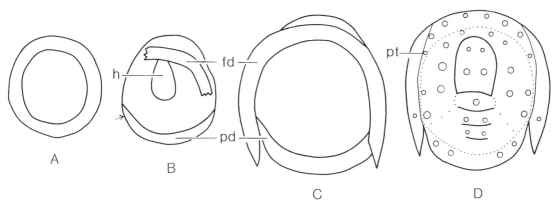

TEXT-FIG. 3. Protaspis development of cheeks and ventral surface of *Dimeropyge virginiensis*, Edinburg Formation, Virginia, U.S.A. A, Stage 1, ventral surface (Pl. 31, fig. 4). Fixed pygidial and free doublures fused. B, Stage 2A, ventral surface (Pl. 31, fig. 5). Free doublure and hypostome displaced. Arrow marks position of posterior outer cranidial tubercle on dorsal surface, which marks the posterior extremity of the cranidium in Stage 3. C, Stage 3A, ventral surface (Pl. 31, fig. 20). Free doublure displaced; note rudimentary genal spine. D, reconstruction of Stage 3A, dorsal surface, with free cheeks in place (Pl. 31, fig. 18). The dotted line indicates the inner margins of the fixed and free doublures. fd, free (anterior) doublure; pd, fixed (pygidial) doublure; pt, palpebral tubercle; h, hypostome. × 50.

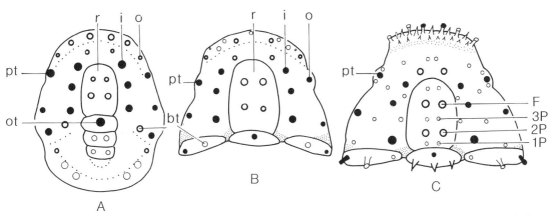

TEXT-FIG. 4. Late protaspis, and early meraspis cranidia of *Dimeropyge virginiensis* compared. A, protaspis Stage 3B (Pl. 31, figs. 23–25), × 60. B, early meraspis cranidium (Pl. 32, fig. 7), × 60. C, meraspis cranidium (Pl. 32, fig. 6), × 50.

cranidium. In Stage 3 the anterior part tapers gently posteriorly, and extends beyond the anterior extremity of the pygidial doublure longitudinally and transversely, thus forming a rudimentary genal spine (text-fig. 3). There is a marked indentation in outline at the front of the pygidium, and the doublure corresponds with the dorsal surface in extent. No visual surface has been detected on the free cheek of the protaspis.

Meraspis development. The transition from protaspis to meraspis is extremely gentle in cranidial development— all the tubercles present in protaspis Stage 3B cranidia can be recognized in the small meraspis (text-fig. 4). The increase in cranidial length between the largest protaspis and the smallest meraspis is about 12 %. However, in other parts, the transition amounts to a metamorphosis—the fused free cheeks become separated by a rostral plate, the hypostome is much altered, and the transitory pygidium incorporates the spine. The smallest meraspis must have been about 0·78 mm in length, an increase of 150 % from the smallest protaspis.

As the cranidium increases in size the protaspis tubercle pattern becomes submerged in the random sculpture of the adult, until by cranidial length 1·1 mm the pattern can no longer be traced. This change probably occurs at the transition from meraspis to holaspis. Chatterton (1980, pl. 22, fig. 11) has illustrated a degree 7 *D. clintonensis* meraspis lacking much of the cephalon; on this basis, the skeletal length of an early holaspis must have been about 2·8 mm, an increase of about 250 % from the smallest meraspis.

SYSTEMATIC DESCRIPTIONS

Family DIMEROPYGIDAE Hupé, 1953
Genus DIMEROPYGE Öpik, 1937
Dimeropyge virginiensis Whittington and Evitt, 1954

Plates 31, 32

1954   *Dimeropyge virginiensis*, Whittington and Evitt, p. 37, pl. 2; pl. 3, figs. 1–30.
1974   *Dimeropyge virginiensis* Whittington and Evitt; Hu, p. 248, pl. 26, figs. 2–21, 24–33.
1980   *Dimeropyge virginiensis* Whittington and Evitt; Chatterton, pl. 8, figs. 30–31.

Localities and horizon. Localities 2, 3, 4 (see Whittington and Evitt 1954, p. 5). Edinburg Formation. Virginia, U.S.A.

Description of protaspides.

Stage 1. 0·32–0·37 mm (Pl. 31, figs. 1–4; text-figs. 2.1, 3A).
Rounded in outline, slightly longer than wide, greatest width posterior to mid-length, convex. Rachis 70 % length of protaspis, extending further posteriorly than anteriorly, tapering more slowly posteriorly than anteriorly, with slight independent convexity. Occipital furrow broad and shallow. Occipital tubercle distinct, placed at 65 % length of protaspis from front. Rachial furrow deeper on some specimens than on others. Facial suture marginal anteriorly, curving upwards anterolaterally and delimiting diminutive free cheek. Three large tubercles spaced alongside rachis (inner series). Three smaller and less distinct tubercles more distantly spaced near margin (outer series); palpebral tubercle (foremost) opposite free cheek. Doublure uniformly narrow, about 15 % length (sag.) of protaspis, flattened, sloping dorsally upwards and inwards. Fixed and free parts apparently fused. Fixed pygidial part lies closer to dorsal surface and extends for about 40 % sagittal length of protaspis from back. Hypostome elongate, globular, narrow anteriorly, greatest width posteriorly, 20 % length (sag.) of protaspis. Borders absent. Surface smooth.

Stage 2. Ovate in outline, broadest posteriorly. Occipital ring faintly defined and slightly depressed; glabella with independent convexity, more strongly defined laterally than anteriorly. An additional (fourth) pair of cranidial tubercles anteriorly in both inner and outer series. One segment of protopygidium represented by one rachial ring bearing a pair of tubercles and one pair of tubercles in the outer series. Surface, including furrows but

---

Silicified protaspides of *Dimeropyge virginiensis* Whittington and Evitt, from locality 4, Edinburg Formation, Virginia, U.S.A. The specimens were whitened before being photographed, and are housed in the National Museum of Natural History, Smithsonian Institution, Washington, D.C.

Figs. 1–4. Stage 1. × 80. 1–2, USNM 258097, length 0·32 mm, dorsal and ventral views (note hypostome). 3, USNM 258095, length 0·35 mm, dorsal view. 4, USNM 258096, length 0·37 mm, ventral view.

Figs. 5–9. Stage 2A. × 70. 5, USNM 258098, length 0·40 mm, ventral view (note hypostome). 6, USNM 258099, length 0·42 mm, dorsal view; anterior pair of glabellar tubercles indistinct. 7–9, USNM 258101, length 0·40 mm, dorsal, anterior, and oblique right anterolateral views.

Figs. 10–15. Stage 2B. × 70. 10, 14–15, USNM 258102, length 0·46 mm, dorsal, oblique left anterolateral, and frontal views. 11–13, USNM 258103, length 0·46 mm, dorsal, ventral, and left lateral views.

Figs. 16–21. Stage 3A. × 60. 16, USNM 258104, length 0·52 mm, dorsal view; note adventitious tubercle alongside second outer tubercle on protopygidium. 17, 21, USNM 258106, length 0·56 mm, dorsal and ventral views. 18–20, USNM 258105, length 0·56 mm, dorsal, oblique left anterolateral, and ventral views.

Figs. 22–25. Stage 3B. × 60. 22, USNM 258107, length 0·54 mm, dorsal view. 23–25, USNM 258108, length 0·60 mm, dorsal, right anterolateral, and ventral views.

PLATE 31

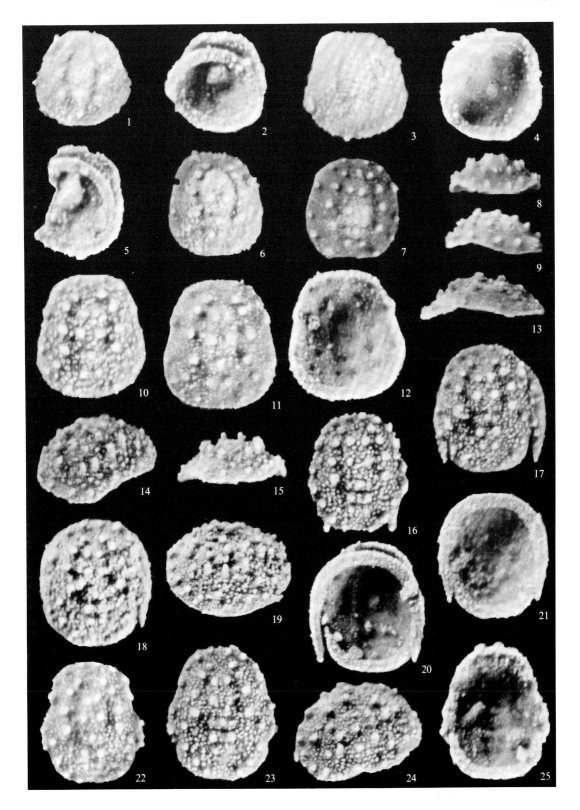

TRIPP and EVITT, *Dimeropyge*

excluding tubercles, finely and closely granular. Substage 2A. 0·38–0·46 mm (Pl. 31, figs. 5–9; text-figs. 2.2A, 3B). Palpebral lobe indistinct. Pair of tubercles at back of glabella distinct, anterior pair faint. Incoming (fifth) pair of cranidial tubercles in outer series may be represented as a pair of granules or small tubercles (Pl. 31, fig. 7). Hypostome similar to that of Stage 1, 0·14 mm in length and approximately equal to glabella in length. Substage 2B. 0·44–0·52 mm (Pl. 31, figs. 10–15; text-fig. 2.2B). More elongate in outline. Palpebral lobe prominent. Two pairs of tubercles on glabella, anterior pair smaller than posterior pair. An additional (fifth) pair of cranidial tubercles anteriorly in outer series. Inner series of tubercles commonly tall (Pl. 31, figs. 14–15).

Stage 3. More elongate in outline. Protopygidium with two rachial rings each bearing a pair of tubercles, posterior pair the smaller; two pairs of tubercles in the outer series. Posterior border of protopygidium faintly marked by change to convexity. Substage 3A. 0·49–0·56 mm (Pl. 31, figs. 16–21; text-figs. 2.3A, 3C, D). Incoming cranidial or pygidial tubercles are present in some specimens. Substage 3B. 0·54–0·63 mm (Pl. 31, figs. 22–25; text-figs. 2.3B, 4A; Hu 1974, pl. 26, figs. 4–6). An additional (sixth) pair of cranidial tubercles in outer series, second from front. Anterior border of cranidium faintly marked by change in convexity. Protopygidium set off by convexity and reduction in width (tr.) behind cranidium; rachis with terminal piece distinguishable; three pairs of tubercles in the outer series.

Postprotaspis development. Smallest meraspis cranidia 0·43–0·55 mm (Pl. 32, fig. 6) have rachial furrow more strongly developed, palpebral lobe swollen, surface coarsely granular, tubercles arranged as in Stage 3B with second tubercle on posterior border always present, paired granules on occipital ring not conspicuous. Small tubercle occurs between the two glabellar tubercles on some specimens. Cranidia 0·55–0·65 mm (Pl. 32, fig. 7) (Whittington and Evitt 1954, pl. 3, figs. 16–17; Chatterton 1980, pl. 8, fig. 30). Glabella with three small tubercles (paired) posterior, between, and anterior to the two large tubercles persisting from protaspis, and placed closer together. We interpret the three posterior pairs of glabellar tubercles as being homologous with tubercles opposite 1L, 2L, and 3L in families with three well-defined lateral glabellar lobes, but in view of Šnajdr's (1981) interpretation of *Scharyia* it is possible that segmental homologies are misleading. Additional tubercles as follows: two tubercles adaxial to inner series on cheek interspaced between first to third, small, anterior (fifth) tubercle in inner series, and five small tubercles abaxial to inner series, also interspaced as shown (text-fig. 4C). Two small tubercles flanking palpebral tubercle. Three aciculate tubercles at back of occipital ring (occipital tubercle reduced), a tubercle at inner extremity of posterior border, numerous tubercles on anterior border. Cranidia 0·65 to 0·75 mm (Whittington and Evitt 1954, pl. 3, figs. 21–23). Five glabellar tubercles (paired) larger than remainder; with further growth tuberculation becomes more dense and more uniform until in holaspis paired tubercles unidentifiable (Whittington and Evitt 1954, pl. 3, figs. 24–26, 28–29; pl. 2, figs. 1–6). No discontinuity in size of meraspis cranidia, and the size/frequency distribution shows no significant clustering. Meraspis free cheeks with suture as in adult, indicating the presence of a rostral plate. About thirty lenses distinguishable in eye of smallest cheek. Hypostome 0·40–0·73 mm; 90% as long as wide, greatest width opposite wing at mid-length, slightly narrower anteriorly, tapering to a broadly rounded extremity posteriorly. Anterior margin transverse in the median 30% width, weakly convex abaxially. Anterior border absent. Anterior lobe of middle body small, convex. Posterior lobe large, with independent convexity. Middle furrow runs backwards and inwards with decreasing strength from anterolateral angle. Posterolateral border

---

EXPLANATION OF PLATE 32

Dimeropyge virginiensis Whittington and Evitt, locality 4 except fig. 7, Edinburg Formation, Virginia, U.S.A. The stereoscan photographs were prepared by the British Museum (Natural History), London, where the specimens are housed.

Figs. 1–2. Protaspis Stage 1, BM IV 132, length 0·33 mm, exterior, dorsal, and left posterolateral views, ×150.

Fig. 3. Holaspis cranidium, BM IV 147, ×25.

Figs. 6–7. Meraspis cranidia. 6, BM IV 148, ×80. 7, BM IV 149, locality 2, ×65.

Figs. 4–5. Meraspis hypostomes. 4, BM IV 156, ×60. 5, BM IV 157, ×60.

Figs. 8–10. Adult hypostome 8, BM IV 159, ventral view, ×85. 9, BM IV 160, oblique right lateral view, ×55. 10, BM IV 161, dorsal view, ×85.

Figs. 11–16. Transitory pygidia. 11, degree 2 with second thoracic segment displaced, BM IV 207, ×50. 12, degree zero, BM IV 206, spine on sixth ring, seventh and eighth segments indicated by marginal spines, ×80. 13, degree 3, BM IV 208, ×55. 14, degree 4, BM IV 209, ×55. 15, degree 5, BM IV 210, ×45. 16, degree 6, BM IV 231, ×50.

PLATE 32

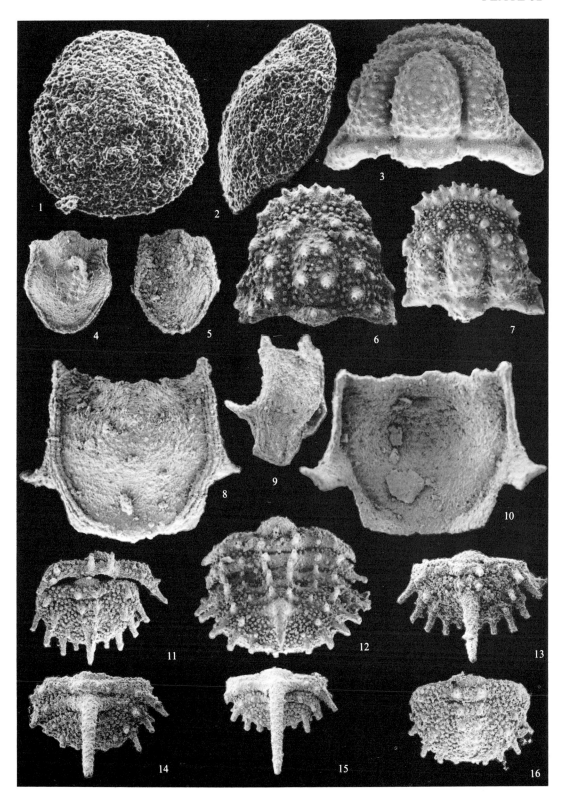

TRIPP and EVITT, *Dimeropyge*

continuous, uniformly narrow, raised, strongly rounded in outline. Doublure narrow, convex. Posterior wing at 40 % length from back, directed straight upwards. Middle body and doublure smooth; posterolateral border with one or two strong raised lines.

Transitory pygidium degree zero 0·35 mm (Pl. 32, fig. 12) consists of eight fused prototothoracic segments, 80 % as long as wide, broadly rounded. Spine of sixth slender, upwardly directed, not extending beyond margin of pygidium in normal view, 40% length of transitory pygidium, tapering steadily to a sharp point. Articulating half-ring long. Segments 1–5 with tall, paired rachial tubercles successively more closely spaced, posterior half anterior width apart. Rachis 35% anterior width, gently convex transversely. Rachial furrow ill defined. Pleural lobe with two columns of tubercles, adaxial smaller than abaxial; pleurae 1–6 with adaxial and abaxial tubercles, pleura 7 with single abaxial tubercle only; no tubercle opposite eighth free point. Segments terminate in short, outwardly directed spines. First segment ends in single broad-based marginal spine. Subsequent spines submarginal, overhanging steep border, bifid; seventh small; eighth minute, post-rachial. Remainder of surface coarsely granular. Articulating half-ring strongly developed, longer (sag.) than first ring. Degrees 2–7 average 0·35 mm in length (excluding rachial spine), with a range of plus or minus 10%; average increase in size, therefore, amounts to length of one thoracic segment. Five segments in all transitory pygidia indicated by five pairs of free points, posterior much smaller than rest, foremost with interpleural furrow stronger than rest, may be incompletely fused. Degrees 2–5 (Pl. 32, figs. 11, 13–16) show the progressive advance of the rachial spine position from fourth to first ring. Spine slender, directed upwards and backwards, longer than sclerite, projecting for about half its length beyond posterior margin. The spine doubles in length/width of pygidium ratio between degrees 0 and 5. Tubercles abaxially on first four pleurae, lacking on fifth. Free point of first segment broad based and usually pleural in appearance; subsequent spines submarginal, short, slender, outwardly directed. Fourth (posterior) segment of degree 7 completes protopygidium; first, second, and third marginal spines with a short spinule in front. Small holaspis pygidium 0·50 mm; adaxial tubercle on first, second, and third pleurae, posterior ring bilobate backwards. With growth rachis becomes relatively broader, smooth median depression develops, bilobate posterior ring merges with marginal spine, tuberculation less conspicuous, single adaxial tubercle on pleurae becomes granular node of decreasing conspicuity.

Holaspis has been described by Whittington and Evitt, but the hypostome is here redescribed and illustrated from more abundant material (Pl. 32, figs. 8–10). Largest hypostomes range from 0·64 to 0·84 mm in width (anterior portion invariably defective, probably because of incomplete original calcification). Differs conspicuously from meraspis hypostome in subquadrate outline and large posterior wing 30 % length from back. Middle furrow absent. Lateral furrow stronger. Posterolateral border widens at posterolateral angles, with three or four strongly raised lines running parallel to margin. Posterior wing large, oblique, parallel-sided, terminating in two small rounded processes; wing projects strongly sideways.

Remarks. Šnajdr (1981) described a series of six protaspis growth stages belonging to *Scharyia micropyga* (Hawle and Corda) from the Silurian Kopanina and Přídolí Formations of Czechoslovakia. There are many points of similarity between the largest *Dimeropyge* and the smallest *Scharyia* protaspides; of particular interest is the occurrence of two morphologically progressive forms with two protopygidial rings (Šnajdr 1981, fig. 1A, B) as is the case in *D. virginiensis*.

Chatterton (1980, p. 33) described protaspis stages of *D. clintonensis* Shaw (1968, p. 40, pl. 11, figs. 12–38) from the Chazy Formation. The development appears to be comparable to that of *D. virginiensis* (Chatterton's text-fig. 5A and pl. 8, fig. 3 resemble Stage 2B; his text-fig. 5B, pl. 8, figs. 1, 2 resemble Stage 3B).

Dimeropyge dorothyae sp. nov.

Plate 33

Derivation of name. After Mrs. H. B. Whittington.

Diagnosis. Glabella ovate in outline, margin of anterior border rounded. Glabella closely granular, larger granules sparsely scattered. Pygidium 35 % as long as wide, furrows shallow, pleural terminations trifid, border subvertical.

Holotype. BM IV 216 (silicified cranidium) (Pl. 33, fig. 6). Locality 8, 275 m north of Greenmount Church, 9 km north of Harrisonburg, Rockingham County, Virginia, U.S.A. Lower 1·5 m of Oranda Formation.

Dimensions of holotype. Length of cranidium (sag.) 3·8 mm; width of cranidium 5·3 mm; length of glabella (sag.) 2·7 mm; maximum width of glabella 2·3 mm.

Other material. Cranidia, free cheeks, thoracic segments, and pygidia common; one rostral plate. Locality and horizon as for holotype.

Description. Cephalon subsemicircular in outline, moderately convex, with short genal spines. Cranidium 70 % as long as wide. Glabella subovate, inflated, widest opposite palpebral lobe, 80 % as wide as long, 70 % length of cranidium. Lateral furrows absent. Occipital ring narrower than maximum width of glabella, comparatively short (sag., exsag.) and narrowing only slightly abaxially, convex in both directions. Occipital furrow broad, more shallow than rachial furrow; small, rounded apodeme at junction with rachial furrow. Rachial and preglabellar furrows continuous, uniformly deep. Preglabellar field 5 % length (sag.) of cranidium, steeply inclined, convex, continuous with convex fixed cheek. Anterior border rounded in outline, 7 % length of cranidium, widest (sag.) mesially, gently convex. Anterior border furrow deep and broad. Posterior border convex, short (exsag.) adaxially, widening strongly abaxially to width of occipital ring; a node adaxially, separated by a smooth depression from a larger node abaxially. Posterior border furrow shallow and narrow. Palpebral lobe 20 % length of cranidium, narrow (tr.), raised, convex; mid-length opposite mid-length of glabella; lobes 160 % width of glabella apart. Palpebral furrow strong. Anterior branch of facial suture runs forwards and inwards, converging more strongly inwards at anterior border, thence running marginally to rostral suture. Posterior branch of facial suture curves posterolaterally backwards to posterior border furrow, curving more strongly outwards across posterior border. Doublure of occipital ring forms an angle at mid-line, narrowing abaxially. Doublure of anterior border gently embayed for median 25 % width. Eye sessile, strongly convex; holochroal visual surface occupies upper part of lobe, lower part granular. Lateral border narrower than field anteriorly, widening strongly backwards and fused with posterior border. Lateral border furrow deep and broad, curving inwards to abut against sutural ridge, which is well developed posteriorly, but weak forwards. Anteroventral projection flexed ventrally, indenting rostral plate deeply at junction of dorsal and ventral surfaces. Genal spine broad-based, tapering rapidly, estimated length 35 % sagittal length of cephalon, inturned at tip. Doublure of free cheek extends to border furrow, curving inwards at base of spine, where crossed by shallow, parallel-sided vincular channel, which runs obliquely backwards. Vincular notch absent. Glabella finely and closely granular with larger granules (basal diameter not exceeding 5 % maximum width of glabella) sparsely and evenly scattered between. Occipital ring, anterior and lateral borders granular with a row of larger granules along margin. Nodes on posterior border granular. Field of fixed cheek tuberculate, with subdued granulation; largest tubercle at posterior inner angle (homologous with hindmost tubercle of inner series of protaspis). Field of free cheek finely tuberculate to granular. Doublure smooth except for a few minute granules adaxially on spine.

Rostral plate small, swollen, subtrapezoidal in outline on exterior (dorsal) surface, narrowing forwards, subtriangular on interior (ventral) surface, widening forwards to twice posterior width. Rostral suture almost transverse; connective sutures strongly convergent dorsally, and even more strongly divergent ventrally. Anterior margin transverse, knife-edged (non-sutural). One (paired) tubercle at posterior outer angle, granules decreasing in size forwards. Ventral surface transversely striate.

Hypostome unknown.

Number of thoracic segments unknown. Rachis 40 % width of segments. Rings of uniform length; articulating rings and furrows strongly developed. Rachial furrow shallow. Articulating process at front, socket at back of each pleura. Pleurae horizontal to fulcrum, downturned beyond, terminations rounded. Successive segments have fulcrum further out and pleura strongly downturned. Pleural furrow shallow, oblique. Axial spine no longer than width of ring. A large pleural node adaxially, smaller nodes near midwidth and abaxially. Doublure extends inwards for 25 % width of pleura; inner margin longitudinal (exsag.) for most of the length, bending obliquely inwards at back without raised anterior margin or vincular notch. Rings, spine, and pleural nodes granular. Doublure smooth.

Pygidium 35 % as long as wide, height 40 % width. Rachis weakly convex, 45 % pygidial width. Three rings transverse and well defined, fourth discontinuous and ill defined. Three ring furrows narrow, shallower mesially than laterally. Median groove posteriorly, widening backwards. Rachial furrow shallow. Articulating half-ring strong. Articulating furrow shallow, apodeme at extremity. Anterior half pleura widens into wide facet laterally. Four pleurae, successively shorter, terminating in trifid (median longest) free point; pleura swollen, curving downwards. Pleural furrow strongly oblique, shallow. Fourth free point immediately posterior to ring. Border subvertical, 30% height of pygidium. Pygidium granular except mesially and in furrows. Doublure convex, recurved, about half height of border, narrowing abaxially, smooth.

Protaspis unknown. Smallest cranidium 0·9 mm in sagittal length (Pl. 33, fig. 1), five (paired) tubercles on glabella equally spaced, five inner series tubercles on fixed cheek alongside rachial furrow, hindmost opposite single adaxial tubercle on posterior border; tuberculate node abaxially on posterior border. Anterior border with a single row of minute granules alongside preglabellar furrow; and an irregular row of spinules along anterior margin. Cranidia 1·3 mm sagittal length (Pl. 33, fig. 2), with small paired tubercles closely spaced between five main pairs present on smaller cranidia; three small tubercles between inner series and rachial furrow. Cranidia 1·4 mm in saggittal length (Pl. 33, fig. 3), with five tubercles on glabella more widely spaced and hardly distinguishable among the densely assorted array; only foremost and hindmost tubercles of inner series distinguishable; tubercles on occipital segment shorter and rounded; anterior border row of granules alongside preglabellar furrow reduced in size and spinules shorter. Appreciably larger than *D. virginiensis* at all stages. With further increase in size tuberculation of glabella, occipital segment, and anterior border, gives way to random granulation and tuberculation of preglabellar field and fixed cheek becomes uniform except that hindmost tubercle of inner series slightly larger. With increase in size of free cheeks, numbers of lenses in eye increases and lenses become less prominent, number of tubercles and granules on field increases, and field increases in size; tubercles on thoracic segments give way to granules; sculpture of pygidium becomes suppressed, and posterior axial groove develops.

TABLE 2. Summary of differential characters in the better-known North American species of *Dimeropyge*.

Species	dorothyae	spinifera	virginiensis	clintonensis
CRANIDIUM				
Glabellar sculpture	Bimodally granular	Tuberculate	Tuberculate	Coarsely granular
Glabellar shape	Ovate	Ovate	Parallel sided	Barrel shaped
Outline of anterior border	Rounded	Rounded	Rounded	Subangular
Aciculate tubercles on anterior margin	Absent	Present	Absent	Absent
Longitudinal depression on preglabellar field	Absent	Slight	Slight	Absent
PYGIDIUM				
Length as % width (max.)	35	40	50	50
Width of rachis as % width (max.)	35	30	25	30
Rachial furrow	Faint	Strong	Strong	Strong
Spinules on pleural tips	3	3	2	3
Border	Subvertical	Rolled	Rolled	Subvertical

Remarks. D. dorothyae differs from all other species in the sculpture of the glabella—closely granular with larger granules sparsely scattered between. In this feature it is closest to *D. raymondi* (Roy 1941, p. 155, fig. 113; Whittington 1954, p. 145, pl. 63, figs. 7, 8, 11, 12, 15, 16) from Silliman's Fossil Mount

---

EXPLANATION OF PLATE 33

Dimeropyge dorothyae sp. nov., locality 8, Oranda Formation, Virginia, U.S.A. The stereoscan photographs were prepared by the British Museum (Natural History), London, where the specimens are housed.

Figs. 1–6. Cranidia. 1, BM IV 211, note absence of tubercles adaxial to inner series of protaspis, × 37. 2, BM IV 212, three small tubercles adaxial to inner series of protaspis, × 25. 3, BM IV 213, × 20. 4, BM IV 214, ventral view, × 10. 5, BM IV 215, × 7. 6, BM IV 216 (holotype), × 10·5.

Figs. 7–10. Free cheeks. BM IV 217, × 30. 8, BM IV 218, oblique lateral view, × 18. 9, BM IV 219, ventral view, × 8. 10, BM IV 220, × 5.

Fig. 11. Rostral plate, BM IV 221, × 60.

Figs. 12–14. Thoracic segments. 12, BM IV 222, showing rachial spine, oblique anterior view, × 17·5. 13, BM IV 223, posterior view, × 10. 14, BM IV 224, anterior view, × 17·5.

Figs. 15–20. Pygidia. 15, BM IV 225, × 42. 16, BM IV 226, anterior view, × 44. 17, BM IV 227, posterior view, × 40. 18, BM IV 228, × 32. 19, BM IV 229, × 23. 20, BM IV 230, × 23.

PLATE 33

TRIPP and EVITT, *Dimeropyge*

but the granulation is closer and the coarser granules are more sparsely scattered. In other features *D. dorothyae* most closely resembles *D. spinifera* (Whittington and Evitt 1954, p. 42, pls. 22, 23) from the Lower Lincolnshire Limestone, but the differences are considerable; for instance, *D. dorothyae* lacks the vincular notch on the free cheek and thoracic segments. The differential characters of the better-known North American species are summarized on Table 2.

The rostral plate of *D. clintonensis* Shaw, 1968 has been described by Chatterton and Ludvigsen (1976, p. 48, fig. 8) and by Chatterton (1980, p. 33, pl. 8, figs. 28–29). That of *D. dorothyae* differs markedly in lacking the posteromedian projection; this margin is knife-edged, and is not a hypostomal suture. It seems certain that, as in the Proetidae (e.g. Šnajdr 1980, p. 22), there was a membrane between the rostral plate and the hypostome.

Acknowledgement. We are most grateful to the staff of the SEM Department, British Museum (Natural History) for preparing the stereoscan photographs.

REFERENCES

CHATTERTON, B. D. E. 1980. Ontogenetic studies of Middle Ordovician trilobites from the Esbataottine Formation, Mackenzie Mountains, Canada. *Palaeontographica,* **(A) 171,** 1–74.

—— and LUDVIGSEN, R. 1976. Silicified Middle Ordovician trilobites from the South Nahanni River area, District of Mackenzie, Canada. *Ibid.* **(A) 154,** 1–106.

HU, CHUNG-HUNG. 1976. Ontogenies of three silicified Middle Ordovician Trilobites from Virginia *Trans. Proc. Palaeont. Soc.* Japan, **101,** 247–262.

HUPÉ, P. 1953. *Classe des Trilobites in* Traité de Paléontologie, **3,** 44–246, edited by Piveteau, J. Paris.

ÖPIK, A. 1937. Trilobiten aus Estland. *Acta Comment. Univ. Tartu.* A, **32,** 1–163.

ROY, S. K. 1941. The Upper Ordovician fauna of Frobisher Bay, Baffinland. *Field Mus. Natur. Hist., Geol.* Mem. no. 2, 1–212.

ŠNAJDR, M. 1980. Bohemian Silurian and Devonian Proetidae (Trilobita). *Rozprzy üst. ust. geol.* **45,** 1–324.

—— 1981. Ontogeny of some representatives of the trilobite genus *Scharyia. Sbornik geologickych ved, paleontologie,* **24,** 7–34.

WHITTINGTON, H. B. 1954. Ordovician trilobites from Silliman's Fossil Mount. *In* MILLER, A. K. Ordovician cephalopod fauna of Baffin Island. *Mem. Geol. Soc. Am.* no. 62, 119–149.

—— and EVITT, W. R. 1954. Silicified Middle Ordovician Trilobites. *Ibid.,* no. 59, 1–137.

R. P. TRIPP
British Museum (Natural History)
Cromwell Road
London SW7 5BD

W. R. EVITT
Stanford University
Stanford
California, U.S.A.

ABNORMAL CEPHALIC FRINGES IN THE TRINUCLEIDAE AND HARPETIDAE (TRILOBITA)

by A. W. OWEN

ABSTRACT. Examples of asymmetry in the outline of the cephalic fringe of both trinucleids and harpetids fall into two groups: those showing indented margins and those in which the posterior part is tapered or abruptly terminated. Most of these abnormalities probably resulted from injury during moulting and they commonly exhibit a degree of healing with the development of a normal marginal band along the edge of the irregularity. Small pits developed close to the affected area on some of the trinucleid fringes may represent an attempt to regain the original pit number. Local asymmetry in pit distribution in otherwise symmetrical fringes is considered a part of normal variation in several trinucleids but in some cases may represent an advanced stage of repair of injury.

RECORDS of teratology and injury are scattered throughout the trilobite literature but only Šnajdr (1978, 1981) has documented a wide variety of abnormalities within individual families. A review of trilobite abnormalities is being compiled by the writer and it is clear that the largest group of cephalic irregularities involves the fringes of trinucleids and harpetids. Whilst it is commonly difficult to distinguish between the effects of genetic mutation and injury (either to the embryo or occurring later in life) in trilobites, many abnormal fringes have features in common and thus probably had a similar origin. All the examples known to the writer, including some recorded for the first time, are reviewed herein and consideration is given to their possible origin and functional consequences. The present study does not include consideration of the neoplasms on fringes of *Bohemoharpes ungula* described by Šnajdr (1978a) and also illustrated by Přibyl and Vaněk (1981, pl. 1, fig. 1). Šnajdr argued that most, if not all, of these swellings probably resulted from infection or parasitic infestation affecting the hypodermis. A review of such structures in trilobites was given by Conway Morris (1981, p. 497).

The trinucleid terminology used here is that of Hughes *et al.* (1975). Other terms are those used in Moore (1959). See Rheder (1973 and references therein) for a discussion of the family-group name Harpetidae. The illustrated specimens are housed in the Hunterian Museum, Glasgow (HM); Sedgwick Museum, Cambridge (SM); Department of Geology, University of Hull (HUD); Department of Geology, University of Birmingham (BU); National Museum, Prague; Paleontologisk Museum, Oslo (PMO); Paleontologiska Institutionen, Uppsala; and Museum of Comparative Zoology, Harvard (MCZ).

ASYMMETRICAL TRINUCLEID FRINGES

Examples of asymmetry in the overall outline of the trinucleid fringe can be divided into two groups; those involving indentations at the fringe margin and those in which the posterolateral part of the fringe is tapered or abnormally truncated. Associated with both these conditions is the disruption of the regular fringe pit distribution. Irregular pit development in otherwise bilaterally symmetrical fringes has been widely documented and is discussed below in the interpretation of the asymmetrical fringes.

(i) *Indented fringes*. The degree of indentation and associated pit disruption varies considerably but in all known cases the marginal band maintains its normal width along the affected area.

In a cephalon of *Onnia cobboldi* illustrated by Hughes *et al.* (1975, pl. 9, fig. 104; Pl. 34, fig. 1) a gentle indentation extends along, and cuts out, three to four radii of the E_1 arc. The E_1 pit immediately adaxial to the indentation is considerably reduced in size but the adjacent I_1 arc is unaffected. A broad (tr.) indentation illustrated in a ventral view of a lower lamella of *Stapeleyella inconstans* by Whittard (1956, pl. 5, fig. 1) also shows the truncation of E arcs. The three E arcs are progressively cut out as the indentation approaches the

[Special Papers in Palaeontology No. 30, pp. 241–247, pl. 34]

girder. Pits in the arcs I_1 and I_n are reduced in size adjacent to where the indentation lies along the outer edge of the girder and at least one I_1 pit may be missing here. A specimen of *O. gracilis* (Pl. 34, fig. 2) shows a V-shaped notch cutting into the E_1 and I_1 arcs and thus approximately five E_1 and three I_1 radii are missing. Three small pits on each side of the indentation may represent a continuation of the E_1 arc. Thus there is a reduction of only about two pits in the total pit number in the affected area.

Considerably more disruption is seen in the cranidium of *Salterolithus caractaci paucus* illustrated by Cave (1957, pl. 10, fig. 2) where a broad indentation cuts across arcs E_{1-3} and, locally I_1 (Pl. 34, fig. 6). An arc of very small pits extends along the edge of the indentation and the adjacent I_1 and I_n pits are reduced in size and irregularly distributed. The F pits are unaffected. Similar extreme disruption of the pitting is seen in the specimen of *Deanaspis goldfussi* described by Přibyl and Vaněk (1969, pl. 8, fig. 5). A photograph kindly provided by Dr. Přibyl shows that although the specimen is poorly preserved, one side of the cranidium has an anterolateral indentation of unknown adaxial extent along which pits in arc E_1 can be traced. These pits are reduced in size and slightly irregularly distributed but clearly cut across the courses of arcs I_1 and I_2. Arc I_3 is deflected forwards slightly and the remaining I arcs are unaffected.

The most extreme example of trinucleid fringe indentation was described by Šnajdr (1979) in *D. seftenbergi* (ascribed to *Marrolithus* by Šnajdr). This shows an irregular anterolateral indentation cutting out approximately eighteen E_1 and ten I_1 pits. Some I_2 pits are also missing and pits in both this arc and I_1 adjacent to the affected area are reduced in size. Šnajdr's illustration (1979, pl. 1, fig. 1) appears to show that a marginal band is not developed along the indentation. The same side of the specimen also shows considerable reduction in size of the pygidial pleural lobe and a disruption in the ribbing of the posterior part of the thorax.

Whittington (1966, p. 88, pl. 26, fig. 20) described a cranidium of *Broeggerolithus nicholsoni* which appears to show the absence of E pits along an anterior indentation (Pl. 34, fig. 3). Detailed examination indicates, however, that the pit development is complete and the apparent indentation is a taphonomic effect. Part of the fringe has been pushed rearwards and upwards along fractures approximately parallel to the pit radii. The affected area of fringe has been overridden slightly by the glabella but overlaps the left gena where its inner edge has been broken off subsequently.

EXPLANATION OF PLATE 34

Figs. 1, 5. *Onnia cobboldi* (Bancroft). Lower part of Onny Shale Formation (Onnian), bed of River Onny, Shropshire, England. 1, HM A10695/2, lateral view of cephalon, × 4; figured Hughes *et al.* (1975, pl. 9, fig. 104), photograph kindly provided by Dr. J. K. Ingham. 5, HM A15150, oblique lateral view of partially exfoliated cranidium, × 2·6.

Fig. 2. *Onnia gracilis* (Bancroft). HM A12031, oblique anterolateral view of partially exfoliated cephalon, middle part of Onny Shale Formation (Onnian), same locality as fig. 1, × 2·5.

Figs. 3, 4, 7, 8. *Broeggerolithus nicholsoni* (Reed). Gelli Grîn Group (middle Caradoc), Bala area, North Wales. 3, BU950, oblique anterolateral view of internal mould of cranidium, × 3·4; figured Whittington (1966, pl. 26, fig. 20). 4, BU952, lateral view of cranidium, × 2·3; figured Whittington (1966, pl. 26, fig. 19). 7, BU953, oblique anterolateral view of cranidium lacking upper lamella of fringe, × 2·4; figured Whittington (1966, pl. 26, fig. 21). 8, BU954, lateral view of internal mould of cranidium, × 2·1; figured Whittington (1966, pl. 26, fig. 22).

Fig. 6. *Salterolithus caractaci* (Murchison) *paucus* Cave. SM A46414, dorsal view of complete individual. Harnagian strata, Trilobite Dingle, Welshpool, Powys, Wales, × 2·8; figured Cave (1957, pl. 10, fig. 2).

Fig. 9. *Onnia superba* (Bancroft) *pusgillensis* Dean. HU D5.52, oblique lateral view of internal mould of cephalon lacking upper lamella of fringe, Onnian, Murthwaite Inlier, northern England, × 2·0; figured Ingham (1974, pl. 10, fig. 15).

Fig. 10. *Paraharpes inghami* Owen. PMO14661, dorsal view of lower lamella of fringe, Bønsnes Formation (probably Rawtheyan), Ringerike, Norway, × 1·5; figured Owen (1981, pl. 7, fig. 21).

Figs. 11, 13. *Wegelinia wegelini* (Angelin). Paleontologiska Institutionen, Uppsala collection, Boda Limestone (Ashgill), Dalarne, Sweden. 11, dorsal view of partially exfoliated cephalon, × 1·0. 13, detail of left genal prolongation, × 2·0; figured Warburg (1925, pl. 4, figs. 22–24).

Fig. 12. *Lioharpes* (*L.*) *venulosus venulosus* (Hawle and Corda). National Museum, Prague collection, dorsal view of cephalon, Koneprusy Limestone (Lower Devonian), Koneprusy, near Beroun, Czechoslovakia, × 2·9; figured Přibyl and Vaněk (in press). Photograph kindly provided by Dr. A. Přibyl.

Fig. 14. *Lioharpes* (*Fritschaspis*) *crassimargo* (Vaněk). National Museum, Prague collection, dorsal view of cephalon, Zlichor Formation (Lower Devonian), Prague, Czechoslovakia, × 3·0; figured Přibyl and Vaněk (in press). Photograph kindly provided by Dr. A. Přibyl.

PLATE 34

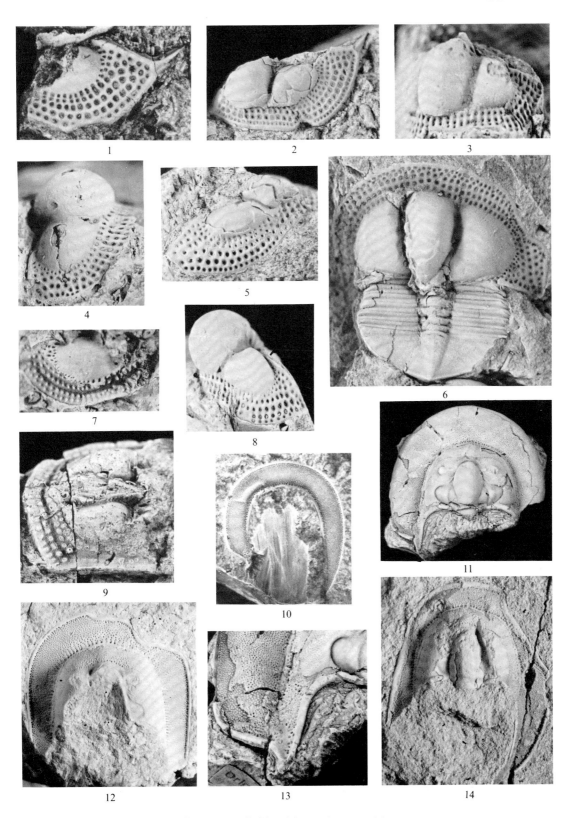

1   2   3

4   5   6

7   8

9   10   11

12   13   14

OWEN, Trilobite fringe abnormalities

(ii) *Fringes truncated or tapered posterolaterally*. Abnormalities affecting the posterolateral part of one side of the fringe include cases where the short dorsal portion of the facial suture cuts forwards obliquely across the course of the outer pit arcs and examples where the outer margin of the fringe is deflected markedly adaxially, again cutting across the outer arcs.

Two cranidia with a forwardly directed suture are known. The specimen of *B. nicholsoni* described by Whittington (1966, p. 90, pl. 26, fig. 19; Pl. 34, fig. 4) also shows an irregular E pit development between R12 and the dorsal suture while E_2 is restricted to three small pits. Some disruption of the I_1 arc is also evident. In contrast, a cranidium of *O. cobboldi* (Pl. 34, fig. 5) shows that while the suture is directed obliquely forwards and the E_1 and I_1 arcs terminate in front of their normal positions, there is no irregularity in pit distribution and the posterior part of the I_1 arc shows the outward curve towards the end of E_1 characteristic of a normal genal angle.

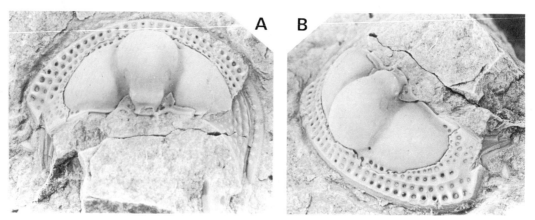

TEXT-FIG. 1. *Cryptolithus tessellatus* Green. MCZ 7719/8. A, dorsal and B, oblique anterolateral view of cephalon. Trenton Group (Middle Ordovician), falls of Montmorency River, near Quebec City, Quebec, Canada, both ×4·5; figured Whittington 1968, pl. 89, fig. 11. Photographs kindly provided by Professor H. B. Whittington.

The best example of oblique adaxial deflection of the posterolateral fringe margin was illustrated by Whittington (1968, pl. 89, fig. 11; text-fig. 1 herein) in *Cryptolithus tessellatus* where the resultant half-fringe pit number in both E_1 and I_1 is eight fewer than on the normal half of the fringe. There also seems to be a slight disturbance of the posterior part of I_2 although this arc does extend to the posterior border. A poorly preserved specimen of *O. suberba pusgillensis* described by Ingham (1974, p. 62, pl. 10, fig. 15; Pl. 34, fig. 9) may represent a similar condition to that seen in *C. tessellatus*. The lower lamella is preserved as an external mould and shows the lateral margin curving sharply adaxially behind the eleventh E_1 pit and apparently cutting across the courses of arcs I_1 and I_2 before becoming parallel to the sagittal line. Details of the fringe in the affected area are difficult to discern but an arc of five large pits ('tubercles' as this is a dorsal view of the lower lamella) is developed close to the outer edge of the fringe here and may be continuous with I_1. A shallow adaxial depression alongside these pits may thus represent a continuation of the pseudogirder between I_1 and I_2 on the unaffected fringe. If this is the case it may be that only the E_1 pits were completely missing from the abnormal part of the fringe and the I arcs were developed, albeit irregularly, parallel to the sagittal line.

A cephalon of *C. tessellatus* illustrated by Kay (1937, pl. 10) shows the posterolateral border directed adaxially across all but the F pits and thus cutting out as many as eight pits from each of E_1 and I_1. The affected border appears to be swollen as a ridge-like callus. There is no genal spine on this side of the cephalon.

ASYMMETRICAL HARPETID FRINGES

Examples of abnormalities affecting the shape of the harpetid fringe have been illustrated by Owen (1981) in *Paraharpes inghami*, Prantl and Přibyl (1954), in *Lioharpes* (*L.*) *venulosus* and *Eoharpes benignensis*, Přibyl and Vaněk (in press), in *Bohemoharpes* (*Unguloharpes*) *ungula viator*, *Lioharpes* (*L.*) *venulosus venulosus*, and *L.* (*Fritschaspis*) *crassimargo*, Sinclair (1947), in *Paraharpes* cf. *ottawaensis* and Warburg (1925) in *Wegelinia wegelini* (see Přibyl and Vaněk 1981, for generic placement). These examples are from Ordovician to Devonian horizons and all show indentations of the fringe (e.g. Pl. 34, figs. 12, 14) or the abrupt termination of one of the genal prolongations (e.g. Pl. 34, figs. 10, 11, 13). In some the indentation extends as far as the girder and in all instances the marginal band is continuous along the edge of the affected area although its width may be slightly increased (e.g. Sinclair 1947). The large pits along the inner edge of the marginal band in most of the specimens continue in this position along the affected area.

INTERPRETATION

All the abnormalities discussed above involve a departure from the normal bilateral symmetry of the fringe outline and individual types of irregularity can be recognized in more than one species. Thus it seems most likely that they are the result of injury and not genetic in origin. Some asymmetry in pit distribution in otherwise symmetrical trinucleid fringes is common in various genera including *Trinucleus* (Hughes 1970, 1971), *Cryptolithus* (Whittington 1968; Lespérance and Bertrand 1976), and *Tretaspis* (Owen 1980) and is considered to be a normal part of phenotypic variation. A small specimen of *L. (L.) venulosus* illustrated by Prantl and Přibyl (1954, pl. 2, fig. 23 after Barrande 1852) shows an anterior indentation on each side of the sagital line and thus is perfectly symmetrical. This probably had a genetic origin but injury early in embryological development cannot be totally dismissed.

Some of the asymmetrical fringes may be the result of attack by predators. The specimen of *D. seftenbergi* described by Šnajdr (1979) certainly belongs in this category with the associated thoracic and pygidial abnormalities indicating predatory attack on an enrolled individual. The majority of examples, however, were probably damaged during moulting. The trilobite moult cycle was reviewed by Henningsmoen (1975) and it seems reasonable to assume that, as in modern arthropods, not only was the animal particularly vulnerable to predation during and immediately after moulting, but the very act of exuviation could involve severe or even fatal injury. Withdrawal from very narrow spaces and 'dead ends' in the exoskeleton was probably a common cause of injury (Šnajdr 1978, p. 7). The bilamellar fringes of trinucleids and harpetids would be likely sites for such moulting problems if the suture between the lamellae opened unevenly or locally failed to open. Tearing of both the new elastic exoskeleton and, more importantly, the underlying soft tissue, could result in permanent deformity in overall fringe shape and in the number and distribution of fringe pits. Thus there would be important consequences for the hydrodynamic stability of the animal and for the efficient functioning of the fringe pits whether they were sensory (e.g. Campbell 1975) or respiratory (Chatterton 1980, p. 21).

Repair of injury (cicatrization) in the trilobite exoskeleton could be initiated soon after damage was sustained (Ludvigsen 1977) with callus development on the affected area. Only one of the abnormal fringes, that illustrated by Kay (1937), shows any such thickening suggesting that the others may reflect much later moult stages. Šnajdr (1981, p. 50) argued that injured proetid pygidia were progressively repaired over several moult periods with scars being healed and some sort of 'morphological equilibrium' eventually being reached. In some instances this closely approached the original morphology. Clearly it is not possible to determine whether a fossil specimen represents a transitional stage in this process of regeneration or whether it marks the final condition beyond which further repair was not possible. Nevertheless, with the exception of Kay's specimen and the *D. seftenbergi* of Šnajdr (1979), all the examples cited above have a marginal band of normal, or very slightly increased, width. Except in the examples of forwardly placed sutures, this band is continuous along the edge of the affected area. This suggests a degree of stabilization, if not regeneration.

In many of the abnormal trinucleids there are small pits present on the affected area perhaps compensating to some extent for the loss of pits at the actual site of the supposed injury. Thus the 'normal' pit number may well have been an important objective in the repair process, even at the expense of losing regularity in pit distribution. Some examples of extreme breakdown in pit arrangement in otherwise symmetrical fringes may represent very advanced repair of injury in which the original outline of the fringe has been re-attained. One such case may be that illustrated by Whittington (1966, pl. 26, fig. 21; Pl. 34, fig. 7) in *Broeggerolithus nicholsoni*. The lower lamella in this specimen shows the anterolateral distribution of E_1 and E_2 pits to be very irregular with one pit even transgressing on to the girder. This could, however, represent a teratological condition as could the development of a single large pit in place of E_1 and E_2 adjacent to an unpitted area on the outer part of the fringe of another specimen of *B. nicholsoni* (Whittington 1966, pl. 26, fig. 21; Pl. 34, fig. 8). A similar 'undeveloped pit' is present in a specimen of *O. superba pusgillensis* illustrated by Ingham (1974, pl. 10, fig. 9). It is tempting, however, to view these last two specimens as representing advanced repair of small injuries at the fringe margin. The same may also apply to specimens of *Trinucleus fimbriatus* illustrated by Hughes (1971, pl. 1, figs. 6, 9).

Dean (1967, p. 107) described two specimens of *Deanaspis* (as *Cryptolithus?*) from the Caradoc of Turkey showing irregular anterior and anterolateral pit distribution in fringes with normal outlines. One of these (1967, pl. 4, fig. 1), termed *C.?* cf. *bedinanensis*, shows an irregular I pit arrangement with I_1 cutting across the course of E_1 in front of the right dorsal furrow and thus the latter arc is absent anteriorly. This specimen may represent late stage repair of a mesial indentation. A second specimen (1967, pl. 2, fig. 10), termed *C.?* sp., shows the anterior breakdown of pit regularity. The left side of the fringe is normal beyond the dorsal furrow, the right side is not preserved. Dean noted that in addition to the fringe irregularity the lateral margins and the shape of the glabella in this specimen differ from those of the associated *D. bedinanensis*. This suggests that the specimen may be a teratological individual belonging to that species. Šnajdr (1979, pl. 1, fig. 2) illustrated a specimen of another species of *Deanaspis*, *D. goldfussi*, in which there is a breakdown in regular pit distribution anteriorly. He also noted that such local irregularities are common in *D. goldfussi* and thus probably were not the result of physical injury.

The individual trinculeids and harpetids described here survived with the fringe abnormalities, at least for some time. The instability consequent on the change of shape must have been a surmountable problem and in this context it is worth noting a specimen of the modern king crab *Limulus polyphemus* shown to the writer by Dr. D. L. Bruton of Oslo. The prosoma is markedly asymmetrical with the posterior part of the left gena spatulate in outline and distinctly wrinkled. The right side shows a normal genal angle. The origin of the abnormality is not known but it certainly did not constitute a fatal condition.

Acknowledgements. I thank Drs. E. N. K. Clarkson, D. A. T. Harper, and Mr. D. M. Rudkin for their comments on earlier drafts of this paper, Mrs. J. Orr for typing the final manuscript and Drs. D. L. Bruton, J. K. Ingham, P. J. Osborne, D. Price, Derek J. Siveter, and Professor R. A. Reyment for access to specimens in their care. I also thank Dr A. Přibyl for kindly providing me with photographs of abnormal harpetids prior to the publication of his joint study of the family with Dr. J. Vaněk.

REFERENCES

BARRANDE, J. 1852. *Système Silurien du centre de la Bohême. 1 ère partie. Crustacés:* Trilobites. i–xxx, 935 pp., pls. 1–51. Prague and Paris.

CAMPBELL, K. S. W. 1975. The functional morphology of *Cryptolithus*. *Fossils and Strata*, **4**, 65–86, pls. 1–2.

CAVE, R. 1957. *Salterolithus caractaci* (Murchison) from Caradoc strata near Welshpool, Montgomeryshire. *Geol. Mag.* **94**, 281–290, pl. 10.

CHATTERTON, B. D. E. 1980. Ontogenetic studies of Middle Ordovician trilobites from the Esbataottine Formation, Mackenzie Mountains, Canada. *Palaeontographica,* **(A) 171**, 1–74, pls. 1–19.

CONWAY MORRIS, S. 1981. Parasites and the fossil record. *Parasitology,* **82**, 489–509, pls. 1–2.

DEAN, W. T. 1967. The correlation and trilobite fauna of the Bedinan Formation (Ordovician) in south-eastern Turkey. *Bull. Br. Mus. Nat. Hist. (Geol.),* **15**, 81–123, pls. 1–10.

HENNINGSMOEN, G. 1975. Moulting in trilobites. *Fossils and Strata*, **4**, 179–200.

HUGHES, C. P. 1970. Statistical analysis and presentation of trinucleid (Trilobita) fringe data. *Palaeontology,* **13**, 1–9.

——1971. The Ordovician trilobite faunas of the Builth–Llandrindod inlier, Central Wales. Part II. *Bull. Br. Mus. Nat. Hist. (Geol.),* **20**, 115–182, pls. 1–16.

——INGHAM, J. K. and ADDISON, R. 1975. The morphology, classification and evolution of the Trinucleidae (Trilobita). *Phil. Trans. R. Soc. Lond.* **B 272**, 537–607, pls. 1–10.

INGHAM, J. K. 1974. A monograph of the upper Ordovician trilobites from the Cautley and Dent districts of Westmorland and Yorkshire. *Palaeontogr. Soc.* [Monogr.], **2**, 59–87, pls. 10–18.

KAY, M. 1937. Stratigraphy of the Trenton Group. *Bull Geol. Soc. Am.* **48**, 233–302.

LESPÉRANCE, P. J. and BERTRAND, R. 1976. Population systematics of the Middle and Upper Ordovician trilobite *Cryptolithus* from the St. Lawrence Lowlands and adjacent areas of Quebec. *J. Paleont.* **50**, 598–613.

LUDVIGSEN, R. 1977. Rapid repair of traumatic injury by an Ordovician trilobite. *Lethaia*, **10**, 205–207.

MOORE, R. C. (ed.). 1959. *Treatise on Invertebrate Paleontology, Part O, Arthropoda* 1. i–xix, 560 pp. Geol. Soc. Amer. and Univ. of Kansas Press (Lawrence).

OWEN, A. W. 1980. The trilobite *Tretaspis* from the upper Ordovician of the Oslo region, Norway. *Palaeontology*, **23**, 715–747, pls. 89–93.

—— 1981. The Ashgill trilobites of the Oslo Region, Norway. *Palaeontographica*, **(A) 175**, 1–88, pls. 1–17.

PRANTL, F. and PŘIBYL, A. 1954. O českých zástupcich čeledi Harpedidae (Hawle et Corda) (Trilobitae). *Rozpr. Ústr. úst. geol.* **18**, 1–170, pls. 1–10.

PŘIBYL, A. and VANĚK, J. 1969. Trilobites of the family Trinucleidae Hawle et Corda, 1847 from the Ordovician of Bohemia. *Sb. geol. věd. Paleontologie*, **11**, 85–137.

—— —— 1981. Preliminary report on some new trilobites of the family Harpetidae Hawle et Corda. *Cas. pro min. geol.* **26**, 187–193. pls. 1–2.

—— —— (in press). A study of the morphology and phylogeny of the family Harpetidae Hawle et Corda Trilobita). *Sborn. Nár. Muz. v. Praze*.

RHEDER, H. A., 1973. Comment on the proposals concerning family names Cassididae and Harpidae. *Z.N.* (S.) 1938. *Bull. Zool. Nom.* **30**, 3.

SINCLAIR, G. W. 1947. Two examples of injury in Ordovician trilobites. *Am. J. Sci.* **245**, 250–257, pl. 1.

ŠNAJDR, M. 1978. Anomalous carapaces of Bohemian paradoxid trilobites. *Sb. geol. ved. Paleontologie* **20**, 1–31, pls. 1–8.

—— 1978a. Pathological neoplasms in the fringe of *Bohemoharpes* (Trilobita). *Vestn. Úst. úst. geol.* **53**, 301–304, pl. 1.

—— 1979. Two trinucleid trilobites with repair of traumatic injury. Ibid. **54**, 49–50, pl. 1.

—— 1981. Bohemian Proetidae with malformed exoskeletons. *Sb. geol. ved. Paleontologie* **24**, 37–61, pls. 1–8.

WARBURG, E. 1925. The trilobites of the Leptaena Limestone in Dalarne. *Bull. geol. Instn. Univ. Upps.* **17**, 1–446, pls. 1–11.

WHITTARD, W. F. 1956. The Ordovician trilobites of the Shelve inlier, West Shropshire. *Palaeontogr. Soc.* [Monogr.], Pt. 2, 41–70, pls. v–ix.

WHITTINGTON, H. B. 1966. A monograph of the Ordovician trilobites of the Bala area, Merioneth. Ibid. Pt. 3, 63–92, pls. 19–28.

—— 1968. *Cryptolithus* (Trilobita); specific characters and occurrence in Ordovician of eastern North America. *J. Paleont.* **42**, 702–714, pls. 87–89.

A. W. OWEN

Department of Geology
The University
Dundee DD1 4HN

NEW CONCAVICARIDA (NEW ORDER: ?CRUSTACEA) FROM THE UPPER DEVONIAN OF GOGO, WESTERN AUSTRALIA, AND THE PALAEOECOLOGY AND AFFINITIES OF THE GROUP

by D. E. G. BRIGGS *and* W. D. I. ROLFE

ABSTRACT. New evidence from the Frasnian Gogo Formation indicates that the notch, characteristic of the bivalved arthropod *Concavicaris*, is situated anteriorly. This notch is occluded in the concavicarid *Harrycaris whittingtoni*, new genus and species, and there appears to be insufficient posterior gape for the egress of an abdomen. Four new species of *Concavicaris* are also established: *C. milesi*, *C. playfordi*, *C. glenisteri*, and *C. campi*. In all these the notch bears a flange that probably served as a rowlock to a presumed antenna. There is evidence of at least ten homonomous segments in *H. whittingtoni* and *C. milesi* (as well as two non-Australian species), while isolated trunks in the gastric residues of fish have up to fourteen segments. Proximally the segments bear dorsal nodes, and were probably fused to the hinge line by an endoskeletal rod. Although the cephalon, appendages, and posterior of the trunk are unknown, the new features and a review of previous evidence eliminate *Concavicaris* from the Phyllocarida. A new order is therefore established for the Austriocarididae, which is extended to include *Ostenia* from the Jurassic and ?Triassic. The affinities of the order remain uncertain, but probably lie with the Crustacea. A consistent carapace damage pattern may result from fish predation, also known from other concavicarid localities.

THE Gogo Formation is one of a suite of formations associated with the celebrated Devonian barrier-reef complex of the Canning Basin. These reefs have been extensively studied (see Playford 1980 for a recent review) as the reef is exceptionally well-exposed and relatively undeformed structurally. The inter-reef Gogo Formation consists mainly of dark-grey shales and siltstones and is poorly exposed (Playford and Lowry 1966). It also includes thin limestone lenses, however, and some intervals with abundant hard calcareous concretions. These weather out in large numbers on the surface, and over half of them contain fossils.

The Gogo Formation is best known for the rich and exceptionally well-preserved fauna of fish which the calcareous concretions have yielded (Gardiner and Miles 1975). Less well-known are the remarkable bivalved arthropods from the area reported by Rolfe (1966) and Brunton *et al.* (1969). This, the first detailed account of any of these arthropods to be published, is devoted to the enigmatic genus *Concavicaris*, classified in the Austriocarididae by Rolfe (1969), but here removed from the Phyllocarida. Interest in the fauna was kindled by the collection by H. A. Toombs of the British Museum (Natural History), in conjunction with the Western Australian Museum, of several hundreds of specimens in 1963. This prompted the organization of a larger joint expedition in 1967 by the British Museum (Natural History), Western Australian Museum, and the Hunterian Museum. This study is based on material from these two collections. The 1967 material is divided between the following museums (abbreviations in parentheses): Western Australian Museum, Perth (WAM); British Museum (Natural History) (In); Hunterian Museum, Glasgow University (HM A); and the National Museum of Natural History, Washington, D.C. (USNM). Additional material has been studied from the Geology Department, University of Western Australia and the University of California Museum of Paleontology, Berkeley. Abbreviations for repositories of comparative material are: Museum of Comparative Zoology, Harvard University (MCZ), Field Museum of

[Special Papers in Palaeontology No. 30, pp. 249–276, pls. 35–38]

Natural History, Chicago (FM). Material from the 1963 expedition came from the Stromatoporoid Camp area and from Bugle Gap (Miles 1971, fig. 1), and is distinguishable by BM catalogue numbers smaller than In62270.

Orientation of carapace. Discussions of *Concavicaris* since its original description by Meek (as *Colpocaris*; 1872, 1875) have reflected uncertainty as to whether the characteristic notch is anterior or posterior. Meek designated the notch posterior. From 1900, however, most authors considered it to be anterior (Clarke 1900; Gürich 1929; Cooper 1932; Shimer and Shrock 1944), but this 'about face' does not appear to have been justified. Chlupáč (1963) following Rolfe (1961), however, argued that 'by analogy in many other phyllocarids the posterior thorny projections and incisions are more prominent than the anterior ones, and the conspicuous broadening of the carapace towards the sinus in *Concavicaris* suggest that Meek's original orientation is correct', an interpretation he considered confirmed by Rolfe's (1962a, p. 918) discovery of a rostral plate at the opposite end to the notch (in *C. rostellata* Rolfe 1969, fig. 140, 1b). The discovery at Gogo of a form with completely occluded notch, preventing the protrusion of an abdomen from that end, prompted reconsideration of this question. We argue here that the indentation which gave *Concavicaris* its name is indeed anterior, a possibility suggested by Rolfe (*in* Brunton *et al.* 1969, p. 81). The rostral plate of *C. rostellata* must therefore be the posterior carapace horn of the opposite valve, broken off; the rostral plate of *Austriocaris* has not been re-evaluated.

Preservation. Fifty-seven per cent of the Gogo concretions contain fossils and 22 % of these are bivalved arthropods, including *Concavicaris* (Brunton *et al.* 1969). Some concretions contain a number of individuals or fragments, possibly as a result of sedimentary processes, but there is no apparent alignment or obvious sorting. The concretions are composed mainly of light-coloured homogeneous, finely bedded, calcareous siltstone, but on splitting they commonly also reveal a concentric or more rarely dendritic or patchy distribution of a darker mineral. The arrangement of the concentric layers does not appear to reflect the position of the fossil which may be eccentric, but rather the margins of the nodule. The fine black granular mineral also occasionally grows in the calcite which replaced the cuticle of the arthropod.

The arthropods are relatively uncompacted. The concretions as a whole tend to split roughly parallel to bedding (possibly due to the influence of the contained fossil), but the lithology shows no tendency to flake off along bedding planes during preparation. Traces of the finely spaced laminations are picked out by weathering of the surface of the nodules, and it is evident that the majority of Gogo concavicarids are preserved with the vertical sagittal plane roughly parallel to the bedding. There is usually some degree of tilting, however, so that opposing valves are commonly preserved offset. In a few cases the carapace is almost normal to the bedding (when it may be possible to prepare the fossil from both sides). Examination with the scanning electron microscope reveals that the original cuticle structure of the concavicarids is not preserved. The cuticle is delicate and in many cases has been lost on splitting of the slab. The carapace thus occurs in four different guises depending on how the concretions split, as outer and inner surfaces of the cuticle, and as external and internal moulds. The surface of the mould may take on a similar lustre and colour to the cuticle it replicates. There is no evidence, however, of the 'transfer replication' described by Rolfe (1962a) in specimens of the phyllocarid *Ceratiocaris papilio* preserved in concretions from the beds close to the Llandovery/Wenlock boundary in Scotland. The cuticle of at least two of the Gogo species is terraced, and although the terraces are external structures they are none the less also evident on the internal surface.

Occurrence. All the new species here described come from the Gogo Formation, *Polygnathus asymmetricus* Zone, Frasnian *to* 1α–*to* 1β, Upper Devonian; Gogo Station, Fitzroy Basin, Kimberley Division, Western Australia. Locality numbers refer to map by Rolfe (*in* Miles 1971, fig. 1).

Dimensions of carapace. These are presented in the Systematic section as follows: observed range, mean (N̄) followed by standard deviation, size of sample in parentheses.

SYSTEMATIC PALAEONTOLOGY

Class Uncertain
Order CONCAVICARIDA order nov.

Diagnosis. Medium to large bivalved carapaces, usually with a prominent, concave, anterior notch; dorsal margin produced into a fused rostrum which may curve down and occlude the notch. At least fourteen homonomous, prominently dorsally lobed trunk segments, diminishing in height anteriorly and posteriorly, attached dorsally to a longitudinal, endoskeletal bar.

Family AUSTRIOCARIDIDAE Glaessner, 1931

Diagnosis (emended from Rolfe 1969). Carapace with anterior notch, which may be shallow or occluded; adductor? muscle scar anteriorly situated; mesolateral or juxtadorsal ridges commonly present.

Genus HARRYCARIS gen. nov.

Etymology. After Professor Harry Whittington, F.R.S.

Diagnosis. Hinge line fused. Notch sub-circular, occluded anteriorly by ventrally curved rostrum, embracing dorsally curved prolongations of anteroventral region of each valve.

Type species. Harrycaris whittingtoni sp. nov.

Harrycaris whittingtoni sp. nov.

Plate 35, figs. 1, 2, 5, 7; Plate 36, fig. 12; Plate 37, fig. 12; text-fig. 1

1969   *Concavicaris* sp. nov. 3, Rolfe *in* Brunton, Miles and Rolfe, p. 81.

Etymology. After Professor H. B. Whittington, who was instrumental in obtaining the National Science Foundation grant which enabled W. D. I. R. to take part in the 1967 Gogo Expedition that collected most of the material described in this paper.

Diagnosis. As for genus.

Holotype. WAM 82.3166 (Pl. 35, fig. 1; Pl. 36, fig. 12).

Material. 166 specimens, from the following localities: loc. 2 (2 specimens), 3(3), 4(2), 5(1), 8(3), 9(2), 12(1), 13(2), 14(19), 15(2), 16(7), 17(1), 18(1), 20(33), 21(4), 22(2), 30(2), 31(3), 34(2), 35(3), 40(3), 46(2), 47(4), 50(1), 51(12), 52(13), 53(15), 58(1), 60(1), 62(6), 63(4), 65(3), 66(2), 74(1), 94(2), and 102(1).

Measurements. Length 9–30 mm, $\bar{N} = 20 \cdot 19 \pm 5 \cdot 16$ mm (39); height 4–18 mm, $\bar{N} = 10 \cdot 78 \pm 3 \cdot 30$ mm (34) (text-fig. 2).

Description. Hinge line straight for most of its length, sloping gently ventrally beyond a slight break of slope near posterior end, and hooked anteriorly, curving downward and then posteroventrally to occlude the carapace gape. Maximum height at about the mid-length of valves. Ventral margin runs posteriorly from this point in an almost straight line, converging towards hinge line into which it curves through a short posterior margin, roughly normal to longitudinal axis of carapace. A narrow ridge follows ventral margin, which anterior to mid-length runs in a straight line converging very gradually with the hinge, to where it curves to meet the dorsal hook at a point just anterior to posterior margin of occluded notch. Notch thus completely enclosed, forming a sub-circular aperture, which is bordered by a narrow flange dorsally and posteriorly which slopes inwards at a steep angle to the plane of symmetry of the carapace. Edge of flange marked by a sharp ridge. There is, in addition, a pronounced lateral ridge on the dorsal hook which marks outer limit of a concave area concentric with aperture. It is complemented by a similar but less-pronounced ridge on anterodorsal extension of ventral part of carapace which is overlapped by the hook. A narrow ridge traverses lower part of valves converging posteriorly with ventral margin, which it parallels from a point not far posterior of mid-length of carapace. Anterior of mid-length, this ridge runs parallel to ventral margin before dying out some distance posterior of notch. Relief on this narrow ridge varies between individuals and it may be virtually absent in some cases. Valves convex in cross-section, outline straightening out towards ridge and becoming slightly concave ventral to it. The pronounced downward curvature of the anterior hook and the way in which it interlocks with the projections of ventral margin of valves renders movement about the hinge impossible. This is borne out by the absence of isolated valves, and continuation of terrace lines across hinge line. The gape is confined to a narrow split separating the valves ventrally and posteriorly.

The cuticle of *H. whittingtoni* is terraced. The terraces are usually spaced *c.* 1 mm apart, but the distance varies over the surface of the carapace, and is less in juveniles. The scarp slopes face posteriorly. Terraces run obliquely across posterior part of valves, anteroventrally from hinge, but curve posteroventrally to run roughly normal to the hinge across most of carapace. Ventral of lateral ridge, terraces form small concentric, acuminate curves, convex anteriorly. Very fine cross striations exist between major terracing.

Genus CONCAVICARIS Rolfe, 1961

Diagnosis. Hinge line usually fused, and produced anteriorly into rostrum; pronounced anterior excavation of carapace valves; up to three lateral ridges present.

Concavicaris milesi sp. nov.

Plate 35, figs. 3, 4, 14–18; Plate 36, figs. 2, 3; Plate 37, figs. 7, 10, 11; text-fig. 1

1966   *Concavicaris* sp. nov. 1 aff. *C. elytroides* (Meek), Rolfe, p. 192.

Etymology. After R. S. Miles, for his work on the Gogo fish.

Diagnosis. Notch U-shaped with inner and outer flanges, rostrum short, curving down anteriorly, one lateral ridge, cuticle terraced.

Holotype. In60161a (Pl. 35, figs. 15, 16).

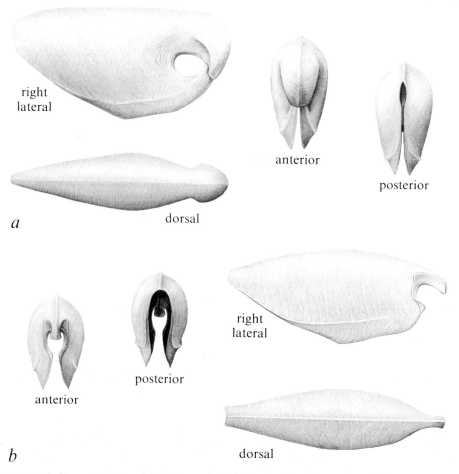

TEXT-FIG. 1. Reconstructions of *a, Harrycaris whittingtoni. b, Concavicaris milesi.* Both × 3.

Material. 84 specimens: loc. 2 (2 specimens), 3(1), 8(7), 14(4), 16(1), 20(2), 21(9), 22(7), 23(4), 31(1), 35(7), 42(1), 46(1), 47(1), 53(2), 58(6), 60(1), 61(1), 62(2), 63(5), 74(5), 75(1), 87(2), 94(2), 99(2), 103(3); specimens unlocalized (4).

Measurements. Total sample: length 5–25 mm, N̄ = 11·58 ± 6·58 mm (29); height 2–11 mm, N̄ = 4·26 ± 3·04 mm (39). Adults, length 9–25 mm, N̄ = 18·15 ± 3.98 mm (13); height 4–11 mm, N̄ = 8·5 ± 2.64 mm (11). Juveniles, length 5–7·3 mm, N̄ = 6·24 ± 0·65 mm (16); height 2–4 mm, N̄ = 2·60 ± 0·48 mm (28) (text-fig. 2).

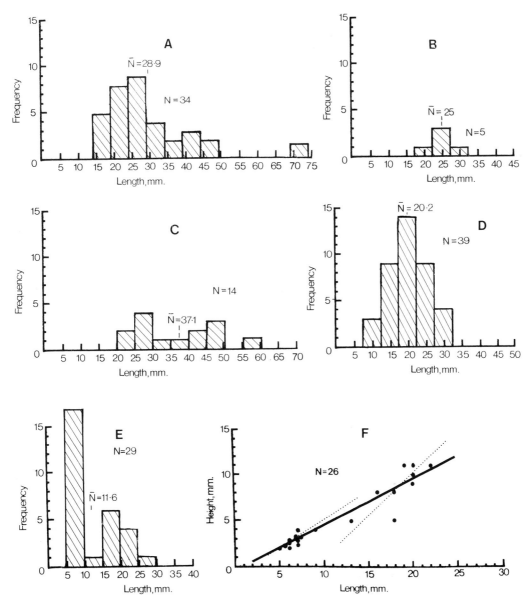

TEXT-FIG. 2. Histograms of length. A, *Concavicaris playfordi.* B, *C. campi.* C, *C. glenisteri.* D, *Harrycaris whittingtoni.* E, F, *C. milesi.* F, plot of length/height, showing cluster at juvenile instar(s). Lines are interpolated by Bartlett's Best Fit method: continuous line for all data ($y = 0·50x - 0·43$); dotted lines show difference in proportion of juveniles ($y = 0·61x - 1·15$, $N = 16$) and adults ($y = 0·9x - 7·88$, $N = 9$; unclustered point from 1963 collection, $x = 9$, $y = 4$, omitted).

Description. Hinge line evenly convex dorsally, except where it curves through an oblique angle at its anterior extremity into downward-facing hook which forms margin of notch. Maximum height occurs at about mid-length of valves. Ventral margin similar to that of *H. whittingtoni*, although posteriorly it converges more rapidly with hinge line. Notch U-shaped, and bordered by two inwardly sloping concentric flanges delimited by slight ridges; the inner narrow, subparallel sided, and steep (i.e. almost transverse to plane of symmetry of carapace); outer is wide, expands dorsally and is less steep. Ridge defining outer margin bears small anterolaterally directed spine. Viewed from anterior hook defining notch expands slightly distally and terminates in a gently convex margin roughly normal to hinge line. A single longitudinal ridge, gently concave dorsally, traverses lower part of valves, converging posteriorly with a less prominent ridge along ventral margin, which it parallels in posterior part of carapace. Longitudinal ridge dies out anteriorly, posterior of notch. Valves evenly convex in cross-section, outline straightening out toward ventral margin. In dorsal view outline of carapace flask-shaped. Width greatest about mid-length, and narrows abruptly just posterior to notch. Curvature of hinge line and anterior hook, and lack of evidence for a functional hinge (the terrace lines are continuous across it—see below), indicate that no movement of the valves about the hinge was possible.

Cuticle of *C. milesi* terraced. 'Primary' terraces, spaced up to 1 mm apart, describe roughly concentric curves, concave anteriorly, roughly symmetrical about long axis of each valve. Relief on these terraces slight, scarp slope, often poorly defined, facing posteriorly. They curve more normal to long axis near hinge and ventral margin of valves, in some cases turning slightly posteriorly. They are more closely spaced near notch. More closely spaced 'secondary' terraces cover most of the carapace. They appear similar to primary terraces under the light microscope but are of such low relief that it is difficult to identify a scarp slope. They run roughly normal to hinge line and therefore oblique to primary terraces in dorsal part of valve, but curve slightly anteriorly to run subparallel to primary terraces in ventral part. Primary terraces are not traversed by secondary ones which form a discrete system between them. Terrace lines continuous across hinge line—in some cases primary terraces become secondary on opposing valve (In60184).

Juveniles. Thirty-one juveniles of this species are present in the 1967 collection, 39 % of the total sample of eighty collected, yielding a bimodal histogram (text-fig. 2). It is unknown why the sample of this species should so predominate in juveniles, although there is some possibility of inclusion of young *H. whittingtoni*, due to the difficulty of identifying fragments of juveniles from ornament alone. Juveniles (Pl. 36, figs. 2, 3), up to *c.* 7 mm in length, differ markedly from adults in having a relatively large anterior notch occupying almost the entire anterior height of carapace. In adults this aperture occupies only *c.* one-third of anterior height. This suggests that young instars had proportionately larger appendages emerging from the notch, to meet the locomotor needs of the larva, as in living nauplii. Even allowing for scaling effects, it implies a different mode of life for the adult from the permanently swimming(?) larva. Posteriorly, juveniles are concavely truncated, making the carapace slightly shorter than in adults. Although some juveniles are broken off along the major terrace lines, which are also concave posteriorly, there is no doubt that other complete juveniles possess this concave border, with its own rim (Pl. 36, fig. 2).

EXPLANATION OF PLATE 35

New Concavicarida from Gogo area, Western Australia; Gogo Formation, Frasnian, except fig 19. All × 2, except where stated, and whitened with ammonium chloride.

Figs. 1, 2, 5, 7. *Harrycaris whittingtoni* sp. nov. 1, lateral, holotype, WAM 82.3166, loc. 66. 2, In62270, loc. 35. 5, In62271, dorsal, anterior to right, loc. 47. 7, WAM 82.3167, dorsal, loc. 47.

Figs. 3, 4, 14–18. *Concavicaris milesi* sp. nov. 3, 4, In62272, crushed anteriorly, but showing terrace lines by different lighting, loc. 63. 14, In62274, posterior of left valve. 15, 16, holotype, In60161a; 15, latex cast, × 4; 16, anterior, × 8. 17, In60184, dorsal, anterior to left, × 8. 18, In 62273, posteroventral region of right valve, showing terrace ornament, loc. 58, × 8.

Figs. 6, 9, 10, 11. *Concavicaris playfordi* sp. nov. 6, In62275, dorsal, anterior to right, loc. 35. 9, holotype, WAM 82.3168, loc. 22. 10, In62276, anterior, loc. 67, × 5. 11, In60188, anterodorsal spines, × 6.

Figs. 8, 12. *Concavicaris glenisteri* sp. nov. 8, holotype, WAM 82.3170, anterior to right, loc. 55, × 1. 12, In62277, 'bitten'?, anterior to left, loc. 53, × 1.

Fig. 13. *Concavicaris campi* sp. nov. Holotype. WAM 82.3171, loc. 14, × 1.

Fig. 19. *Concavicaris sinuata* (Meek and Worthen). HM A2363, latex cast from Herdina Collection FM specimen, originally Caro 185; north Illinois, Mecca Quarry Shale, Westphalian C, D, × 2.

PLATE 35

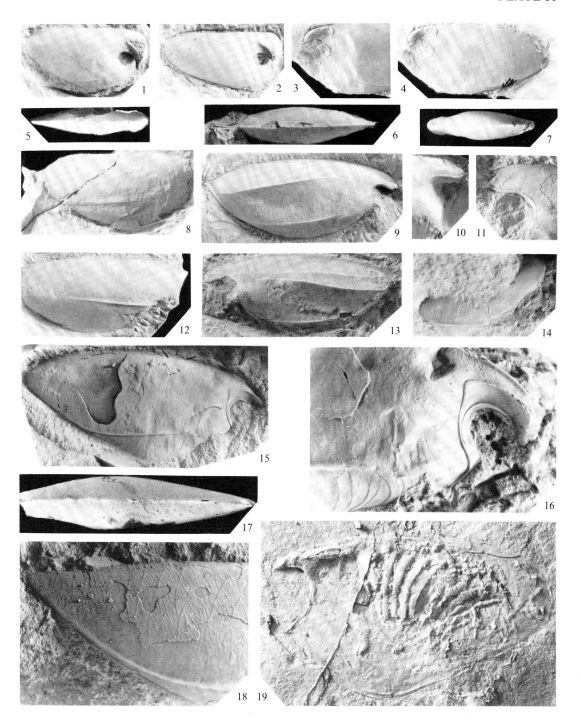

BRIGGS and ROLFE, Concavicarida

Remarks. This species resembles the Mississippian *C. elytroides* (Meek) in its elongation, and in having a rostrum. Plate 38 illustrates the specimen described by Cooper (1932), from which it can be seen that there are many points of specific difference. None of the Gogo species has the distinctive serrate ridge of *C. elytroides*.

Concavicaris playfordi sp. nov.

Plate 35, figs. 6, 9, 10, 11; Plate 36, fig. 10; text-fig. 3

1966   *Concavicaris* sp. nov. 2, Rolfe, p. 192.

Etymology. After P. E. Playford, for his work on the Canning Basin reef.

Diagnosis. Notch shallow, with single flange, hinge line projecting anteriorly as small median spine flanked by larger anterodorsal spine on each valve; rostrum short and stout, two lateral ridges, cuticle smooth with regularly spaced pits.

Holotype. WAM 82.3170 (Pl. 35, fig. 9).

Material. 112 specimens: loc. 2 (2 specimens), 5(2), 8(2), 12(18), 13(2), 14(5), 16(2), 17(2), 20(18), 20/21(1), 21(1), 22(13), 35(3), 40(10), 41(1), 43(1), 45(1), 49(1), 53(4), 57(1), 58(1), 60(1), 61(1), 62(2), 63(2), 65(2), 66(2), 71(1), 74(1), 76(1), 80(1), 97(2), 98(1), 102(2); specimen unlocalized (1).

Measurements. Length 16–73 mm, $\bar{N} = 28·88 \pm 11·48$ mm (34); height 5–29 mm, $\bar{N} = 13·35 \pm 6·00$ mm (43) (text-fig. 2).

Description. Hinge line gently convex dorsally, curving more strongly toward anterior and posterior ends. Commonly it projects only slightly at its posterior extremity, but rarely it is produced into a prominent spine (text-fig. 3). Ventral margin of valves more strongly convex than dorsal; evenly curved outline broken only by anterior notch. Maximum height occurs at about mid-length of valves. A narrow ridge follows ventral margin. Notch semicircular in outline and flanked dorsally and posteriorly by an inwardly directed flange, which is widest dorsally and separated from rest of valve by a slight ridge. Dorsal margin of notch extends anteriorly into a slender spine which projects anterolaterally. This spine is separated from its counterpart on the other valve by a median indentation which accommodates a much smaller spine which forms the termination of the hinge line, and projects dorsal to the other two. Valves traversed by two gently curving longitudinal ridges, the dorsal sinuous, the ventral gently concave dorsally. These delimit three subequal facets or areas of carapace, dorsalmost convex in section, those ventral to it flat or gently concave. Lower of the two ridges runs parallel and adjacent to margin of valve posteriorly but, like dorsal ridge, dies out anteriorly just behind notch. Upper of the two ridges usually terminates just before posterior border but, rarely, continues right up to it. In dorsal view outline of valves almost symmetrical about mid-length. Width greatest a little posterior of this point, however, and abruptly reduced at anterior end coinciding with position of notch. Slight adjustment of gape between valves may have been possible by movement about the hinge. Internal structures are unknown.

The cuticle of *C. playfordi* preserves no evidence of terracing, although there appear to have been corrugations of low relief. Most specimens preserve traces of a fairly regularly spaced pitting.

Concavicaris glenisteri sp. nov.

Plate 35, figs. 8, 12; Plate 36, figs. 6, 7; text-fig. 3

Etymology. After B. G. Glenister, for his work on the Gogo ammonoids and conodonts.

Diagnosis. Notch shallow with single flange, rostrum of medium length, slender; one lateral ridge, cuticle with regularly spaced pits.

Holotype. WAM 82.3170 (Pl. 35, fig. 8).

Material. 14 specimens: loc. 5 (1 specimen), 14(2), 15(1), 20(2), 40(1), 51(1), 53(4), 55(1); specimen unlocalized (1).

Measurements. Length 22–60 mm, $\bar{N} = 37·14 \pm 11·16$ mm (14); height 13–26 mm, $\bar{N} = 17·85 \pm 4·10$ mm (13) (text-fig. 2).

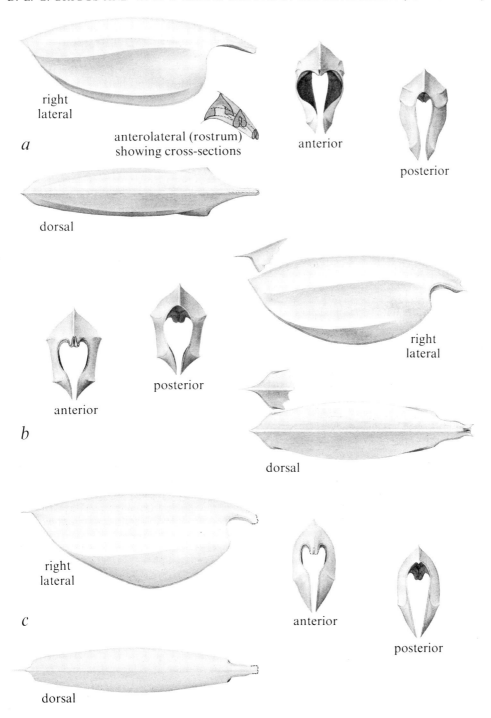

TEXT-FIG. 3. Reconstructions of *a*, *Concavicaris campi*. *b*, *C. playfordi*. *c*, *C. glenisteri*. All ×2.

Description. Hinge line evenly convex dorsally and produced posteriorly into a spine, ventral floor of which in the holotype is obtusely infolded longitudinally along midline, like the rostrum of *C. elytroides* (Pl. 38, fig. 10). Ventral margin shows more pronounced convexity, valves expanding slightly anteroventrally. Highest point just anterior of mid-length. A narrow ridge and interior marginal doublure follows ventral margin. Anterior notch shallow; its dorsal margin parallels hinge line to form a rostrum, the tip of which is unknown. A single ridge traverses valves obliquely below mid-height, sloping posteroventrally from a point some distance behind notch, and curving to run almost parallel to ventral margin posteriorly. This ridge shows considerable variation in position and amplitude. In smaller specimens (In62278, WAM 82.3172), it is situated much more ventrally, and is almost completely effaced, indicated largely by a groove that runs below it. No terrace lines are known, but regularly spaced pits are present.

Remarks. This form closely resembles *C. playfordi*. A similar form is present in the Famennian Woodruff Shale of Nevada (Smith and Ketner 1975), although it is still uncertain whether the longitudinal ridge is present in that less well-preserved material.

Concavicaris campi sp. nov.

Plate 35, fig. 13; Plate 36, figs. 1, 4, 5; text-fig. 3

Etymology. After C. L. Camp, who probably collected the first Gogo concavicarids, on his 1960 expedition (Camp 1962).

Diagnosis. Notch very shallow, with single flange; rostrum long and slender; three lateral ridges, cuticle smooth, lacking pits.

Holotype. WAM 82.3171 (Pl. 35, fig. 13; Pl. 36, fig. 1).

Material. 8 specimens: loc. 14 (4 specimens), 52(1), 53(1), 60(1), 61(1).

Measurements. Length 21–29 mm, $\bar{N} = 25\cdot0 \pm 2\cdot92$ mm (5); height 11–12 mm, $\bar{N} = 11\cdot50 \pm 0\cdot71$ mm (2) (text-fig. 2).

EXPLANATION OF PLATE 36

Concavicarida from the Frasnian Gogo Formation, Western Australia and the Devonian and Carboniferous of U.S.A. Figs. 1–6, 8, 10–13, 15 whitened with ammonium chloride.

Figs. 1, 4, 5. *Concavicaris campi* sp. nov. 1, detail of Plate 35, fig. 13, denticulate rostrum, × 8. 4, In62279, right rostrum *in situ* and 5, detached, showing vertical doublure, loc. 14, × 4.

Figs. 2, 3. *Concavicaris milesi* sp. nov., juveniles, × 8. 2, In62280, posterior concave truncation, loc 87. 3, WAM 82.3173, right valve, loc. 23.

Fig. 6. *Concavicaris glenisteri* sp. nov., WAM 82.3172, medium-sized carapace, left valve, loc. 20, × 1·3.

Figs. 7, 10. Damaged carapaces, with anterior portions missing, × 1·3. 7, *C. glenisteri*, WAM 82.3174, loc. 51. 10, *C. playfordi*, In62281, loc. 62.

Fig. 8. *Concavicaris sinuata* (Meek and Worthen), FM 35 A 3–4, showing cuticle ornament; Mecca Quarry, Indiana, Mecca Quarry Shale, Westphalian C, D, × 10.

Fig. 9. *Ostenia*? sp. nov., FM specimen, Heebner Shale, Stephanian B, Hamm Quarry, Kansas, × 0·66. Photograph by Dr. E. S. Richardson.

Fig. 11. Concavicarida indet., In62286, typical preservation of trunk segments in gastric residues, loc. 54, × 4.

Fig. 12. *Harrycaris whittingtoni* gen. et sp. nov., holotype, Plate 35, fig. 1, × 4.

Figs. 13, 15. *Concavicaris bradleyi* (Meek), Blue Lick, Junction City, Danville, Kentucky, upper New Albany Shale, basal Mississippian. 13, MCZ 6913/1, left valve, × 1. 15, USNM 27128/7a, latex cast of interior of right valve, showing doublure; anterior missing, × 1·3.

Fig. 14. *Concavicaris* cf. *sinuata* (Meek and Worthen), R. and W. Lund Collection, Adelphi University, New York, 78-70204-a, Bear Gulch, Montana, Bear Gulch Limestone, late Mississippian, × 0·7.

Fig. 16. *Concavicaris* aff. *bradleyi* (Meek), USNM 353932, Woodruff Creek, Nevada, Woodruff Formation, Famennian, collected by K. B. and V. J. Ketner, 1982. Trunk segments and appendages protruding from posteroventral region of carapace, × 0·66.

PLATE 36

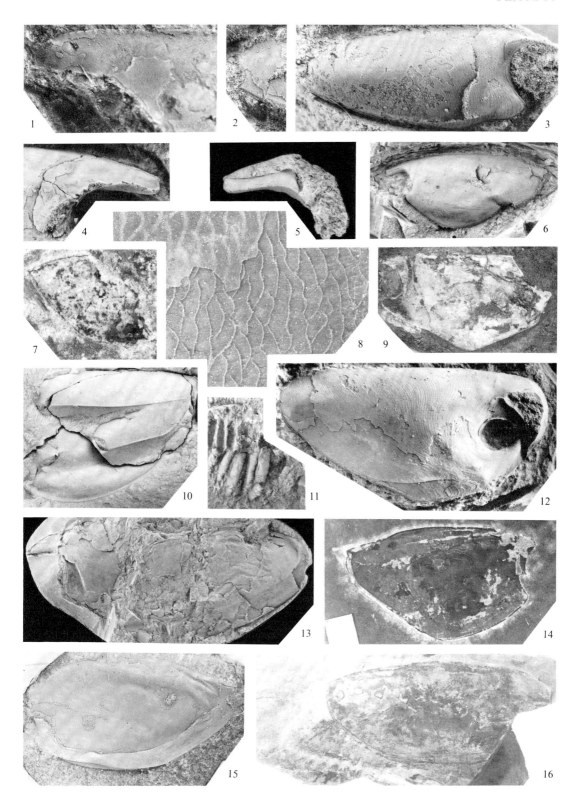

BRIGGS and ROLFE, Concavicarida

Description. Hinge line, as in *C. playfordi*, gently convex dorsally, curving more strongly toward anterior and posterior ends. It projects slightly at its posterior extremity. Ventral margin of valves more strongly convex than dorsal, valves expanding slightly anteroventrally. Maximum height of valves lies at about 0·4 of length of carapace from anterior, in contrast to *C. playfordi* where it is at about mid-length. A narrow ridge follows ventral margin. Valves traversed by three gently curved longitudinal ridges. Lower two similar in position and form to those of *C. playfordi*. Additional dorsal ridge also gently sinuous, converging anteriorly with, but not meeting, median ridge. All these ridges die out posterior to notch. They divide the carapace into distinct facets which are almost straight or slightly concave in transverse section. A long, dorsally notched rostrum is present, floored by the flanges, which are transversely concave. Mesially the flanges deflect dorsally to form continuation of doublure (text-fig. 3). They thereby form a longitudinal tube, narrowing distally to what may have been an opening, although preservation makes it impossible to be certain about this. Between reflexed doublure and carapace surface, anterior rostrum bears a crescentic flange (or wall—Rolfe 1969, fig. 124), which is widest at angle between rostrum and carapace. Dorsal outline of carapace similar to that of *C. playfordi* except for a greater increase in width at anterior end just behind notch. Anterior gape appears to have been much wider than in *C. playfordi*. Cuticle shows no evidence of terracing or pitting.

Sexual dimorphism. The similarity of carapace morphology among the Gogo concavicarids suggests that they might include sexual dimorphs. *C. campi* might represent the rarer male, and *C. playfordi* the female, of a single species, for example. Glaessner (1931) interpreted variation in one species of *Austriocaris* as the result of sexual dimorphism. Rolfe (1969, p. 305) noted that all Recent Leptostraca show pronounced sexual dimorphism, although there are no unequivocal examples among fossil phyllocarids. We have described the different Gogo concavicarids as separate species in the absence of conclusive evidence of dimorphism.

INTERNAL STRUCTURES OF CONCAVICARIDS

(i) *Gogo specimens.* The Gogo Formation is unusual in preserving structures hitherto virtually unknown within the carapace of Concavicarida. Although what we suspect are identical structures were illustrated by Böhm (1935), these were misidentified and perhaps misinterpreted by him in Carboniferous material from the Montagne Noire.

Such structures have been found inside carapaces of *H. whittingtoni* and *C. milesi*, but their occurrence in other species elsewhere (see below) suggests that they were present in all Gogo species. Their preservation in greatest numbers in *H. whittingtoni* is probably due to the enclosed nature of its carapace. When found inside carapaces the internal structures lie directly against one of the valves, indicating post-mortem displacement.

Internal structures of Gogo concavicarids occur much more commonly in association with a mass of white phosphatic material, which may be bounded by a trace of the valve outline. This phosphatic material includes fragments of palaeoniscoid fish (P. L. Forey, pers. comm.) (Pl. 37, figs. 4, 8). Such specimens may represent arthropods which have been ingested by fish, the carapace softened or dissolved by the digestive juices, and the cuticle of the internal hard parts regurgitated with other indigestible debris. The absence of a carapace, or its preservation only in outline, associated with wrinkling of the cuticle in such internal structures, suggests decalcification during digestion or regurgitation.

EXPLANATION OF PLATE 37

Trunks of Concavicarida from Gogo area, Western Australia; Gogo Formation, Frasnian. All from gastric residues except as indicated and all, except fig. 7, whitened with ammonium chloride; lateral views.

Figs. 1–6, 8, 9, 13, 14. Type B trunks. 1, 3, WAM 82.3175, loc. 14, × 10, × 4. 2, WAM 82.3176, fourteen-segment trunk (anterior to left), loc. 37, × 2. 4, 8, In62282, gastric residue with *c.* thirteen segments and palaeoniscoid fish bones, loc. 93, × 2. 8, detail of left end of fig. 4, showing bones, some inside pleurites, × 5. 5, In62283, offset pleurites, loc. 30, × 3. 6, In62284, pleurites of both sides almost completely displaced, showing overlap only at right, loc. 98, × 1·5. 9, WAM 82.3178, divided dorsal rod, loc. 30, × 4. 13, 14, In62285, gut infill, loc. 30, × 4, × 8.

Figs. 7, 10–12. Type A trunks. 7, 10, 11, *Concavicaris milesi* sp. nov., WAM 82.3177, loc. 63. 7, pleurites showing internal lineations, × 10. 10, divided dorsal rod, × 10. 11, trunk *in situ* within carapace, but inverted, × 6. 12, *Harrycaris whittingtoni* gen. et sp. nov., In60185, trunk *in situ* within carapace, × 6.

PLATE 37

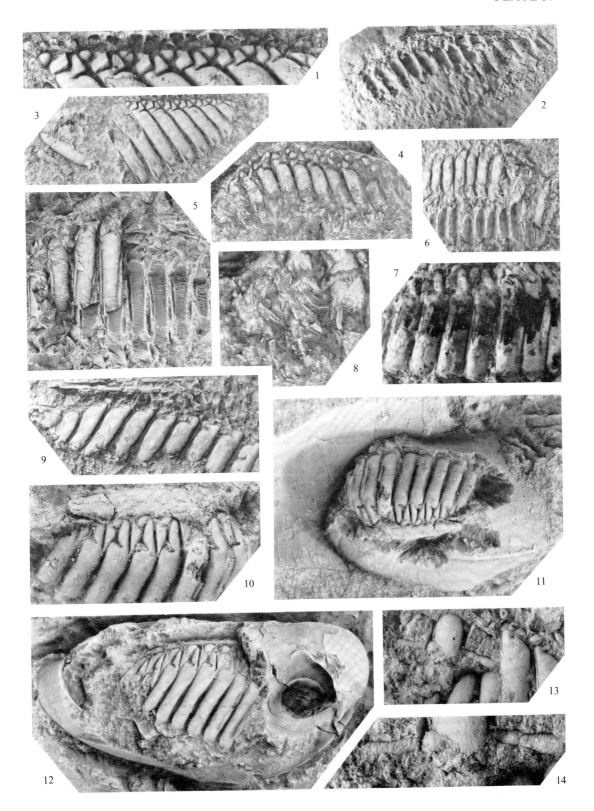

BRIGGS and ROLFE, Concavicarida

Concavicarids represented by these gastric residues tend to be large relative to those commonly found as carapaces, suggesting that predators (see below) may have selected, or at least regurgitated, the larger individuals. Variations in the morphology of the internal structures preserved in this way and described below, may represent the different concavicarid species present in the fauna.

Internal structures are rarely preserved in association with the carapace of *Concavicaris*. They are known in about 5 % of the carapaces of *H. whittingtoni* (Pl. 37, fig. 12) and with one carapace of *C. milesi* (Pl. 37, fig. 11), but have not been found with those of the other Gogo species. A much larger number (fifty internal structures), however, is known from gastric residues, but as these do not include the carapace it is impossible to determine to which species they belong. Two similar but distinct types of internal structures have been identified: 'type A' occurring in specimens of *H. whittingtoni* and *C. milesi* and in some gastric residues, 'type B' in other gastric residues. Types A and B differ in their proximal (dorsal) and distal morphology. These areas are often not preserved in gastric residues, however, many of which are therefore impossible to identify. Type B may represent *C. playfordi*, *C. glenisteri*, or *C. campi*, the internal structures of which are unknown in association with the carapace.

Description. The internal structures of *Concavicaris* consist of a series of similar divisions which were laterally flattened in life, the general outline in transverse section similar to that of the carapace. The divisions appear to represent trunk segments. Cutile of these segments consists of two lateral halves or sides which are symmetrical about the median (sag.) axis. These sides or pleurites are connected dorsally to a longitudinal rod-like structure. They appear to have been separated in the mid-line ventrally, and may have been joined by a band of unmineralized cuticle which bore appendages. Internal structures normally preserved with sagittal plane parallel to bedding and either left or right side exposed (Pl. 36, fig. 11). Where segments are preserved in a continuous series it is clear that their anterior and posterior margins were joined. Some isolated segments, however, appear to be enclosed anteriorly and posteriorly, presumably due to an incurving of the cuticle along these margins when the segments became separated. Most internal structures associated with carapaces appear to retain their original position but are not complete, and some have been rotated with respect to the valves (Pl. 37, fig. 11). In gastric residues opposing sides of segments are often offset vertically, i.e. with respect to sagittal axis (Pl. 36, fig. 11; Pl. 37, fig. 5), in rare cases to such an extent that they no longer overlap (Pl. 37, fig. 6).

Maximum number of segments known within a carapace is ten (Pl. 37, fig. 12). Gastric residues, however, include up to fourteen (Pl. 37, fig. 2), and there may have been more. Segments gradually increase in height posteriorly as far as the sixth or seventh, but behind the seventh, height gradually decreases (Pl. 37, fig. 2). Posterior segments are smaller than those at the anterior. Cuticle of each pleurite consists of a lobed dorsal (proximal) area, which curves sharply into a long extension or lateral area that slopes posteroventrally and parallels the convexity of the carapace.

Type A. Dorsal part of each pleurite consists of four lobes bounded by margins of segment and by grooves which traverse it (Pl. 37, figs. 10, 12). Most proximal lobe subcircular and bounded by a shallow transverse groove. Ventral to this, two elongate lobes, the first tapering ventrally, the other dorsally, are separated by a deep groove which slopes steeply anteroventrally but does not reach anterior part of segment. Second of these lobes bounded ventrally by shallow transverse groove which traverses posterior half of segment. A similar additional groove defines a small lobe, often poorly developed in more posterior segments, distal to the rest. Height of this lobed area decreases only slightly through the series of segments. Most of this reduction in height of the segments is a result of the more pronounced gradation in the height of the lateral area which varies from about 50 % to 67 % of the total.

Lateral area parallel-sided and convex outward in horizontal section. A pronounced groove runs parallel to most of length of anterior margin in more anterior segments (Pl. 37, fig. 12). About five regularly spaced fine lineations run most of length of lateral area parallel to its longitudinal axis (Pl. 37, fig. 7). They are not evident on outer surface of cuticle, but appear to have provided a template for the growth of diagenetic calcite within segments. A small tubercle of low relief is evident in some specimens near distal end of pleurite, situated more proximally in successively posterior segments (Pl. 37, figs. 7, 10). Some specimens (e.g. Pl. 37, fig. 7) preserve evidence of perforations of the cuticle near distal extremity, just dorsal of where convexity diminishes abruptly. Ventral of this half-tergites terminate in a short flat subtriangular projection which expands slightly anteroventrally.

Type B. Dorsal part of each pleurite consists of only three lobes (Pl. 37, fig. 1), apparently equivalent to all but the most distal in type A. Most proximal lobe similar to that in type A. The other two, however, are much less elongate than their type A equivalents. Both are V-shaped due to a pronounced angle in the groove which separates them. As in type A, groove separating lobed area from lateral part of pleurite is transverse and does not reach anterior margin. Lobed area smaller than in type A; lateral area makes up 70 to 80% of total height of segments.

Lateral area only differs from that in type A in its distal morphology. A shallower groove parallels posterior margin in larger specimens (Pl. 37, fig. 5), in addition to that along anterior margin. In type B, convexity of pleurites decreases gradually to distal extremity where posterior margin curves forward to meet straight anterior margin in an angle. Abrupt change in convexity and flat terminal projection in type A are absent. Type B also lacks the small tubercle evident on the pleurite of some specimens of type A, but is characterized by the presence of a small elongate depression or pit near anterior margin at the distal end (Pl. 37, fig. 3). This pit, which corresponds to a prominent projection on inner surface of the pleurite, is bounded by a ridge anteriorly. Fine lineations or wrinkles of the cuticle radiate from it posteriorly (WAM 82.3175). Minor granulation and rugosity present in larger specimens (Pl. 37, fig. 5).

Rod-like structure. Pleurites of both types A and B are attached dorsally to a rod-like structure which appears to extend the length of the trunk, curving gently to follow dorsal configuration of carapace (Pl. 37, fig. 12, WAM 82.3175). The rod is made up of bundles of calcite parallel to the longitudinal axis. It is a double structure, the two halves defined by a medium dorsal trough or groove (Pl. 37, fig. 10), and is only rarely present in gastric residues (Pl. 37, fig. 9). It is not clear whether the rod was originally solid, or has been infilled diagenetically with calcite. A similar structure preserved between the pleurites in In62285 is probably a gut infill (Pl. 37, figs. 13, 14).

The rod may have been an endotergite, fusing the two pleurites together (although the separation of the latter doubtless only occurred to facilitate moulting). Its lateral profile (Pl. 37, fig. 12) suggests that it fused the trunk to the interior of the carapace hinge line, i.e. it is an endoskeletal bar. Such fusion has been suggested for some true fossil phyllocarids (Rolfe 1981, p. 18), and occurs more extensively in ostracods and eumalacostracans, for example. It resulted in a rigid cephalothorax of functional, but not necessarily phylogenetic, significance. The rod probably transmitted the stresses incurred by such relatively large crustaceans during locomotion, from the trunk to the carapace, and prevented shear between those two components.

(ii) *Non-Gogo specimens.* Trunk segments can now be recognized in *Concavicaris* from two other localities: Indiana, U.S.A., and the Montagne Noire region, Hérault, southern France. *C. sinuata* from the Mecca Quarry Shale, Indiana, shows a trunk of c. ten segments arching posteroventrally (Pl. 35, fig. 19), but this may result from post-mortem displacement.

Böhm (1935, p. 112, pl. 5, fig. 6a–d) illustrated what he thought were 'branches de phyllocaridés', associated with carapaces here reidentified as *C.* aff. *bradleyi*. They are so diagenetically overgrown with mamillated phosphate, like that on the Cretaceous ostracod appendages illustrated by Bate (1972, pl. 67, fig. 1), that little can be distinguished. At least eight segments are present, however, and what may be equivalents of the pleurites are opened out ventrally and spread out laterally (Böhm 1935, pl. 5, fig. 6d; Université . . . du Languedoc specimen 57). They have thus come to overlie the dorsum of a segmented structure, which may be the equivalent of the rod in the Gogo species.

Discussion. No living bivalved arthropod provides a direct comparison with the trunk segments of concavicarids, as the cuticle of the thorax of comparable living forms lacks the mineralization of the dorsal tergites. Trunk segments are not preserved in the majority of fossil phyllocarids (Rolfe 1969, p. 299) but those of *Ceratiocaris papilio* (Silurian, Scotland) show some similarities. Rolfe (1962a, p. 919) noted that in several specimens of *C. papilio* 'the sternum of a segment has split open longitudinally and the two halves displaced relative to one another', and concluded that the sternal cuticle was unmineralized (as is apparently the case in concavicarids). 'A well-marked groove or channel surrounds the anterior end and posterior end of each segment, and probably facilitated articulation' in *Ceratiocaris* (Rolfe 1962a, p. 919), a structure comparable to the grooves of concavicarids. The thoracic segments of *Ceratiocaris* (Rolfe 1962a, pl. 131, fig. 6) show a distal depression or pit, at least superficially similar to that in type B of *Concavicaris*. The pronounced internal projections corresponding to these pits may have functioned in muscle attachment. It is not known whether they were confined to a distinct thorax in concavicarids (as in *Ceratiocaris*), due to inadequate preservation of gastric residues containing larger numbers of segments.

The dorsal lobed area has no parallel in living crustaceans of which we are aware. They are similar to the distinct oblique furrows on the proximal part of the pleurae of cheirurinid trilobites (cf. Whittington and Evitt 1954, text-figs. 3, 4). The function of these structures in concavicarids is unknown: they may have provided attachment sites for the muscles of the trunk, or strengthened the cuticle. Internally, the ridges produced by such lobes form a truss in concavicarids, perhaps to withstand the stresses transmitted from the trunk to the dragging carapace, set up by locomotor antennae acting about a fulcrum in the notch (see below).

DAMAGED CARAPACES

The carapace of Gogo concavicarids is rarely fragmented and no evidence of abrasion of the cuticle has been observed, suggesting that post-mortem transport and disturbance was minimal. Occasional concretions preserve a scatter of fragments suggesting the activity of some external agent, possibly a scavenger. Most of the fracturing of carapaces is clearly post-depositional, however, the result of compaction of the relatively rigid valves during diagenesis. As might be expected the smaller valves are less susceptible to compaction damage than the larger, but there is also a variation between species. Examination of a small sample shows that whereas up to 75 % of specimens of *Concavicaris playfordi* are undamaged, this figure is reduced to 25 % in *C. milesi*, and only 12 % in *H. whittingtoni*. A combination of two factors may be invoked to explain this. First, the carapace of *H. whittingtoni* does not separate into two, apparently even at moulting. It is more enclosed and therefore less likely to be completely filled with sediment during burial than carapaces of the other species, resulting in less support to resist compaction. The valves of 30–40 % of specimens of the other two species are separate, and seem not to have been fused to the same extent along the hinge line. Secondly, the cuticle of *H. whittingtoni* may have been weaker than that of the other species.

A significant proportion of specimens lack the anterior end of the carapace (text-fig. 4). Thus, 15 of the 112 specimens of *C. playfordi*, 4 of the 46 adult *C. milesi*, and as many as 6 of the 13 known *C. glenisteri*, are so damaged. Carapaces are either broken off along a jagged line, suggesting an originally tough, semi-brittle cuticle, or they fade out insensibly, sometimes marked by wad? dendrites, suggesting diagenetic corrosion. This consistent damage pattern is probably mainly the result of predation. The occurrence of only the anterior region of an indeterminate concavicarid in a gastric residue, along with a trunk, at locality 30 (In62292) supports this interpretation.

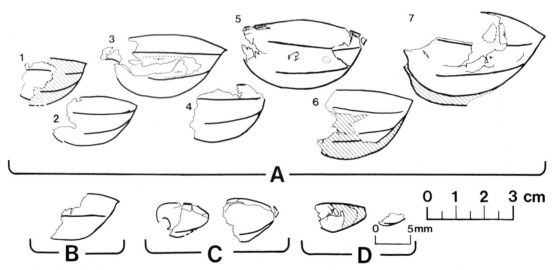

TEXT-FIG. 4. Damage patterns in carapaces of Gogo concavicarids. A, *Concavicaris playfordi*. 1, WAM 82.3179 (loc. 40). 2, USNM 353931 (20). 3, In62287 (53). 4, HM A2421 (14). 5, In62288 (20). 6, In62281 (Pl. 36, fig. 10, loc. 62). 7, WAM 82.3180 (16). B, *C. glenisteri*, WAM 82.3174 (Pl. 36, fig. 7, loc. 51). C, *Harrycaris whittingtoni*, WAM 82.3181 (14); In62289 (52). D, *C. milesi*, In62290 (16); juvenile, In62291 (22). Where it is necessary to show both valves, one is shaded. Damaged edges shown by thin lines; dashed lines indicate broken edges of counterpart; and thick lines show features of original carapace morphology. Dotted lines show crests of major crush folds, with dip symbols on major incrushed planes. Drawings reoriented with anteriors to left, to facilitate comparison.

Similar damage occurs in the true phyllocarids of the Gogo fauna, especially *Montecaris*, which may lack the anterodorsal part of the carapace. Such damage has also been illustrated by Zangerl and Richardson (1963, pl. 22C) in *C. sinuata* from the Pennsylvanian Mecca Quarry Shale, which also yields carapaces chewed to pieces. Zangerl and Richardson (1963, pp. 138, 140, 197) attribute this damage to shark predation and subsequent regurgitation (Zangerl 1981, pp. 9, 36), since *Concavicaris* fragments have been recognized among shark gastric or intestinal contents. This is further confirmed by the discovery by Dr. M. E. Williams of an entire *C.* cf. *bradleyi* carapace within the stomach of *Cladoselache* from the Famennian Cleveland Shale of Cleveland, Ohio (Cleveland Museum 8228). The anterior is similarly missing from *Concavicaris* cf. *sinuata* from the Mississippian of Ada, Pontotoc Co., Oklahoma (HM A2364) and, for example, *Dithyrocaris? chesterensis* (Worthen 1890, fig. 2).

Predation on the Gogo concavicarids is also evidenced by the large numbers of gastric residues containing body segments: fifty residues are known from twenty localities. The predator was presumably one of the numerous Gogo fish (Gardiner and Miles 1975; Miles and Dennis 1979, p. 32). Although pebbles (Miles 1971, p. 177, fig. 82) and arthrodire plates (Dennis and Miles 1981, p. 254, fig. 21) have been found as stomach content in some Gogo arthrodires, no concavicarids have been recorded: they would be rapidly dissolved during the acid extraction of the fish from the concretions.

A rarer type of damage is the longitudinal cracking and incrushing of carapaces, seen on text-figs. 4A3, 4A7. This suggests a different predator from that causing the majority of damage, perhaps a lungfish with crushing tooth-plates, or the shearing jaws of an arthrodire. The occasional slivers of carapace collected may also result from this type of predation, e.g. ventral margins of *C. playfordi* and *C. campi*, and even of one juvenile *C. milesi* (text-fig. 4D2).

Only 3 of the 166 *H. whittingtoni* specimens show this damage pattern, although a fourth (WAM 82.3182, loc. 52) shows lateral crushing of the rostrum that seems to have been repaired during life. This rarity may partly result from difficulty of detection in this small species, especially when so many show the diagenetic crushing mentioned above. If this rarity is real, however, it could mean that *H. whittingtoni* was better able to avoid or withstand predation, and that the remarkable anterior occlusion was an adaptation to that end.

Text-fig. 5 shows the distribution of damaged carapaces. No obvious correlation with fish types is discernible from distribution records of the latter kindly provided by Dr. S. M. Andrews, Mrs. K. Dennis-Bryan, and Dr. P. L. Forey. There may be some suggestion of onychodont and palaeoniscoid concentration in the reef marginal zone, where gastric residues containing concavicarid trunks abound.

DISTRIBUTION OF GOGO CONCAVICARIDS

Concavicarids account for 463 of the *c.* 2300 crustaceans collected by the 1967 Gogo expedition (Brunton *et al.* 1969, p. 81); 379 are specifically identifiable. The relative abundances (gastric residues excluded) are: *H. whittingtoni* 166; *C. playfordi* 112; *C. milesi* 80; *C. glenisteri* 13; *C. campi* 8. Since material was sampled at circumscribed points within the Gogo outcrop, it is possible to plot relative abundances of species, where the sample is sufficiently large to warrant it. The result is shown in text-fig. 5 which can be compared to fig. 1 in Miles (1971), to ascertain the complete number of localities collected.

H. whittingtoni occurs at 36 localities, and preponderates at many of them. As was noted in the field (Brunton *et al.* 1969, p. 81), this species occurs almost to the exclusion of others, particularly *C. milesi*, in the Long's Well outcrop, which was probably an atoll within the Great Devonian Barrier Reef complex. No other equally striking pattern emerges from text-fig. 5. *C. playfordi* is found at 35 localities, associated with all other concavicarids. It commonly occurs with the abundant *H. whittingtoni*, but is also found at 12 localities (41, 43, 45, 49, 57, 61, 67, 71, 76, 80, 97, 98) where that species is lacking.

A plot of species against distance from the reef front shows that *C. campi* appears about 700 yards out from the fore-reef in the Stromatoporoid Camp area (area around loc. 8, 12, 14 of text-fig. 5), and 900 yards out in the Number 10 Bore area (around 40, 62). It occurs much closer to the reef in the Long's Well atoll (52, 53), however, and these discrepancies may only be a reflection of the rarity of this species. Attempts to analyse distributions, particularly in the Stromatoporoid and Long's Well areas, are risky. It is not known how far the Gogo Formation extends to the north-east away from the Stromatoporoid Camp reef due to an inadequate exposure, or whether a further reef exists there, as suggested by an extrapolation from the geological map of Playford and Lowry (1966, pl. 4), and implied by their cross-section E–F. Straightforward extrapolations of distance from the reef front, and of correlated water depth, are therefore not possible in those well-sampled areas.

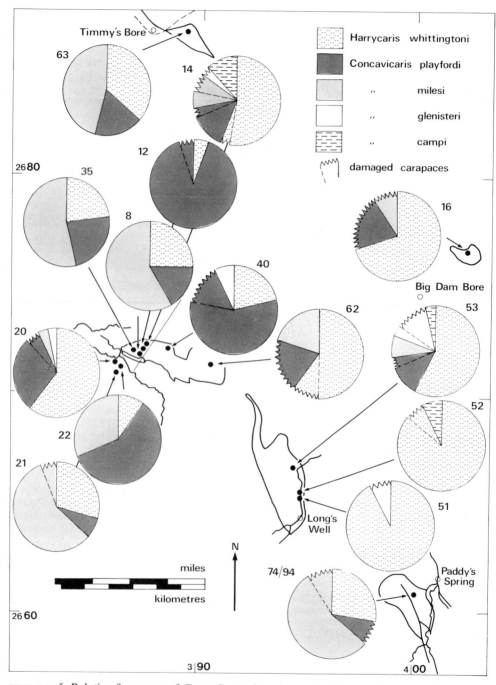

TEXT-FIG. 5. Relative frequency of Gogo Concavicarida, and incidence of carapace damage, at localities yielding more than a few concavicarids to the 1967 Gogo Expedition. Numbers shown beside each pie diagram refer to localities given by Rolfe *in* Miles (1971, fig. 1). Sample sizes: locality 8 (12 specimens), 12 (19), 14 (33), 16 (10), 20 (55), 21 (14), 22 (22), 35 (13), 40 (14), 51 (13), 52 (14), 53 (26), 62 (10), 63 (11), 74 and 94 (7+4 = 11, data pooled).

Distribution of gastric residue trunks. These are found in all the larger areas of Gogo Formation, except that of Big Dam Bore. They are concentrated in strips along the reef margin, as from loc. 36, through 30, 37 to loc. 7, and from 73, through 79, 80 to 92, or in the shallows by the reef from 54, through 55 to 56 (Rolfe *in* Miles 1971, fig. 1), although rare specimens occur further from the reef.

Of 51 localities yielding concavicarids, only 10 yield carapaces as well as gastric residues with trunks. A further 10 yield trunks but no concavicarid carapaces. Furthermore, the most abundant trunks are found in areas that are barren of concavicarid carapaces (e.g. 7 were found at loc. 54, 4 at 56). One other locality (30—unusual in being a reef-front strip, 50 yards *long*), yielded only two *H. whittingtoni*, yet provided thirteen residues with trunks. This suggests that predators were very effective in clearing such localities of living concavicarids, or that such trunks were regurgitated near the reef front, or that the near-reef sea-floor was unsuitable for the preservation of carapaces, as opposed to residues containing concavicarid trunks. Differences in trophic relationships have been invoked to explain similar discrepancies in marginal, as opposed to central basinal, fish predation (Zangerl 1981, p. 36).

MODES OF LIFE

The carapace of *Harrycaris* is almost closed posteriorly, and it is therefore unlikely that an abdomen extended beyond it in this genus, although a small abdomen could have projected in *Concavicaris*. The narrow ventral gape suggests that the trunk appendages may have been confined within the valves as well. It is possible, however, that locomotion was achieved by cephalic appendages projecting from the notch, as in living myodocopid ostracods and cladocerans.

The morphology of the anterior notch varies from the open indentation of *C. playfordi*, *C. campi*, and *C. glenisteri*, through *C. milesi* with its downwardly projecting rostrum, to *H. whittingtoni*, where the anterior of the carapace is completely occluded. This morphological 'gradient', which may reflect a similar gradient in habit or habitat, is accompanied by the elimination of the possibility of movement about the hinge, a reduction in the gape and number of longitudinal ridges on the valves, and a rounding of the carapace. In addition, species with an open notch and anteriorly projecting rostrum lack terrace lines, which are present in *C. milesi* and *H. whittingtoni*.

Gogo concavicarids are distinctive for their development of longitudinal ridges, which presumably served to stiffen the carapace. Similar ridging is known only in *C. rostellata* and *C.* sp. nov. from Nevada, and in the subdued, serrate dorsal ridge of *C. elytroides* (Pl. 38, figs. 1, 2, 6, 7, 8) and *C. woodfordi*.

The flanges around the notch probably served as a bearing surface, or rowlock (Maddocks 1982, p. 227), for an appendage like the locomotor second antenna of myodocopids. A flange of similar form and function is present around the myodocopid incisure (Kornicker 1975, figs. 146b, 154b, 157, 160b). The slight differences in position of the maximum width of this flange in different concavicarids, as in myodocopids, presumably reflects different habits.

The carapace shape and its surface texture might be expected to provide some indication of the mode of life. At present, however, we have not even conclusive evidence that individual species were infaunal or epifaunal!

The relatively large size of concavicarids argues against their being able to swim by antennae alone. The overall resemblance of body shape, especially in *Harrycaris*, to chydorid cladocerans is striking, particularly in anterior view (see Frey 1982). Like large, thick-shelled cladocerans, *Harrycaris* might have been 'a benthic species not given to great activity' (Fryer 1963, p. 337). In the largest Recent chydorid, *Eurycercus*, the broad rostrum is used in head-on pushing through plants or detritus, and the abbreviated trunk is fused to the carapace to transmit thrust from the post-abdomen (Fryer 1963, p. 341).

The carapaces of all the species are reasonably streamlined which might indicate locomotion by swimming. The maximum width (dorsal view) occurs nearer the anterior end in *H. whittingtoni* and *C. campi*, and near the mid-length in the others, the carapace tapering gradually posteriorly. The rostrum distorts the streamlined profile, but the extent to which the cephalic appendage might have increased the width in this area is unknown. The hydrodynamic shape of concavicarids would have facilitated movement through water but, at the same time, would not have inhibited burrowing in sediment, at least in an anterior direction.

'The shells of soft bottom burrowers are either perfectly smooth, or they bear sculptures that keep the shell from sliding back while the actively burrowing part of the body probes into the sediment' (Seilacher 1973). The scale and distribution of the terrace lines in *Harrycaris* and *Concavicaris* do not suggest a strengthening function. Thus both smooth and terraced Gogo concavicarids might have been burrowers. Rare specimens occur with the sagittal plane near vertical with respect to the bedding, and it is tempting to suggest that such an unstable orientation represents a specimen *in situ* in a burrow. Three Gogo *Concavicaris* species have the hinge line produced anteriorly into a fused rostrum, formed by the carapace, flange, and doublure complex. In *C. campi* this seems to have formed an open tube (text-fig. 3), which may have served to canalize an exhalent? water current from the carapace. Together, this evidence suggests some Gogo concavicarids may have lived buried in the substrate, like some living cumaceans (Kaestner 1970, figs. 16-4, 16-5). It is not clear how concavicarids might have burrowed, however, unless antennae were employed. The cumaceans, for example, burrow using the thoracic appendages, but those of concavicarids, particularly *Harrycaris*, would almost certainly have been constrained by, if not confined within, the carapace.

Seilacher (1973) and Schmalfuss (1978) demonstrated that the terraces of living decapods are an adaptation for interaction with the substrate, and Schmalfuss (1978, p. 19) 'assumed that cuticular terraces in arthropods always function as a frictional resistance in interaction with a solid or loose substrate'. In decapods which burrow in sediment the number of terraces increases during ontogeny so that their density relative to the sediment grain size is maintained ('allometric densing' of Seilacher 1973). In Gogo concavicarids, however, the space between the terraces becomes relatively larger in larger individuals. This corresponds to the arrangement in decapods that use rock crevices to avoid predators, the number of friction points rather than their density being the critical factor. There is no evidence, however, that the distribution of Gogo species with terraces (text-fig. 5) is concentrated close to the reef (a rock substrate). Other possible functions for terraces include Miller's (1975) suggestion that in trilobites they provided a site for current monitoring setae, a view disputed by Schmalfuss (1981). Cursory observations with the scanning electron microscope have revealed no concentration of pits along the terrace scarps and this, together with the high density of terraces in the juveniles, suggests that it is unlikely that the terraces functioned as sites for setae.

The terraces do not correspond to the paradigm for burrowing sculpture (Seilacher 1973; Schmalfuss 1978) and the suggestion that they functioned in interaction with a hard substrate is difficult to reconcile with the distribution of terraced species some distance from the reef. It is hoped that further work on the carapace shape and terrace system of the Gogo species will lead to an understanding of their modes of life.

STRATIGRAPHIC DISTRIBUTION OF CONCAVICARIDA

Concavicarida have a much wider distribution in the geological record than the present literature indicates. Many of these forms await description, but a preliminary synopsis of their stratigraphic distribution is given as text-fig. 6. Detailed ranges of the Devonian species were given by Rolfe and Edwards (1979, text-fig. 2).

C. desiderata (Barrande) is the earliest concavicarid known, from upper Emsian, basal *wenkenbachi* Zone (Chlupáč 1963; Chlupáč and Zikmundova 1979, p. 147, fig. 9). The Gogo concavicarids here described are the first known from the Frasnian (*to* 1a–*to* 1β). *C. incola* Chlupáč (1963, p. 113) occurs in the *Clymenia* Zone (*to* Vα), Famennian of Czechoslovakia.

The type species of *Concavicaris*, *C. bradleyi* (Meek) (text-fig. 6i, Pl. 36, figs. 13, 15) is a widely distributed species. Meek's 1872 types came from the New Albany Shale of Linietta Springs, one mile west of Junction City, near Danville, Kentucky. Campbell (1946, p. 870) showed this to be continuous with the Falling Run Bed of Indiana (where *C. bradleyi* also occurs), now taken to be earliest Mississippian (Gray 1979, p. K7). *C.* aff. *bradleyi* occurs in the Woodruff Formation of Nevada, of *Platyclymenia* Zone, Famennian to IIIβ age (Mackenzie Gordon, pers. comm. October 1982; Smith and Ketner 1975, p. A29). It is one of a number of *Concavicaris* species collected from this unit in 1982 by K. B. and V. J. Ketner, and W. D. I. R. One of these is a new species bearing only a short mesolateral ridge. *C.* aff. *bradleyi* also occurs rarely in the Cleveland Shale of Ohio (Cleveland Museum 8228), of Upper Devonian–Mississippian age.

Figs. 1–12. *Concavicaris elytroides* (Meek). 1–3, 8, MCZ 6916/2, Blue Lick, near Junction City, Kentucky, Sanderson Formation, Mississippian, Coll. H. K. Brooks. 1, broken left valve, internal, × 2·8. 2, latex cast from internal mould of that in fig. 1, × 2·8. 3, detail of fig. 1, showing ornament anteroventral to serrate ridge, anterior to right, × 12·5. 8, detail of fig. 2, showing anterior end of serrate ridge, × 12·5. 4–7, 9–12, USNM 112024, original of Cooper 1932, north of Springer, Oklahoma, Woodford Formation, Upper Devonian–Mississippian. 4, muscle? scar of fig. 10, anteroventral of serrate ridge, left valve, anterior to left, × 6. 5, latex cast of external mould of left valve, ventrolateral view, anterior to left, × 2·5. 6, external mould of left valve serrate ridge of fig. 7, anterior to left, × 18. 7, dorsal, × 3. 9, ventral and anterior gapes, × 5. 10, anterolateral view, note doublure, × 4. 11, lateral, anterior of right valve, × 5·5. 12, anterior gape, × 8.

PLATE 38

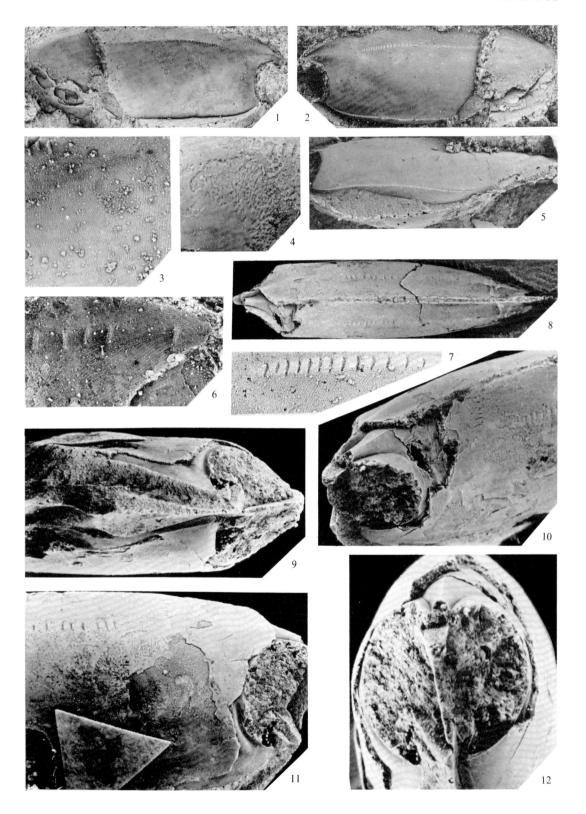

BRIGGS and ROLFE, Concavicarida

Carapaces from the Montagne Noire, previously identified as *Dithyrocaris glabra* and *Ceratiocaris truncata* by Böhm (1935), Genson (1937), and Gèze (1971) from Tournaisian Tn2c–Tn3a (as revised by Jean Galtier and A. Scott, pers. comm., after Coudray *et al.* 1979, p. 132) can be reidentified as *Concavicaris bradleyi* (text-fig. 6*l*). The same two genera recorded by Delépine (1935) from the Pyrénées are also *Concavicaris*. In both these areas, telson-like structures occur, but there is no proof that these articulated with the *Concavicaris* carapaces.

C. woodfordi (Cooper) (text-fig. 6*j*) from the Woodford Formation, uppermost Devonian–basal Mississippian (Kinderhook—Fay *et al.* 1979, fig. 4) resembles *C. bradleyi* in outline, but differs (according to Cooper 1932) in having a lateral serrated ridge, like that of *C. elytroides* (Pl. 38, text-fig. 6*m*) from the same locality (Cooper 1932).

C. sinuata (Meek and Worthen; text-fig. 6*d*, Pl. 35, fig. 19) is widespread throughout the Pennsylvanian (*c.* Westphalian C/D) of the central United States. It shows considerable variation in outline and ornament, although some of the latter may be intra-cuticular (Pl. 36, fig. 8). Abundant material exists (e.g. *c.*1000 specimens in Dr. R. K. Pabian's collection at the University of Nebraska State Museum) for a regional study of variation in this species. Much of the material from the Westphalian Mecca and Logan Quarry Shales of Indiana and Illinois (Zangerl and Richardson 1963, pl. 22A–C, figs. 31, 32; Gray 1979, p. K14, fig. 3) is found fragmented, although the Field Museum's holdings of 28 complete and 264 fragmented carapaces is probably a fair reflection of the field occurrence. The same strata yield rare juveniles (FM PE14690 is *c.* 12 mm long) that show a deep incision at the back of the anterior notch, with an outline similar to the larger *Concavicaris* sp. nov. shown (text-fig. 6*e*) from the Stephanian Kansas City Group, Sarpy Formation.

C. cf. *sinuata* (text-fig. 6*h*, Pl. 36, fig. 14) forms a rare member of the Bear Gulch Limestone crustacean fauna from Montana (Schram 1978), of late Chesterian, late Mississippian (= lower Namurian) age (Smith and Gilmour 1979, p. X24). *C. sinuata*? occurs with other crustaceans collected by Dr. R. Mapes from the Moorfield Shale, Meramecian, early late Mississippian of Buffalo Wallow, Batesville, Arkansas. *C.* cf. *sinuata* is associated with *Archaeocaris graffhami* Brooks (Schram 1979) in the Delaware Creek Shale, lower Chesterian, Mississippian, near Ada and Clarita in the Ardmore Basin, Oklahoma (Fay *et al.* 1979, figs. 1, 2, 4). It differs in having a shallow, slightly more ventrally situated, notch (Oklahoma University 4610, MCZ 6915/1, HM A2364). *C. rostellata* Rolfe (text-fig. 6*g*) is found in the Carbondale Formation, Kewanee Group, of Westphalian C/D age, Illinois.

An unusual form (text-fig. 6*f*, Pl. 36, fig. 9), from the Stephanian B Shawnee Group, Oread Limestone Formation, Heebner Shale, north of Lawrence, Kansas, is here tentatively referred to *Ostenia*, pending its description. The posterodorsally truncated, and denticulate, margin is reminiscent of the condition here described in juvenile *C. milesi*. This species was illustrated by Case (1982) who also misidentified a *Dithyrocaris* tailpiece as *Concavicaris*. A similarly angulated carapace, lacking the posterodorsal truncation but otherwise close to *C. sinuata*, is known from the Pennsylvanian of Ace Hill Quarry, Nebraska (photograph of specimen 1936/1937 kindly supplied by W. Rushau of Omaha, Nebraska).

The upper Triassic *Austriocaris striata* Glaessner is here reassigned to *Ostenia* (text-fig. 6*b*), since its carapace form is quite different from that of the type species of *Austriocaris*, *A. carinata* Glaessner (text-fig. 6*c*). The carapace outline, juxta-dorsal rib, and position of adductor muscle scar is almost identical to some specimens figured as *O. cypriformis* by Arduini *et al.* (1980, p. 12, fig. 1, pl. 13). These differ considerably from the holotype of *Ostenia* figured by those authors (pl. 9, text-figs. 1, 2), however, which is the largest concavicarid known, at up to 20 cm long.

AFFINITIES

Concavicarids have been accepted, until recently, as phyllocarid crustaceans based on the similarity of the carapace to that of other phyllocarids, and on the assumption that they possessed the phyllocarid abdomen and tail. The lack of the latter in association with the Gogo concavicarids is significant, particularly as the abdomen and tail are usually preserved in specimens of the true phyllocarids that occur at Gogo. *Harrycaris* shows that a normal phyllocarid abdomen could not have emerged from the carapace (text-fig. 1), and the evidence suggests that the trunk was contained within it. The trunks described here do not appear to have been divided into tagmata and this, together with the enclosure of the body within the carapace, forms the basis for erecting the new Order Concavicarida.

The previous assumption that *Concavicaris* bore a phyllocarid abdomen and tailpiece was based on a few localities where these occur, detached, in association with the carapace (e.g. Chlupáč 1963, p. 113). None of these associations withstands scrutiny; in all cases the abdomen and telson can be identified as those of known phyllocarids, although the carapaces and thoraces are absent at the particular localities. Thus the two para-type tailpieces found at the same locality as carapaces of the type species of *Concavicaris*, *C. Bradleyi*

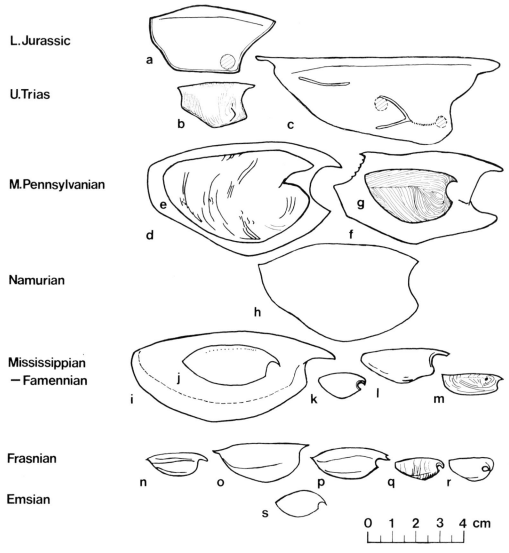

L.Jurassic

U.Trias

M.Pennsylvanian

Namurian

Mississippian
— Famennian

Frasnian

Emsian

0 1 2 3 4 cm

TEXT-FIG. 6. Stratigraphic occurrence of *Concavicaris* species, and related genera, all ×0·7, except *a*, which is ×0·3. *a, Ostenia cypriformis* Arduini, Pinna and Teruzzi, large individual, *bucklandi* Zone, Sinemurian, Osteno, Como, Italy (Arduini *et al.* 1980, fig. 1). *b, c, Ostenia striata* and *Austriocaris carinata* Glaessner, Lunzer Schichten, Lunz, Austria (Glaessner 1931, figs. 1, 8). *d. Concavicaris sinuata* (Meek and Worthen), holotype, Coal Measures, Grundy Co., Illinois (Rolfe 1969, fig. 140, 1a). *e, C.* sp. nov., Wea Shale, Sarpy Co., Nebraska (University of Nebraska State Museum, IP17221, Coll. R. K. Pabian). *f, Ostenia*? sp. nov., Heebner Shale, Hamm Quarry, Kansas (FM no number). *g, C. rostellata* Rolfe (1969, fig. 140, 1b), Coal 3, Henry Co., Illinois (Illinois State Geol. Survey 2955b). *h, C.* cf. *sinuata*, Bear Gulch Limestone, Bear Gulch, Montana (R. and W. Lund Collection, Adelphi University, Garden City, New York 78-70204-A). *i, C. bradleyi* (Meek) (MCZ 6913/1), upper New Albany Shale, basal Mississippian, Blue Lick, Junction City, Danville, Kentucky. *j, C. woodfordi* (Cooper), Woodford Formation, Arbuckle Mountains, Oklahoma (Cooper 1932, pl. 52, fig. 6). *k, C. incola* Chlupáč, Hády Limestones, Famennian, Hády, near Brno, Czechoslovakia (Chlupáč 1963, pl. 12, fig. 1). *l, C.* cf. *bradleyi*, Lydiennes à nodules phosphatés, Tournaisian Tn 2c–3a, Mont Peyroux, Montagne Noire, Hérault, France (Université des Sciences et Techniques du Languedoc = *Ceratiocaris* cf. *truncata* Woodward: Böhm 1935, pl. V, fig. 3). *m, C. elytroides* (Meek), same data as i (MCZ 6916/2a; pl. 38, figs. 1–3, 8). *n–r*, all Gogo Formation, Gogo, Fitzroy, Western Australia. *n, C. campi, o, C. glenisteri, p, C. playfordi, q, C. milesi, r, H. whittingtoni. s, C. desiderata* (Barrande), Daleje Shales, Choteč, Czechoslovakia (Chlupáč 1963, pl. 12, fig. 4).

(USNM 27 128/6d, 4; Meek 1875, pl. 18, fig. 6d, e) show the shorter telson and the pattern of ridges and setal bases on the furca that characterize *Dithyrocaris*. This phyllocarid genus must therefore form part of the larger New Albany Shale crustacean fauna, which is already known to include *Archaeocaris* and *Palaeopalaemon* (Schram 1979, p. 60; 1981, p. 133), but which has not yet been adequately sampled. The converse mis-identification has also occurred: what Packard (1883, p. 212) identified as the carapace of *D. venosa* (Scudder 1878) is *C. sinuata* (FM UC60884). Yet no tailpiece, other than that of *D. venosa* (of which the abdomen and carapace *are* known), occurs at that locality.

The Concavicarida therefore lack the typically large phyllocarid abdomen, telson, and furca. On this, and the morphological grounds discussed above, we remove the new Order Concavicarida from the Phyllocarida, a possibility suggested by Rolfe (in Brunton *et al.* 1969, p. 81). This solves Zangerl and Richardson's (1963, p. 131) 'mystery . . . of the complete absence of limbs, abdomina and rostral plates' from the otherwise perfectly preserved Mecca Shale *Concavicaris* (Pl. 35, fig. 19).

The possibility that concavicarids are larval forms merits consideration, although the large size of some individuals is difficult to reconcile with this interpretation. The absence of an appropriate adult might reflect a pelagic mode of life in such larvae, resulting in a different distribution to that of a benthonic adult. The Hoplocarida (Schram 1982, p. 118) are possible candidates for the adult as their stratigraphic range coincides for the most part with that of the Concavicarida. However, although the pelagic larvae of recent stomatopods (Hoplocarida) have carapaces similar to those of concavicarids (Gurney 1946), and a practically limbless, homonomous thorax, the segment number never approaches that in Concavicarida. The stomatopods also lack the anterior and posterior diminution of the trunk, and this suggestion may therefore be excluded.

The possible affinity of concavicarids to other major groups is discussed below.

Malacostraca. The relatively large number of somites (at least fourteen) and lack of tagmosis indicate that concavicarids are not malacostracans. Only in exceptional cases, such as *Canadaspis* (Briggs 1978) and *Perspicaris* (Briggs 1977) which lack abdominal limbs, are the malacostracan pleomeres similar in length to the thoracomeres.

Cirripedia. Arduini *et al.* (1980) interpret the Jurassic arthropod *Ostenia cypriformis* as showing a mixture of adult and cypris larval characters of Cirripedia. W. A. Newman (pers. comm.) rejects *Ostenia* from the Cirripedia because the three appendages illustrated by Arduini *et al.* (1980, fig. 1, pl. 11) are unlike those of any maxillopodan, the number of trunk segments is too large for cirripedes, which never exceed six, and because of the large size. *Ostenia* (text-fig. 6a) is here tentatively referred to the Concavicarida because of its similar carapace and the presence of eight to ten internal segments (Arduini *et al.* 1980, p. 366, fig. 2), which resemble those in Concavicarida. In particular a specimen of *Concavicaris* aff. *bradleyi* (Pl. 36, fig. 16) shows at least ten apparently identical structures, which we presume to be body segments, drifted out from within the carapace. The fine cusped ornament on some specimens of *Ostenia* (HM A2139/3) is also like that in some concavicarids.

Arduini *et al.* (1980, p. 369) speculate whether *Pseuderichthus* and *Protozoe* (Dames 1886), from the Cretaceous of Lebanon, are related to *Ostenia*, since there are difficulties in regarding at least *Protozoe* as a stomatopod erichthus larva (Roger 1946). These arthropods also require consideration as possible Concavi-carida, but restudy is required before extending the range of the order to the Cretaceous.

Ostracoda. Concavicarida show superficial similarities to Ostracoda, but reach a much larger size. The lack of hinge structures or of ventral valve overlap, however, rule out a comparison with most ostracods. The similarity between the concavicarid notch and the myodocopid incisure is probably convergent. Even within the myodocopids the incisure is thought to have been convergently acquired at least twice (Kornicker and Sohn 1976, p. 5). Among the Palaeozoic Ostracoda, supposed myodocopid groups such as Entomozacea and Entomo-conchacea show certain similarities to concavicarids in carapace form, ornament, and incisure. Kornicker and Sohn (1976, p. 115) have questioned the orientation and affinity of the onchotechmonine Entomoconchacea, and have suggested that they may not be true ostracods. All these differ from concavicarids, however, in the characters available for comparison.

Trunk segmentation is known in few Palaeozoic ostracods, is usually undiscernible in living forms, and absent from those that are completely cephalized. Vestiges of up to ten post-cephalic segments have been detected in the living *Cytherella* (Maddocks 1982, p. 222), however, indicating that a larger number was formerly present. The trunk of Concavicarida, however, included at least fourteen clearly defined segments. In addition, the Cambrian bradoriids already had a reduced thorax (McKenzie and Jones 1979; Müller 1979). The assignment of the bradoriids to the ostracods by these authors appears vindicated (Briggs 1983) and this feature of ostracods had thus evolved by at least the middle Cambrian.

Branchiopoda. Of the branchiopods the Concavicarida show greatest similarity to the Conchostraca, particularly the Lynceidae. Conchostraca are non-marine, however, but some occur in brackish water, and the possibility of fossil marine representatives cannot be entirely ruled out (Tasch 1969). The Lynceidae have a trunk of eleven segments in the male, and thirteen in the female, thus differing from other conchostracans (Linder 1945, p. 9). Lynceids also retain at most one moult and therefore lack the characteristic growth lines of the group. In detail, however, the carapace differs from that in concavicarids in the lack of a well-developed incisure and rostrum, the situation of the hinge in a furrow and the low degree of mineralization.

Conclusion.

Carapace form is known to be a highly convergent character and the grouping of species in the Concavicarida on this basis alone (in the absence of additional criteria) may prove in some cases artificial. In the absence of information on the cephalon, the appendages, and the posterior of the trunk and its termination, the affinities of the order remain obscure, but probably lie with the Crustacea.

Acknowledgements. We are indebted to many colleagues for assistance, but particularly to Dr. J. K. Ingham, for the painstaking accuracy of his reconstructions (text-figs. 1, 3). The material was prepared by a succession of preparators employed at the Hunterian Museum under Manpower Services Commission auspices, supervised by Christina Heywood, Valerie Boa, and Stan Wood. We also thank: D. Barbour, J. Berdan, members of the British Micropalaeontological Group for discussion, S. M. Andrews, K. Bryan-Dennis, P. Forey, J. Galtier, B. Gardiner, J. Holmes and A. Scott, J. Herman, K. B. Ketner, K. G. McKenzie, R. Mapes, S. F. Morris, W. A. Newman, R. Pabian, P. E. Playford, E. S. Richardson, G. Pinna and G. Teruzzi, S. Rolfe, H. Schmalfuss, F. Schram, A. Seilacher, P. Selden, and M. E. Williams. W. D. I. R. is indebted for support: to the National Science Foundation which enabled him to partake in the 1967 Gogo Expedition; to the Field Museum, Chicago, where some of this work was carried out in 1981 under the visiting scientist programme, and to the Carnegie Trust for the Universities of Scotland. D. E. G. B.'s research was supported by Goldsmiths' College Research Fund.

REFERENCES

Phyllocarid references before 1934 are given by Van Straelen and Schmitz (1934).

ARDUINI, P., PINNA, G. and TERUZZI, G. 1980. A new and unusual Lower Jurassic cirriped from Osteno in Lombardy: *Ostenia cypriformis* n.g. n.sp. *Atti Soc. ital. Sci. nat. Museo civ. Stor. nat. Milano*, **121**, 360–370.

BATE, R. H. 1972. Phosphatized ostracods with appendages from the Lower Cretaceous of Brazil. *Palaeontology*, **15**, 379A–393.

BÖHM, R. 1935. Études sur les faunes du Dévonien supérieur et du Carbonifère inférieur de la Montagne Noire. *Ire. Thèse Faculté Sci. Univ. Montpellier, D. ès Sci. nat.* 203 pp.

BRIGGS, D. E. G. 1977. Bivalved arthropods from the Cambrian Burgess Shale of British Columbia. *Palaeontology*, **20**, 595–621.

—— 1978. The morphology, mode of life and affinities of *Canadaspis perfecta* (Crustacea: Phyllocarida), Middle Cambrian Burgess Shale, British Columbia. *Phil. Trans. R. Soc.* **B 281**, 439–487.

—— 1983. Affinities and early evolution of the Crustacea: the evidence of the Cambrian fossils. *In* SCHRAM, F. R. (ed.). *Crustacean Phylogeny*. A. A. Balkema, Rotterdam (in press).

BRUNTON, C. H. C., MILES, R. S. and ROLFE, W. D. I. 1969. Gogo Expedition 1967. *Proc. geol. Soc. Lond.* no. 1655, 79–83.

CAMP, C. L. 1962. Journey through north western Australia and central Australia in search of fossil vertebrates. *Ann. Rep. West. Aust. Mus.* **1960–1961,** 27–32.

CAMPBELL, G. 1946. New Albany Shale. *Bull. geol. Soc. Am.* no. 57, 829–908.

CASE, G. R. 1982. *A pictorial guide to fossils*. Van Nostrand Reinhold, New York.

CHLUPÁČ, I. 1963. Phyllocarid crustaceans from the Silurian and Devonian of Czechoslovakia. *Palaeontology*, **6**, 97–118.

—— and ZIKMUNDOVA, J. 1979. The Lower/Middle Devonian boundary beds in the Barrandian area, Czechoslovakia. *Geol. Palaeont.* **13**, 125–156.

COOPER, C. L. 1932. A crustacean fauna from the Woodford Formation of Oklahoma. *J. Paleont.* **6**, 346–352.

COUDRAY, J., FEIST, R., GALTIER, J. and MICHEL, D. 1979. Nouvelles précisions sur l'âge et le paléoenvironnement des lydiennes à nodules phosphatés du Mont Peyroux (Montagne Noire) et de leur encaissant carbonaté. *Réunion ann. Sci. Terre*, **7**, 123.

DAMES, W. B. 1886. Ueber einige Crustaceen aus den Kreideablagerungen des Libanon. *Zeitschr. deutsch. Geol. Gesell.* **38**, 551–575.

DELÉPINE, G. 1935. Contribution a l'étude de la faune du Dinantien des Pyrenées. *Bull. Soc. géol. France*, (5) **5**, 65–75, 171–191.

DENNIS, K. and MILES, R. S. 1981. A pachyosteomorph arthrodire from Gogo, Western Australia. *Zool. J. Linn. Soc.* **73**, 213–258.

FAY, R. O., FRIEDMAN, S. A., JOHNSON, K. S., ROBERTS, J. F., ROSE, W. D. and SUTHERLAND, P. K. 1979. The Mississippian and Pennsylvanian (Carboniferous) Systems in the United States—Oklahoma. *Prof. Pap. U.S. geol. Surv.* no. 1110, R1–35.

FREY, D. G. 1982. Honeycombing of the carapace in the chydorid Cladocera: the elusive male of *Chydorus faviformis*. *J. Crust. Biol.* **2**, 469–476.

FRYER, G. 1963. The functional morphology and feeding mechanism of the chydorid cladoceran 'Eurycercus lamellatus' (O. F. Müller). *Trans. R. Soc. Edinb.* **65**, 335–381.

GARDINER, B. and MILES, R. S. 1975. Devonian fishes of the Gogo Formation, Western Australia. *Colloques int. Cent. natn. Rech. scient.* **218**, 73–79.

GENSON, E. 1937. Découverte d'une faune et d'une flore dans l'horizon à lydiennes de la base du Carbonifère du Massif du Foulon. *Bull. Soc. Etud. Sci. Nat. Béziers*, **40**, 81–84.

GÈZE, B. 1971. *Notice explicative, carte géologique de la France, Bédarieux*, 3rd edn. B.R.G.M., Orléans-la-Source. Pp. 1–23.

GLAESSNER, M. F. 1931. Eine Crustaceen fauna aus den Lunzer Schichten Niederösterreichs. *Jahrb. geol. Bundesanst.* **81**, 467–486.

GRAY, H. H. 1979. The Mississippian and Pennsylvanian (Carboniferous) Systems in the United States—Indiana. *Prof. Pap. U.S. geol. Surv.* no. 1110, K1–20.

GURNEY, R. 1946. Notes on stomatopod larvae. *Proc. Zool. Soc. Lond.* **116**, 113–175.

KAESTNER, A. 1970. *Invertebrate Zoology*, **3**. Interscience, New York.

KORNICKER, L. S. 1975. Antarctic Ostracoda (Myodocopina). *Smithson. Contrib. Zool.* no. 163, 1–720.

——and SOHN, I. G. 1976. Phylogeny and morphology of living and fossil Thaumatocypridacea (Myodocopa: Ostracoda). Ibid. no. 219, 1–124.

LINDER, F. 1945. Affinities within the Branchiopoda, with notes on some dubious fossils. *Ark. Zool.* **37A**, 1–28.

MCKENZIE, K. G. and JONES, P. J. 1979. Partially preserved soft anatomy of a Middle Cambrian bradoriid (Ostracoda) from Queensland. *Search*, **10**, 444–445.

MADDOCKS, R. F. 1982. Ostracoda. *In* ABELE, L. G. (ed.). *The biology of Crustacea*, **1**, 221–239. Academic Press, New York.

MILES, R. S. 1971. The Holonematidae (placoderm fishes), a review based on new specimens of *Holonema* from the Upper Devonian of Western Australia. *Phil. Trans. R. Soc. Lond.* **B 263**, 101–234.

——and DENNIS, K. 1979. A primitive eubrachythoracid arthrodire from Gogo, Western Australia. *Zool. Jl. Linn. Soc.* **66**, 31–62.

MILLER, J. 1975. Structure and function of trilobite terrace lines. *Fossils and Strata*, **4**, 155–178.

MÜLLER, K. J. 1979. Phosphatocopine ostracodes with preserved appendages from the Upper Cambrian of Sweden. *Lethaia*, **12**, 1–27.

PLAYFORD, P. E. P. 1980. Devonian 'Great Barrier Reef' of Canning Basin, Western Australia. *Bull. Am. Assoc. Petrol. Geol.* **64**, 814–840.

——and LOWRY, D. C. 1966. Devonian reef complexes of the Canning Basin, Western Australia. *Bull. geol. Surv. West. Aust.* **118**, 1–150.

ROGER, J. 1946. Les invertébrés des couches à poissons du Cretacé Supérieur du Liban. *Mém. Soc. géol. France*, N.S. **51**, 1–92.

ROLFE, W. D. I. 1961. *Concavicaris* and *Quasicaris*, substitute names for *Colpocaris* Meek, 1872 and *Pterocaris* Barrande, 1872. *J. Paleont.* **35**, 1243.

——1962*a*. Grosser morphology of the Scottish Silurian phyllocarid crustacean, *Ceratiocaris papilio* Salter in Murchison. Ibid. **36**, 912–932.

——1962*b*. The cuticle of some middle Silurian ceratiocaridid Crustacea from Lanarkshire, Scotland. *Palaeontology*, **5**, 30–51.

——1966. Phyllocarid fauna of European aspect from the Devonian of Western Australia. *Nature, Lond.* **209**, 192.

——1969. Phyllocarida. *In* MOORE, R. C. (ed.). *Treatise on Invertebrate Paleontology*. Geol. Soc. Am., New York and Lawrence. Pt. R, Arthropoda 4, 1, R296–331.

——1981. Phyllocarida and the origin of the Malacostraca. *Géobios*, **14**, 17–27.

——and EDWARDS, V. A. 1979. Devonian Arthropoda (Trilobita and Ostracoda excluded). *Spec. Pap. Palaeont.* **23**, 325–329.

SCHMALFUSS, H. 1978. Structure, patterns and function of cuticular terraces in Recent and fossil arthropods, 1. Decapod crustaceans. *Zoomorphologie*, **90**, 19–40.

—— 1981. Structure, patterns and function of cuticular terraces in trilobites. *Lethaia*, **14**, 331–341.

SCHRAM, F. R. 1978. Crustacea of the Mississippian Bear Gulch Limestone of Central Montana. *J. Paleont.* **52**, 394–406.

—— 1979. The genus *Archaeocaris*, and a general review of the Palaeostomatopoda (Hoplocarida: Malacostraca). *Trans. San Diego Soc. nat. Hist.* **19**, 57–66.

—— 1981. Late Paleozoic crustacean communities. *J. Paleont.* **55**, 126–137.

—— 1982. The fossil record and evolution of Crustacea. *In* ABELE, L. G. (ed.). *The biology of Crustacea*, **1**, 93–147. Academic Press, New York.

SEILACHER, A. 1973. Fabricational noise in adaptive morphology. *Syst. Zool.* **22**, 451–465.

SHIMER, H. W. and SHROCK, R. R. 1944. *Index fossils of North America*. Wiley, New York and London, 837 pp.

SMITH, D. L. and GILMOUR, E. H. 1979. The Mississippian and Pennsylvanian (Carboniferous) Systems in the United States—Montana. *Prof. Pap. U.S. geol. Surv.* no. 1110, X1–32.

SMITH, J. F. and KETNER, K. B. 1975. Stratigraphy of Paleozoic rocks in the Carlin-Pinon Range area, Nevada. Ibid. no. 867-A, 1–87.

TASCH, P. 1969. Branchiopoda. *In* MOORE, R. C. (ed.). *Treatise on Invertebrate Paleontology*. Geol. Soc. Am., New York and Lawrence. Pt. R, Arthropoda **4**, 1, R128–191.

VAN STRAELEN, V. and SCHMITZ, G. 1934. Crustacea Phyllocarida (= Archaeostraca). *Foss. Cat.* **64**, 1–246.

WHITTINGTON, H. B. and EVITT, W. R. 1954. Silicified Middle Ordovician trilobites. *Mem. geol. Soc. Amer.* no. 59, 1–137, pls. 1–33.

WORTHEN, A. H. 1890. Palaeontology of Illinois. *Geol. Surv. Illinois*, **8** (2).

ZANGERL, R. 1981. Chondrichthyes I, Paleozoic Elasmobranchii. *In* SCHULTZE, H. P. (ed.). *Handbook of Paleoichthyology*, **3A**, 1–117. G. Fischer Verlag, Stuttgart and New York.

—— and RICHARDSON, E. S. 1963. The paleoecological history of two Pennsylvanian black shales. *Fieldiana Geol. Mem.* **4**, 1–352.

D. E. G. BRIGGS

Department of Geology
Goldsmiths' College
University of London
Creek Road
London SE8 3BU

W. D. I. ROLFE

Hunterian Museum
University of Glasgow
Glasgow W2

NOTE ADDED IN PROOF

Since this paper was written, we have received new information on *Ostenia cypriformis* from Pinna *et al.* (1982). Those authors now accept that the cirripede features are convergently acquired (cf. p. 272 above). They therefore establish a new crustacean Class Thylacocephala, characterized by what they interpret as an anteriorly developed cephalic sac, covered with cuticular sclerites. This distinctive structure is also present in *Concavicaris* aff. *bradleyi* collected in June 1983 by W. D. I. R. and K. B. Ketner, from the Famennian of Nevada (p. 268 above). This latter discovery provides further evidence of a relationship between these forms (p. 272 above; Pinna *et al.* 1982, p. 481). This reticulate structure may represent a huge compound eye, in which case the notch of *Concavicaris* accommodated an eye peduncle, rather than an antenna (Rolfe in Brunton *et al.* 1969; cf. p. 267 above). This might explain the large size of the notch in juvenile *C. milesi*, as negative differential growth of the eyes occurs in many arthropods. If the 'cephalic sac' represents an eye it is unlikely that it probed the substrate, as shown by Pinna *et al.* (1982, fig. 4). We have argued above (p. 273) that there are insufficient data available to classify Concavicarida above ordinal level, but the Class Thylacocephala is now available for the group, even if its relationship to the major arthropod taxa remains uncertain.

Dr. F. Schram has drawn our attention to Secretan's 1983 paper on the Callovian *Dollocaris* from

La-Voulte-sur-Rhône, for which she proposes the Class Conchyliocarida. This is a junior synonym of Thylacocephala; the presence of a large compound eye in *Dollocaris* independently supports our interpretation of the structure in *Concavicaris*.

REFERENCES

PINNA, G., ARDUINI, P., PESARINI, C. and TERUZZI, G. 1982. Thylacocephala: una nuova classe di crostacei fossili. *Atti Soc. ital. Sci. nat. Museo civ. Stor. nat. Milano,* **123,** 469–482.

SECRETAN, S. 1983. Une nouvelle classe fossile dans la super-classe des Crustaces: Conchyliocarida. *C.r. Acad. Sci., Paris,* **296** (III), 437–439.